including West Nipissing

Forest Plants
OF
Central ONTARIO

Written by
Brenda Chambers,
Karen Legasy,
and **Cathy V. Bentley**

Illustrated by
Shayna LaBelle Beadman
and
Emma Thurley

The Publisher
Lone Pine Publishing

206, 10426-81 Avenue	202A, 1110 Seymour Street	16149 Redmond Way, #180
Edmonton, Alberta	Vancouver, British Columbia	Redmond, Washington
Canada T6E 1X5	Canada V6B 3N3	USA 98052

Canadian Cataloguing in Publication Data

Chambers, Brenda.
  Forest plants of Central Ontario

    Includes bibliographical references and index.
    ISBN 1-55105-061-7

    1. Forest plants—Ontario—Identification—Handbooks, manuals, etc.
I. Legasy, Karen L. II. Bentley, Cathy, 1958- III. LaBelle-Beadman, Shayna.
IV. Thurley, Emma. V. Title.
QK203.05C52 1996   581.9713   C95-911181-6

The Authors

Brenda Chambers   *Central Region Science & Technology*
                            Ontario Ministry of Natural Resources

Karen Legasy   *Northeast Science & Technology*
                            Ontario Ministry of Natural Resources

Cathy V. Bentley   *Forest Research Consultant*
                            Churchill, Ontario

Shayna La-Belle Beadman *Northeast Science & Technology*
                            Ontario Ministry of Natural Resources

Emma Thurley   *Consultant*
                            Haliburton, Ontario

Senior Editor: Nancy Foulds
Botanical Editor: Linda Kershaw
Editors: Roland Lines, Jennifer Keane
Design: Bruce Timothy Keith
Cover Design: Carol Dragich
Cover Photo: Brenda Chambers

Layout & Production:
Gregory Brown, Carol Dragich
Maps: Volker Bodegom
Separations:
Elite Lithographers Co. Ltd.,
Edmonton, Alberta, Canada
Printing:
Quality Color Press,
Edmonton, Alberta, Canada

Funding for the development and printing of this publication has been made available in part through the Northern Ontario Development Agreement, Northern Forestry Program (NODA).

This field guide to the forest plants of Central Ontario is a co-publication of Central Region Science & Technology and Lone Pine Publishing. This guide was developed in partnership with Northeast Science & Technology unit of the Ontario Ministry of Natural Resources (OMNR 50817). Cette publication spécialisée n'est disponible qu'en anglais.

The publisher gratefully acknowledges the assistance of Alberta Community Development and the Department of Canadian Heritage.

TABLE OF CONTENTS

Acknowledgements • 6
Introduction • 7

TREES
13–42

SHRUBS
43–129

HERBS
130–291
Herb Chart • 132

GRASSES
292–309
Grass Key • 294

SEDGES & RUSHES
310–328
Sedge Key • 310

FERNS & ALLIES
329–362
Fern Key • 341

MOSSES & LIVERWORTS
363–414

Mosses and Liverworts Key • 364
Peat Mosses Key • 370
Upright (Acrocarpous) Mosses Key • 380
Trailing (Pleurocarpous) Mosses Key • 398

LICHENS
415–432
Lichens Key • 415

Appendix • 433
Glossary • 435
Pictorial Glossary • 438
References • 440
Index • 443

ACKNOWLEDGEMENTS

The producers of this guide would like to acknowledge the contributions of the following individuals: Steve Newmaster, Consultant, Wawa; Dr. Bill Crins, Regional Ecologist, Central Region, OMNR, Huntsville; Dr. John Morton, Professor Emeritus, Biology Department, University of Waterloo; Dr. Peter Ball, Department of Botany, Erindale College, University of Toronto; Andy MacKinnon, Research Branch, B.C. Forest Service, Victoria, B.C.; Colleen Beckett, Central Region Science and Technology, OMNR, North Bay; Dr. Brian Naylor, Habitat Biologist, Central Region Science and Technology, OMNR, North Bay; Tim Bellhouse, Wildlife Biologist, Central Region Science and Technology, OMNR, North Bay; Trevor Howard, Systems Analyst, Central Region Science and Technology, OMNR, North Bay; Brendan G. MacKey, Senior Lecturer, Department of Geography, Australian National University; Dan McKenney, Chief Forest Resource Economist, Canadian Forest Service, OMNR, Sault Ste. Marie; Ago Lehela, Silviculture Impact Assessor, Ontario Forest Research Institute, Sault Ste. Marie; Trevor Goward, Naturalist and Lichenologist, Clearwater, B.C.

Thanks must also be given to people who have helped along the way and to any we might have missed: Ken Baldwin, Dave Coleman, Wasyl Bakowski, Norm Sczyrek, Kevin Lawrence, Yin-Qian Yang, Dr. Irwin Brodo, Albert Dugal, Dr. Robert Ireland, Neil Mauer, John Thompson, Sylvain Levesque, Peter Uhlig, Lyn Thompson, Kevin Etmanski, John Askitis, Louise Clement and Leith Hunter.

Additional thanks go to Steve Newmaster, who contributed two liverwort, five moss and three lichen illustrations, and to Trevor Goward, who contributed three lichen illustrations (pp. 418, 420 and 421).

We acknowledge the use of plant specimens from the herbaria of Algonquin Park and the Canadian Forestry Service, Sault Ste. Marie, for photographic purposes.

INTRODUCTION

This field guide has been developed for resource managers, naturalists and recreational users of the forest who wish to improve their ability to identify the plants they encounter in Central Ontario. It contains descriptions of about 400 plants of upland and lowland forest ecosystems. The majority of species described are characteristic of forest ecosystems, but some introduced species of roadsides and other disturbed habitats are also included.

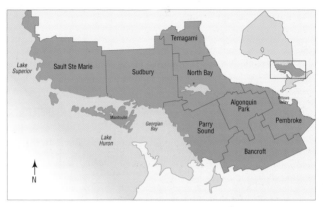

REGIONAL OVERVIEW

The Central Region of Ontario lies within the Great Lakes-St. Lawrence Forest Region, a forest zone that also extends into Quebec and northwestern Ontario. The Central Region is south of the Boreal Forest and north of the Carolinian (Southern Hardwood) Forest of southern Ontario, and it is noted for its diverse mix of conifer and hardwood forest ecosystems.

The northern boundary of the Central Region extends from Lake Superior just south of Wawa, south of Chapleau, Gogama and Kirkland Lake to the Quebec border. The southern boundary stretches from Arnprior in the east to Honey Harbour on Georgian Bay in the west. The limestone-based landscape of southern Ontario lies to the south.

Climate, topography, geology and human history have all played a significant role in determining the pattern and composition of the forest ecosystems in the region. Minimum temperatures generally decline from south to north, and from low to high elevations.

Climatic patterns in the Central Region are affected by its topography and its proximity to the Great Lakes.

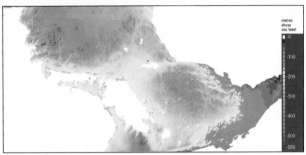

Digital elevation model, illustrating contours (metres above sea level) at 20 m intervals. From Mackay et al. (1994)

Low-elevation areas along the north shore of Lake Huron, the east side of Georgian Bay, Lake Temiskaming (in the Little Clay Belt) and within the Ottawa Valley have moderated climate regimes. Plants that typically occur farther south extend their ranges into these areas, where more growing degree days (see glossary) are available for plant growth than in higher-elevation areas.

Higher-precipitation zones are found on the windward sides of the Algoma and Algonquin Highlands along the eastern shores of Lake Superior and Georgian Bay. The lee side of the Algonquin Highlands experiences a 'rain shadow' effect. The Renfrew County area in the Ottawa Valley has the driest climate in the region.

Geologically, the Central Region is diverse. In the southern half of this region, acidic soils have developed from Precambrian granites, gneisses and quartzites. In the northern part, more base-rich soils are found where intrusions of igneous rocks, such as diabase, exist.

The Precambrian marble belt is a significant geological feature of the southern part of the region, in the Minden, Bancroft, Tweed and south Pembroke areas. Here, diverse upland and lowland plant communities have developed in response to more calcareous soil conditions and a warmer climate. Tolerant hardwood species like red oak, white ash, basswood, ironwood, beech and black cherry are more prevalent on this southern landscape than elsewhere in our region. Bulblet fern, maidenhair fern, leatherwood and downy arrowwood are also found in these floristically rich areas.

The area east of Pembroke, in the Ottawa Valley, also supports a rich flora. Deep beds of calcareous clays, silts and sands are evidence of a Paleozoic marine environment—the Champlain Sea—where sea life abounded. Outcrops of sedimentary limestone appear in this area, as well as in the extreme south of the Central Region and in the northeastern part of

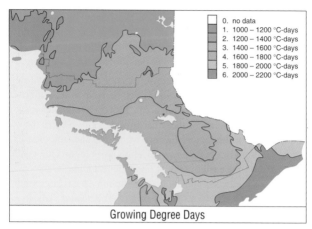

Growing Degree Days

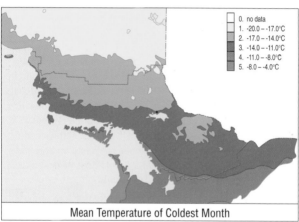

Mean Temperature of Coldest Month

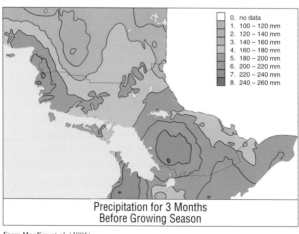

Precipitation for 3 Months
Before Growing Season

From MacKay et al. (1996)

the region, near the Little Clay Belt. These same limestones underlie the soils of Manitoulin and St. Joseph's Island, in the North Channel of Lake Huron. Plants characteristic of both southern and northern calcareous sites flourish in these areas.

Tolerant hardwood mixtures of sugar maple and yellow birch dominate the Algonquin and Algoma Highlands, which are cool, humid, high-elevation areas. Eastern hemlock is a characteristic associate of these hardwoods, especially in the southern half of the region. Eastern white, red and jack pine forests are found in warm, low-elevation areas in the Ottawa Valley and on the eastern side of Georgian Bay. Intolerant hardwood mixedwood stands, with trembling aspen, largetooth aspen, white birch, white spruce, eastern white cedar, balsam fir, red maple, eastern white pine and red pine are widespread across the landscape.

Lowland hardwood forests, found in warm valleys and flats, contain a great diversity of plants. Eastern white cedar is a common companion of white elm, balsam poplar, black ash, yellow birch and red maple in these stands. Occasional pockets of silver maple, green ash and bur oak can be found in the rich deposits of seasonally flooded river banks. Dense mats of sensitive fern and lady fern, tolerant of the fluctuations in water level caused by periodic flooding, are often present. Lowland conifer forests in cool, poorly drained valleys and plains have a distinctive boreal flavour, with an overstorey of black spruce, tamarack, and balsam fir and mats of peat moss on the forest floor.

As one moves north in the Central Region, the climate cools, the growing season shortens, and boreal species become more common. This is particularly noticeable in the northern part of the Sudbury district, where dry sand flats laid down by lakes and glacial streams support extensive stands of jack pine. Black spruce peatlands have developed in cool, poorly drained flats and depressions. Sugar maple and yellow birch stands are restricted to warm slopes. A varied and mixed forest is typical throughout this transition forest zone, and it includes intolerant hardwoods (poplars, white birch), white spruce, balsam fir, eastern white cedar and eastern white and red pine. This bedrock-controlled landscape, with its rolling, often rugged terrain, is home to Ishpatina Ridge near Temagami—this is the highest point in Ontario at 701 metres above sea level.

Natural and human disturbances have also shaped the forests of the Central Region. Fire has played a significant role in the pattern, stability and diversity of conifer and some hardwood ecosystems in this region, as in the boreal forest to the north. From the tree canopy to the mosses on the forest floor, many plant species and their ecosystems have evolved in association with repeated fire. Windfall is another natural disturbance that is important in the regeneration of the tolerant hardwood forest. Human activities that have had a major influence on forest ecosystem patterns include the removal of plants for settlement, logging, mining and agriculture and the introduction of non-native species.

Whether you explore the tolerant hardwood forests, with their distinctive spring flora, the expanses of pine forest or the species-rich hardwood swamps, you can discover a fascinating array of plant species in the Central Region.

ORGANIZATION OF THE GUIDE

This guide is divided into several sections based on plant growth forms—trees, shrubs, herbs, grass-like plants (grasses, sedges and rushes), ferns and fern allies (horsetails and club-mosses), liverworts, mosses and lichens. Several sections include easy-to-use pictorial keys or charts that aid in identification.

Each species account includes a detailed line drawing, a colour photograph (when available) and descriptive text. The text for the plant descriptions is based on a review of available taxonomic literature and is written in plain English, with a minimum of technical terms. Each entry begins with a general description of the plant, followed by detailed descriptions of important plant parts, such as leaves, flowers, fruits and spore-bearing structures.

Habitat information was obtained from the Central Region Forest Ecosystem Classification dataset and existing literature. Similar species and important distinguishing features are discussed in the 'Notes' sections, along with historical, aboriginal, nutritional and medicinal uses of many plants. Information regarding food uses by a number of mammals and birds that live in the region is based on available knowledge and current literature.

Caution: We are NOT recommending the use of any of these plants for medicinal, healing or food value purposes. We also caution that many of the plants in our region are poisonous or harmful and may cause adverse reactions if consumed or used externally. The information about food and medicinal values of plants is provided for interest's sake and historical value only. The information has been compiled from other books and its accuracy has not been tested.

HOW TO USE THIS GUIDE

When trying to identify a plant in the field, first identify its life form (is it a tree, shrub, herb, etc.?) and go to the appropriate section of the guide. When possible, refer to the chart or key at the beginning of the section, and match the characteristics of your specimen (e.g., leaf shape, toothed or toothless margins) with those listed. If the chart or key doesn't lead you to the correct species, try to match your specimen with a photo or line drawing. When you find an illustration that appears to match your specimen, refer to the text for more specific information. When the photo, line drawing and text match your specimen, you have made a field identification.

If you are not able to identify your specimen using this guide, a more comprehensive treatment should be consulted. Check the list of references (pp. 441–43) for some possible sources. A glossary and pictorial glossary (pp. 435–40) are included for users who are unfamiliar with some of the technical terms used in plant descriptions.

PLANT NAMES

The nomenclature for this guide follows *The Ontario Plant List* (Newmaster and Lehela 1995) and *A Checklist of the Flora of Ontario* (Morton and Venn 1990). Common names (English and French) are provided whenever possible.

General: Conifer, evergreen, up to 20 m tall (occasionally over 25 m), symmetrical with a narrowly pyramid-shaped crown; trunk slightly tapered, usually covered with dead branches that persist for years, up to 60 cm in diameter; roots shallow; young bark greyish, smooth, with raised 'blisters' filled with aromatic resin or gum; older bark with irregular, brownish scales; twigs greenish, hairless or slightly hairy; buds waxy, rounded at tips.

Leaves: Needle-like, single, spirally arranged but appear to be in 2 rows (making needled branches or sprays appear flat), flat with rounded or notched tips, 2–4 cm long; upper surface shiny and dark green; underside with 2 white bands or lines; margins toothless.

Fruiting Structures: Cones; male cones small, yellowish-red or purplish-tinged, short-lived, hanging from axils of previous year's leaves; female cones erect, purplish to greenish, about 5–10 cm long; male and female on same tree.

Karen Legasy

Fruits: Winged seeds eventually released from female cones.

Habitat: All moisture regimes and soil textures; a component in many stand types, including intolerant hardwood mixedwoods, pine and tolerant hardwood stands; in conifer swamps in association with black spruce, tamarack and eastern white cedar.

Notes: The resin or gum from the bark blisters is used to make turpentine, and the resin has historically been used in remedies for colds, sores and wounds. See caution in Introduction. • Balsam fir's wood is soft, light, odourless and white. It is used for lumber or pulpwood. Balsam fir has long been a Christmas tree favourite for its balsamic fragrance and long-lasting needles. See notes on ground hemlock (p. 44).

Food Use by Wildlife: bud, leaf, seed, sap, bark, twig.
• **Mammals:** snowshoe hare, red squirrel, beaver, white-footed mouse, porcupine, moose. • **Birds:** spruce and ruffed grouse, yellow-bellied sapsucker, black-capped chickadee, white-winged crossbill.

] 1 cm

TAMARACK • *Larix laricina* • Larch
MÉLÈZE LARICIN

PINE FAMILY (PINACEAE)

Karen Legasy

General: Conifer, deciduous, up to 21 m tall (occasionally over 25 m); crown small, narrow, open and cone-shaped; trunk straight with little taper, up to 60 cm in diameter; roots shallow but wide-spreading; young bark grey, thin and smooth, becomes reddish-brown and scaly; inner bark dark reddish-purple; twigs orange-brown, slender and hairless; buds dark red, small, rounded and smooth.

Leaves: Needle-like, single, spirally arranged on elongated twigs and in clusters of 10–20 on dwarf twigs, triangular in cross-section, slender, 2–3 cm long, soft and flexible, light green, turn golden-yellow and shed in fall; margins toothless.

Fruiting Structures: Cones; male cones small, short-lived; female cones light-brown, upright, rounded, 1–2 cm long; male and female on same tree, appear in May, often stay on tree over winter and throughout following summer.

Fruits: Winged seeds; released from female cones in fall.

Habitat: Wet organic to moist sandy upland sites; an associate of black spruce, eastern white cedar, and balsam fir.

Notes: Tamarack is also known as 'American larch.' • Aboriginal peoples used tamarack roots to sew the edges of birch-bark canoes and to weave baskets. They also used the fresh inner bark as a poultice for wounds or steeped it for a medicinal tea. See caution in Introduction. • Tamarack wood is somewhat hard, heavy and oily, and it is considered decay-resistant even in water. It has been used to make railway ties, poles, fence posts and crates, and the curved roots have been used in shipbuilding.

Food Use by Wildlife: bud, leaf, bark, twig. • **Mammals:** eastern cottontail, snowshoe hare, eastern chipmunk, grey and red squirrels, porcupine. • **Birds:** spruce grouse, black-capped chickadee, red-breasted nuthatch, red crossbill.

1 cm

1 cm

PINE FAMILY (PINACEAE)

General: Conifer, evergreen, up to 24 m tall (occasionally over 30 m); crown cone-shaped, with spreading to slightly drooping branches extending down to base (except in dense stands with heavy shade where lower branches are gradually shed); trunk noticeably tapered, up to 60 cm in diameter (occasionally up to 120 cm); roots shallow but widespread; bark light greyish-brown, thin and scaly; inner bark silvery-white; twigs rough due to persistent woody leaf-bases, whitish-grey to yellowish, usually hairless; buds blunt tipped, with pointed outer scales.

Leaves: Needle-like, single, spirally arranged, straight, pointed, 4-sided in cross-section, around 2 cm long, green to dull bluish-green with a whitish powdery covering (bloom), stiff and thick, often with a skunk-like fragrance when crushed; margins toothless.

Fruiting Structures: Cones; male cones small, pale red, on branch

Karen Legasy

ends; short-lived; female cones slender, cylindrical, reddish to light-brown, 5–7 cm long, open in fall and drop from tree during fall or winter; male and female on same tree.

Fruits: Winged seeds, released from female cones in fall.

Habitat: Moist to dry, clayey to sandy upland sites, occasional in wet organic sites; a component in many stand types, including intolerant hardwood mixedwoods, pine and tolerant hardwood stands and hardwood and conifer swamps.

Notes: Aboriginal peoples used the pliable roots to sew birch-bark canoes and to make baskets and snowshoes. They also used the inner bark to make poultices for wounds or swellings. See caution in Introduction. • White spruce is commercially important for lumber and pulpwood, and it is also used to manufacture musical instruments such as guitars and violins.

Food Use of Spruces (*Picea* spp.) by Wildlife: bud, leaf, seed, bark, twig. • **Mammals:** snowshoe hare (white spruce), grey and red squirrels, deer mouse, porcupine. • **Birds:** spruce grouse (white and black spruce), black-capped chickadee, red-breasted nuthatch, cedar waxwing, pine grosbeak, red and white-winged crossbills, pine siskin.

1 cm [   ] *1 cm*

BLACK SPRUCE • *Picea mariana*
ÉPINETTE NOIRE

Karen Legasy

General: Conifer, evergreen, up to 18 m tall (occasionally over 25 m); trunk straight and without branches for much of its length, up to 25 cm in diameter (occasionally up to 40 cm); crown narrow, often club-shaped at top; branches drooping with ends turned upward; roots shallow; bark dark greyish-brown, thin and scaly; inner bark deep olive-green; twigs dark brown, covered with dense, short, rusty hairs (especially on new growth); buds have finely hairy outer scales with long, slender, hair-like points that extend beyond the bud tip.

Leaves: Needle-like, single, spirally arranged giving branchlets a cylindrical appearance, linear, pointed, 4-sided in cross-section, 0.5–1.5 cm long, dull, dark bluish-green, thick and stiff; margins toothless.

Fruiting Structures: Cones; male cones small, dark red, at branch ends, short-lived; female cones purplish to dark-brown, 2–3 cm long, egg-shaped, become rounded when open, often remain on tree for 20–30 years; male and female on same tree.

Fruits: Winged seeds gradually released from female cones throughout winter, some may remain in cones for a number of years.

Habitat: Wet organic sites, dry to moist, rocky to clayey upland sites; in pure stands or an associate of tamarack, eastern white cedar and balsam fir in lowland sites; an associate of jack pine, eastern white pine and red pine in upland sites.

Notes: When young, black and white spruce may be difficult to distinguish in the field. A key distinguishing feature is the buds. Black spruce buds are hairy at the tips whereas white spruce buds are not. Also, the young twigs of black spruce have short rusty hairs, whereas those of white spruce are hairless. • The roots were used for sewing birch-bark canoes. The gum or resin of black spruce was historically used as a chewing gum. See caution in Introduction. • Black spruce, with its long, strong fibres, is a commercially important pulpwood species used for manufacturing newsprint and other paper products.

Food Use by Wildlife: See notes under white spruce (p. 15).

1 cm

1 cm

General: Conifer, evergreen, up to 23 m tall; trunk straight with little taper, up to 60 cm in diameter; crown cone-shaped, with horizontal-spreading and upturned branches extending down to base (except in dense stands where lower branches are absent); roots shallow, wide-spreading; bark thin, scaly, light reddish-brown; inner bark pale olive-green; twigs with dense short orange-brown hairs; buds hairy.

Leaves: Needle-like, 12–19 mm long, firm, more or less curved above the middle, pointed, 4-angled, green to yellowish-green, shiny.

Fruiting Structures: Cones; male cones oval, nearly stemless, bright red; female cones narrow, egg-shaped, with thin, rounded scales, 3.8–5 cm long, short-stalked; scales rigid, close fitting, with toothed or rough edges, green, turn brown to reddish-brown; open in autumn and stay attached over first winter.

Ron Lee

Fruits: Seeds dark brown, 3 mm long, with short broad wings.

Habitat: Dry to moist upland sites in association with sugar maple and eastern hemlock or other conifers; in wet organic sites with black spruce, tamarack and balsam fir; reaching the eastern limit of its range in this region, uncommon.

Notes: Typically, red spruce is a smaller tree than white spruce. It can interbreed with black spruce, producing intermediate forms.
• Historically, young twigs were used for medicine, and the needles were steeped for tea. See caution in Introduction. • The wood is important for lumber and pulpwood.

Food Use by Wildlife: See notes under white spruce (p. 15).

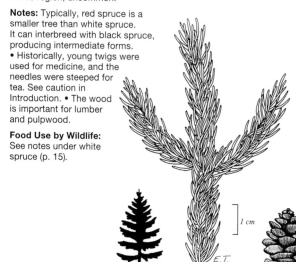

1 cm

E.T.

EASTERN HEMLOCK • *Tsuga canadensis*
PRUCHE DU CANADA

PINE FAMILY (PINACEAE)

Emma Thurley

General: Conifer, evergreen, up to 21 m tall (occasionally over 30 m); trunk tapered, up to 90 cm in diameter; crown pyramid-shaped, extending to the ground; crown tip (leader) usually bent; branches slender, spread horizontally and droop at ends; foliage heavy, often with a silvery cast; roots shallow, wide-spreading; bark dark and rough, deeply furrowed with age; inner bark bright cinnamon-red; twigs flexible, slender, rough-hairy, yellowish-brown; buds blunt and brownish.

Leaves: Needle-like, in flat sprays, rounded or indented at tips, slender-stalked, those on upper side of branch much shorter than others, 8–17 mm long, with 2 white bands on underside.

Fruiting Structures: Cones; male cones rounded, in leaf axils of previous year; female cones at branch tips, oval to oblong, pointed, almost 2 cm long, nearly stemless, with rounded scales, green or rarely purple, turning brown, pendent; remain on tree overwinter.

Fruits: Seeds compressed, nearly surrounded by long oblong wings; shed over first winter.

Habitat: Moist to dry, clayey to rocky upland sites; in pure stands, or an associate of sugar maple, red maple, yellow birch and other hardwoods, including American beech and basswood; also a component of eastern white pine and red oak stands; in the southern half of this region, and extending into the Algoma Highlands as an associate of sugar maple and yellow birch.

Notes: Aboriginal peoples and early white settlers made a tea rich in vitamin C from the leaves. Aboriginal peoples used the leaves as a spice for bear and porcupine meat. The bark and roots were used in a red dye for spoons and other utensils. The needles and bark were used in medicine and the dry, pounded inner bark was applied to wounds. See caution in introduction.
• The wood is hard and low in strength. It is used for coarse lumber and pulp.

Food Use by Wildlife: bud, leaf, seed, sap, bark, twig.
• **Mammals:** eastern cottontail, snowshoe hare, red squirrel, flying squirrels, beaver, white-footed mouse, porcupine, white-tailed deer, moose. • **Birds:** ruffed grouse, yellow-bellied sapsucker, black-capped chickadee, red and white-winged crossbills, pine siskin, American goldfinch.

1 cm

General: Conifer, evergreen, up to 18 m tall (occasionally over 30 m); trunk straight, slender, with little taper, up to 30 cm in diameter (occasionally up to 50 cm); crown shape highly variable; branches spreading to ascending; roots moderately deep and wide-spreading; young bark reddish-brown to grey and thin, turns dark brown and flaky; twigs yellowish-green and slender, turn dark greyish-brown; buds rounded, pale reddish-brown.

Karen Legasy

Leaves: Needle-like, in spirally arranged, spreading pairs, straight or slightly curved, sharp-pointed, with a papery sheath at base, 2–4 cm long, yellowish-green, stiff; margins minutely toothed.

Fruiting Structures: Cones; male cones yellowish, small, about 1 cm long, short-lived, clustered at branch tips; female cones yellowish-brown, varying from oblong to cone-shaped and from straight to strongly curved inward, 2.5–7.5 cm long, often whorled on branches, usually remain closed and stay on tree for a number of years; male and female on same tree.

Fruits: Winged seeds in female cones, black, often ridged.

Habitat: Dry to fresh sandy to coarse loamy upland sites, occasionally on wet organic sites; in pure stands or an associate of black spruce, eastern white pine and/or red pine; also in intolerant hardwood mixedwood stands.

Notes: Aboriginal peoples dug up the long roots, split them in half, rolled them up and soaked them in water to loosen and remove the bark, then used them for sewing things like canoe seams. • Today, jack pine is used for lumber and pulpwood. • Female cones require a temperature of about 50°C or more to open and release their seeds. Therefore, pure stands of natural jack pine usually originate following a fire.

Food Use of Pines (*Pinus* spp.) by Wildlife: bud, leaf, seed, sap, bark, twig. • **Mammals:** eastern cottontail, snowshoe hare, eastern chipmunk, grey and red squirrels, deer and white-footed mice, porcupine, white-tailed deer, moose. • **Birds:** spruce grouse (jack pine), ruffed grouse, wild turkey, yellow-bellied sapsucker, red-bellied wood-pecker, blue jay, black-capped chickadee, white-breasted and red-breasted nuthatches, brown creeper, pine warbler, evening and pine grosbeak, dark-eyed junco, red and white-winged crossbills, pine siskin, American goldfinch.

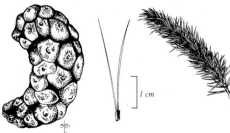

1 cm

RED PINE • *Pinus resinosa*
PIN ROUGE

PINE FAMILY (PINACEAE)

Karen Legasy

General: Conifer, evergreen, up to 24 m tall (occasionally over 35 m); trunk straight, and has little taper; up to 60 cm in diameter (occasionally larger); crown usually oval; roots moderately deep and wide-spreading; bark reddish to pinkish and scaly, older bark is furrowed or grooved with long, flat, scaly ridges; twigs orange to reddish-brown, thick and shiny; buds have sharp, pointed tips and loosely overlapping, hairy scales.

Leaves: Needle-like; in spirally arranged pairs, slender, straight and flexible, flattened on inside of clusters, rounded on outside, with sharp, pointed tips and papery sheaths at base; 10–15 cm long; dark green and shiny; margins minutely toothed.

Fruiting Structures: Cones; male cones small and short-lived; female cones egg-shaped, about 4–7 cm long, release seed in fall and usually drop from tree the following spring; male and female on same tree.

Fruits: Winged seeds enclosed in female cones.

Habitat: Dry to fresh, sandy to loamy upland sites; in pure stands or an associate of eastern white pine and jack pine; a component in intolerant hardwood mixedwood stands.

Notes: Red pine has 2 needles in every cluster; this distinguishes it from eastern white pine (p. 21), which has 5. • Aboriginal peoples used the sap or resin to make a tar for waterproofing the seams on birch-bark canoes and roofs. • Red pine is often used for lumber, telephone poles and railway ties. • Its reddish bark helps to explain the common name.

Food Use by Wildlife: See notes under jack pine (p. 19).

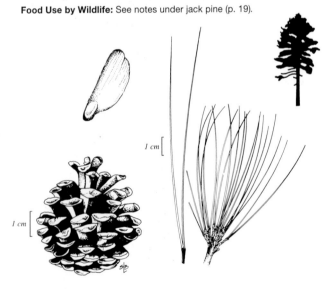

1 cm

1 cm

General: Conifer, evergreen, up to 30 m tall (occasionally over 35 m); trunk little tapered, often branchless for over 1/2 its height, up to 90 cm in diameter (occasionally over 100 cm); branches wide-spreading, often at right angles to trunk near midsection, upward-growing near top; crown widely oval to irregular; roots moderately deep, wide-spreading; young bark greyish-green, thin, smooth; older bark dark greyish-brown with deep longitudinal furrows; twigs green and downy, become orange-brown and hairless; buds reddish-brown, slender, with pointed tips and overlapping scales.

Leaves: Needle-like, about 7–12 cm long, slender, in clusters of 5, soft and flexible, bluish-green, sheathless at base; margins finely toothed.

Fruiting Structures: Cones; male cones small, short-lived; female cones

Karen Legasy

cylindrical, often curved, about 8–20 cm long, on stalks about 1 cm long, green in summer, dark brown in fall when they open, drop from tree during late fall or winter; male and female on same tree.

Fruits: Winged seeds enclosed in female cones; released in fall.

Habitat: Dry to moist, rocky to fine loamy upland sites, occasional in wet organic sites; an associate in many stand types, including pure conifer (with red pine, jack pine and black spruce), tolerant hardwoods and other hardwoods (including red oak) and intolerant hardwood mixedwoods.

Notes: Eastern white pine is the tallest conifer in eastern Canada. When growing in the open, branches on mature trees are often angled east because of prevailing westerly winds. • Aboriginal peoples used eastern white pine gum in a sore-throat remedy and the bark in a cold remedy. See caution in Introduction. • Eastern white pine is valued for its lumber, which is used to make furniture, panelling, window frames and doors, among other things. The British Navy used eastern white pine for shipbuilding. The tall, straight trunks were used for ship masts. • Large, tall eastern white pine are sanctuary trees for female black bears with cubs. They also are important nesting or perching trees when located along shorelines.

Food Use by Wildlife: See notes under jack pine (p. 19).

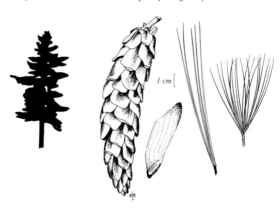

1 cm

Karen Legasy

General: Conifer, evergreen, up to 15 m tall (occasionally over 25 m); trunk rapidly tapered, often twisted and rugged-looking, up to 60 cm in diameter (occasionally over 80 cm); crown narrow, cone-shaped, often extends to ground; roots shallow, wide-spreading; bark reddish-brown, thin, shreds; older bark forms narrow, flat ridges; twigs yellowish-green, form wide, fan-shaped sprays; buds green, minute.

Leaves: Scale-like, tightly overlapping, opposite, dull yellowish-green, 2–4 mm long; upper and lower leaves flat, pointed at tips, with a prominent yellowish resin spot; side leaves folded, clasping bases of upper and lower leaves; margins toothless.

Fruiting Structures: Cones; male cones tiny, short-lived, at branch tips; female cones oval, woody, about 1 cm long, with 10–12 scales in opposite pairs; male and female on same tree.

Fruits: Seeds 2-winged, enclosed by scales of female cone; released a year after cone develops.

Habitat: Wet organic sites, moist to dry, fine loamy to sandy uplands; in pure stands, or an associate in many stand types, including hardwood and conifer swamps, tolerant hardwoods, intolerant hardwood mixedwoods and pine stands; in areas of moderate to low soil acidity.

Notes: If eastern white cedar oil is ingested in quantity, it can cause abnormally low blood pressure, convulsions and death. Eastern white cedar has historically been taken internally as a diuretic and used externally for skin diseases and as an insect repellent. See caution in Introduction. • The soft wood is light and not overly strong, but it is somewhat decay-resistant and has been used for items such as posts, poles and shingles, and in canoe- and boat-building. The cedar boughs were used as brooms, and the sweeping action deodorized the house with cedar fragrance.

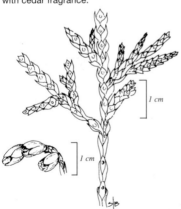

1 cm

1 cm

Food Use by Wildlife: bud, leaf, seed, bark, twig. • **Mammals:** snowshoe hare, red squirrel, beaver, porcupine, white-tailed deer, moose. • **Birds:** common redpoll, pine siskin.

WILLOW FAMILY (SALICACEAE)

General: Hardwood, deciduous, up to 25 m tall (occasionally over 30 m); trunk long, straight, cylindrical, up to 60 cm in diameter; crown narrow, open, with a few stout, upward-growing branches; roots shallow; young bark greenish-brown, smooth; older bark dark greyish, furrowed with irregular, V-shaped crevices; twigs reddish-brown, round in cross-section, stout, hairless; buds reddish-brown, slender, long, pointed at tips, sticky, fragrant.

Leaves: Alternate, stalked, simple, egg-shaped, tapering to point at tip, tapered or rounded at base, 5–12 cm long; upper surface shiny, dark green; underside paler, often has rusty or brownish resin blotches; margins with fine, rounded teeth that turn inward at the tips; stalks round in cross-section.

Brenda Chambers

Flowers: Greenish or reddish, minute, male and female on separate trees; numerous in hanging catkins; appear in early spring before leaves.

Fruits: Capsules in catkins, egg-shaped, hairless, thick-skinned; open to release many minute seeds with long, white, cottony hairs; mature in spring or early summer.

Habitat: Moist clayey to sandy uplands to wet organic sites; an associate of eastern white cedar, black ash, white elm and trembling aspen.

Notes: Aboriginal peoples used the buds to make an ointment for cuts and wounds and rubbed the buds inside their nostrils to help relieve congestion. See caution in Introduction. • Balsam poplar is used for plywood and pulp. Pieces of the lower bark were used as floats for fishing nets or lines.

Food Use by Wildlife: See notes under largetooth aspen (p. 24).

1 cm

LARGETOOTH ASPEN • *Populus grandidentata*
PEUPLIER À GRANDES DENTS

WILLOW FAMILY (SALICACEAE)

General: Hardwood, deciduous, up to 30 m tall (occasionally over 35 m); trunk short, tapered, up to 60 cm in diameter; crown oval but often uneven with a few coarse, irregular branches; roots shallow; young bark pale green to yellowish-grey, usually orange-tinged; older bark dark grey, furrowed; twigs brownish-grey, dull, often slightly hairy and stout; buds have greyish, downy hairs.

Leaves: Alternate, on flattish stalks, simple, egg-shaped to widely oval or almost round, blunt to pointed at tip, wedge-shaped to rounded or straight at base, 4–12 cm long; upper surface dark green; underside paler, downy when leaves unfolding; margins have large, irregular, wavy or rounded teeth.

Brenda Chambers

Flowers: Minute, numerous in catkins, male and female on separate trees; male catkins 5–10 cm long; female catkins 7.5–12.5 cm long in fruit; appear before leaves.

Fruits: Capsules in catkins, about 6 mm long, split open to release tiny seeds with silky hairs; May–June, while leaves are expanding.

Habitat: Dry to fresh sandy to clayey upland sites; an associate of eastern white pine, red oak and other hardwoods including white ash and ironwood.

Notes: Largetooth aspen can be recognized in the spring by the downy hair covering its twigs and buds and later in the season by the large teeth on its leaf margins. • The wood is used to make pulp, plywood, veneer and matches. • The scientific name *Populus* means 'people' and most likely was chosen because poplar species were often planted in public squares where people congregated.

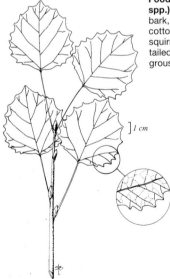

Food Use of Poplars (*Populus* spp.) by Wildlife: bud, leaf, seed, bark, twig. • **Mammals:** eastern cottontail, snowshoe hare, red squirrel, beaver, porcupine, white-tailed deer, moose. • **Birds:** ruffed grouse, purple finch.

] *1 cm*

WILLOW FAMILY (SALICACEAE)

General: Hardwood, deciduous, up to 30 m tall (occasionally over 35 m); trunk cylindrical, little tapered, branch-free for most of its length, up to 60 cm in diameter (occasionally over 80 cm); crown short, rounded; roots shallow, wide-spreading; young bark pale green to almost white, smooth, waxy; older bark grey with long furrows or ridges; twigs brownish-grey, round in cross-section, slender, shiny; buds small, shiny, dark reddish-brown, with pointed tips, not sticky or fragrant.

Leaves: Alternate, stalked, simple, widely egg-shaped to rounded, pointed at tip, flat to rounded or slightly heart-shaped at base, 2–8 cm long, 1.8–7 cm wide; upper surface deep green; underside paler, hairless; margins have fine, irregular teeth; stalks flat, usually longer than blade.

Brenda Chambers

Flowers: Minute, male and female on separate trees; numerous in hanging catkins; appear in early spring before leaves.

Fruits: Capsules in catkins, narrowly cone-shaped, hairless, split open to release tiny seeds with long, white, cottony hairs.

Habitat: Dry to moist, sandy to clayey upland sites, occasional in wet organic sites; in pure stands and an associate in many stand types, including pine, tolerant hardwoods and intolerant hardwood mixedwoods.

Notes: Trembling aspen is considered an invasive species after a forest disturbance such as fire or clearcutting. Aspen reproduces mainly by root suckers. • As a member of the willow family, trembling aspen contains salicin, a compound similar to aspirin. See caution in Introduction. • This tree is used to make pulp, plywood, veneer, wooden matches and chopsticks. • The common name refers to the way the leaves tremble or quake in the slightest breeze.

Food Use by Wildlife: See notes under largetooth aspen (p. 24).

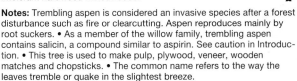

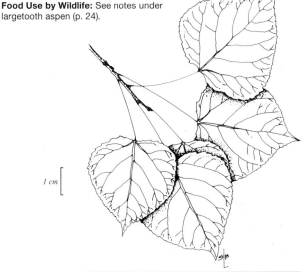

1 cm

General: Hardwood, deciduous, up to 18 m tall (occasionally over 25 m); trunk long, straight, little tapered, up to 45 cm in diameter; crown short, rounded, with slender, ascending branches; branchlets slender, drooping, smooth, greenish to greyish-brown; bark smooth and greenish-grey when young, remaining tight and developing a fine network of ridges with age; buds slender, orange-yellow.

Leaves: Alternate, up to 23 cm long, compound with 7–9 (occasionally 5) leaflets; leaflets long-pointed, elliptic, narrow, finely toothed, dark green above, paler green below.

Flowers: Small, greenish, male and female on same tree; male catkins slender, hanging, in groups of 3 at base of new growth or from lateral buds; female spikes 2–10-flowered, at end of new twig growth; appear in spring with the leaves.

Brenda Chambers

Fruits: Nuts, egg-shaped to round, hard, 2–3 cm long, slender-pointed at tip, with a thin-shelled husk enclosing a single seed; mature and fall in autumn.

Habitat: Fresh to moist upland sites as an associate of sugar maple and other hardwoods; occasional in southern part of this region.

Notes: Butternut (*Juglans cinerea*), another member of the walnut family, is also found occasionally in our region. It also has alternate, compound leaves, but its leaves have 11–17 leaflets (rather than 7–9) and the middle leaflets are largest (rather than those nearest the tip). • Historically, these nuts were used in medicine, dye and food. The leaves and bark were also used in medicine. See caution in Introduction. • The wood was used for handles and fuel, and the bark was used to make chair backs or seats. The wood is still valued for tool handles and sporting goods, and it gives the best flavour to smoked ham and bacon. • The name *cordiformis* means 'heart-formed,' in reference to the nut.

Food Use by Wildlife: bud, leaf, seed, sap, bark, twig. • **Mammals:** eastern cottontail, eastern chipmunk, grey, red and flying squirrels, black bear, raccoon, white-tailed deer. • **Birds:** ring-necked pheasant, wild turkey, yellow-bellied sapsucker, red-bellied woodpecker, blue jay, white-breasted nuthatch, rose-breasted grosbeak.

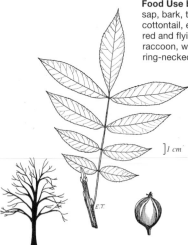

]*1 cm*

E.T.

YELLOW BIRCH • *Betula alleghaniensis*
BOULEAU JAUNE • (merisier)

Brenda Chambers

General: Hardwood, deciduous, up to 22 m tall (occasionally over 30 m); trunks in forest branch-free for more than half their height, with very little taper, up to 60 cm in diameter (occasionally over 90 cm); crown in the forest is short and irregularly rounded; crown in the open is long and wide-spreading; roots deep, wide-spreading; young bark dark reddish and shiny to yellowish or bronze; older bark darker with age, breaks into large, raggedy pieces, does not peel easily; twigs brown, slightly hairy, have a wintergreen taste when broken; buds chestnut brown, sharply pointed at tips.

Leaves: Alternate, stalked, simple, oval, gradually tapering to a sharp, pointed tip, narrow, often slightly heart-shaped at base, 8–12 cm long, 2.5–5 cm wide; upper surface deep yellowish-green, turns yellow in fall; underside paler, soft-downy on veins; 12 or more pairs of veins, each extends to 1 of the large teeth on margin; margins finely to sharply toothed.

Flowers: Tiny, in catkins, male and female on same tree; male catkins 7.5–12.5 cm long, present during winter; female catkins about 1.5 cm long, develop from buds in spring.

Fruits: Small, winged seeds in cone-like female catkins; catkins oval, about 3 cm long, 1.5 cm wide, erect on branches.

Habitat: Moist to fresh upland sites, wet organic sites, all soil textures; an associate of sugar maple, red maple, basswood and eastern hemlock in upland sites; an associate of eastern white cedar, red maple and white elm in lowland sites.

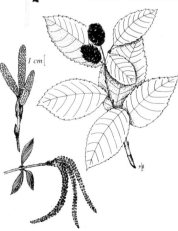

1 cm

Notes: One way to distinguish yellow birch from white birch (p. 27) is to taste the twigs. Yellow birch twigs have a distinctive wintergreen taste, whereas white birch twigs do not. • The inner bark has historically been considered edible, and a tea was made with the twigs. Aboriginal peoples mixed yellow birch sap with maple sap. See caution in Introduction. • Yellow birch saplings were used for wigwam poles. The wood is used for furniture, cabinets, moulding, flooring, doors, veneer, plywood and other products.

Food Use by Wildlife: See notes under white birch (p. 27).

General: Hardwood, deciduous, up to 24 m tall (occasionally over 30 m); trunk slender, often curves and extends almost to top of crown, up to 60 cm in diameter; crown in forest consists of many upward-growing branches ending in clusters of fine branchlets; crown in the open is pyramid-shaped, irregularly rounded in outline; young bark reddish-brown, thin and smooth; older bark creamy-white, papery, peels easily to expose reddish-orange inner layer that turns black with age; twigs dark reddish-brown, slender, sometimes hairy, without wintergreen taste; buds greenish-brown, darker on tips of scales, blunt or occasionally pointed at tips, often sticky.

Karen Legasy

Leaves: Alternate, slender-stalked, simple, egg-shaped to triangular, sharply pointed at tip, straight to rounded or slightly heart-shaped at base, 5–10 cm long; upper surface dull green; underside paler, slightly hairy; margins singly to doubly shallowly toothed from tip to about 1.5 cm from leafstalk.

Flowers: Tiny, in catkins, male and female on same tree; male catkins cylindrical, in hanging clusters, up to 10 cm long; female catkins single or in pairs, erect, to about 5 cm long; in May before leaves.

Fruits: Rounded, winged nutlets in cone-like female catkins; shed in early fall.

Habitat: All moisture regimes and soil textures; in pure stands and an associate in many stand types, including pine, tolerant hardwoods and intolerant hardwood mixedwoods.

Notes: One of the best-known traditional uses of birch by aboriginal peoples is the construction of birch-bark canoes. These peoples also used birch bark as a waterproof layer on wigwams and to make items such as torches, funnels, splints, spoons, dishes and artwork. Birch bark was often wrapped around foodstuffs and sometimes even bodies to prevent their decay. • In emergency situations, birch bark can be used to protect eyes from damage that may be caused by the extreme brightness of the sun reflecting on snow. Placing strips of birch bark over the eyes protects them while allowing some visibilty through the natural openings in the bark.

Food Use of Birches (*Betula* spp.) by Wildlife: bud, leaf, seed, sap, bark, twig. • **Mammals:** eastern cottontail, snowshoe hare (white birch), eastern chipmunk, red squirrel, beaver, porcupine, white-tailed deer, moose. • **Birds:** ruffed grouse, yellow-bellied sapsucker, black-capped chickadee, purple finch, common redpoll, pine siskin (yellow birch), American goldfinch (yellow birch)

1 cm

BIRCH FAMILY (BETULCEAE)

General: Hardwood, deciduous, up to 15 m tall (occasionally over 20 m); trunk straight, up to 25 cm in diameter; crown wide-spreading with slender branches; branchlets slender, dark reddish-brown, hairy when young, becoming hairless, drooping at ends; bark light brown, shredding in long loose strips.

Leaves: Alternate, stalked, simple, tapered at both ends, sharp-pointed at tip, wedge-shaped at base, 6–13 cm long, 2.5–3 cm wide, dark yellowish-green above and below, soft-hairy beneath; margins sharply toothed, with 2 sizes of teeth.

Flowers: Tiny, in catkins, male and female on same tree; male catkins densely flowered, slender, hanging; female catkins cylindrical, 3–5 cm long; appear in spring before leaves.

Emma Thurley

Fruits: Flat nutlets, 5–6 mm long, enclosed in inflated bracts that form a hop-like structure, whitish.

Habitat: Dry to moist, sandy to fine loamy upland sites; an associate of sugar maple and other hardwoods, including largetooth aspen, red oak, white ash and basswood; also a component of eastern white pine, largetooth aspen and red oak stands.

Notes: Also known as 'hop hornbeam.' • Aboriginal peoples used this wood to build frames for their dwellings. A liquid made from the inner bark of choke cherry (p. 66), the root of beaked hazel (p. 54) and the heartwood of ironwood was used to treat bleeding of the lungs. Ironwood was combined with spruce (pp. 15–17) and pine (pp. 19–21) to treat stiff joints, and with eastern white cedar (p. 22) to make cough syrup. See caution in Introduction. • Historically, ironwood was used for sleigh runners, and the strong, tough wood is used for tool handles.

Food Use by Wildlife: bud, leaf, seed. • **Mammals:** grey and red squirrels, white-footed mouse, white-tailed deer. • **Birds:** ruffed grouse, downy woodpecker, rose-breasted grosbeak, purple finch.

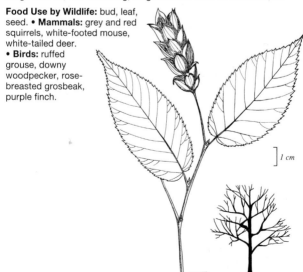

] *1 cm*

E.T.

AMERICAN BEECH • *Fagus grandifolia*
HÊTRE À GRANDES FEUILLES

Emma Thurley

General: Hardwood, deciduous, up to 25 m tall (occasionally over 35 m); trunk straight, clear, up to 60 cm in diameter (occasionally over 100 cm); crown small to wide-spreading; branches slender, somewhat zigzag, with long, thin, pointed buds; wide-spreading shallow root system; bark thin, smooth, grey.

Leaves: Alternate, short-stalked, simple, egg-shaped to elliptic, 5–13 cm long, pointed at tip, tapering to base, leathery, dark bluish-green, turn golden bronze and often persist into winter, very silky when young, silky on the midvein when old; veins straight, each ending in a tooth; no teeth on margin between veins.

Flowers: Tiny, in small clusters, male and female flowers on same tree; male clusters ball-like, on long hanging stalks; female clusters 2–4-flowered, short-stalked, bur-like; after leaves unfold in spring.

Fruits: Sharp-pointed, 3-angled nuts, 1.3 cm long, in pairs, surrounded by a prickly reddish-brown husk which splits into 4 parts to release nuts.

Habitat: Moist to fresh, sandy to loamy upland sites; occasionally in pure stands, but more typically associated with sugar maple, eastern hemlock and other hardwoods, including red oak, white ash, ironwood and basswood.

Notes: The fresh nuts are very high in protein and were an important food for aboriginal peoples. Historically, American beech nuts were roasted and used as a coffee substitute, and the inner bark was used in bread making. The bark, fruit and leaves were used in medicines. See caution in Introduction. • American beech is used for flooring, furniture, containers, handles and wooden ware. The leaves have been used to fill mattresses. • The species name *grandifolia* means 'large-leaved.'

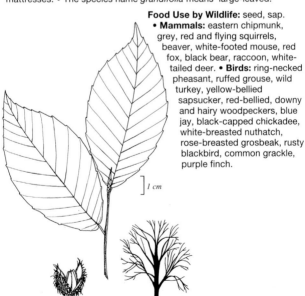

Food Use by Wildlife: seed, sap. • **Mammals:** eastern chipmunk, grey, red and flying squirrels, beaver, white-footed mouse, red fox, black bear, raccoon, white-tailed deer. • **Birds:** ring-necked pheasant, ruffed grouse, wild turkey, yellow-bellied sapsucker, red-bellied, downy and hairy woodpeckers, blue jay, black-capped chickadee, white-breasted nuthatch, rose-breasted grosbeak, rusty blackbird, common grackle, purple finch.

1 cm

General: Hardwood, deciduous, up to 20 m tall; trunk long, straight, or with a few large branches near base when open grown, up to 60 cm in diameter; wide-spreading branches, often with a rugged appearance; deep taproot with deep lateral roots; bark light grey, scaly to slightly furrowed; twigs mostly hairless; buds red-brown, small, blunt.

Leaves: Alternate, stalked, simple, oblong to egg-shaped, with 7–9 ascending narrow lobes, deeply notched between lobes, tapered to base, 15 cm long, 7.5 cm wide, downy when young, hairless at maturity.

Flowers: Male and female flowers separate on same tree; male flowers in slender naked catkins; female flowers solitary or in small spikes, each with a bract and surrounded by many scales; appear with the leaves.

Brenda Chambers

Fruits: Rounded nuts (acorns) in bowl-shaped cups which cover up to 1/3 of the nut, pointed at tips, stalkless or short-stalked, green turning brown, 15–20 mm long, sweet, edible; mature in 1 year.

Habitat: Dry to fresh shallow upland sites as an associate of eastern white pine and red oak; occasional in southern part of this region.

Notes: Aboriginal peoples used the inner bark of young white oak trees as an antiseptic and for the treatment of diarrhea and piles. Historically, the acorns were used for food. See caution in Introduction. • White oak is a valuable species for wood; its uses include furniture, flooring and barrels for liquid. The bark was used historically to produce a brown dye.

Food Use of Oaks (*Quercus* spp.) by Wildlife: bud, leaf, seed, sap, bark, twig. • **Mammals:** opossum, eastern cottontail, snowshoe hare, eastern chipmunk, grey and red squirrels, flying squirrels (white oak), beaver, white-footed mouse, red fox, black bear, raccoon, white-tailed deer, moose. • **Birds:** ring-necked pheasant, ruffed grouse, yellow-bellied sapsucker, red-bellied and downy wood-peckers, northern flicker, blue jay, American crow, white-breasted nuthatch, European starling, rose-breasted grosbeak, rusty blackbird.

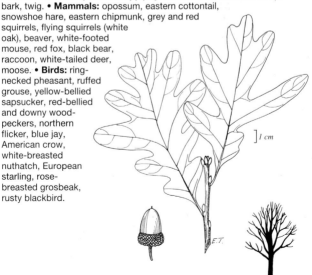

]1 cm

E.T.

BEECH FAMILY (FAGACEAE)

Emma Thurley

General: Hardwood, deciduous, up to 15 m tall (occasionally over 20 m); trunk straight, up to 60 cm in diameter; crown cone-shaped, rounded, regular to irregular with crooked, gnarled branches; deep tap root and deep lateral roots, or shallow spreading roots; bark light grey, rough, with deep furrows and dark scaly ridges; twigs yellow-brown, hairless to hairy; buds blunt, hairy.

Leaves: Alternate, stalked, simple, inversely egg-shaped to oblong, tapered to base, up to 18 cm long, 9 cm wide, fine white-hairy beneath, variable, with wide upper part round-toothed and narrow lower part lobed, or deep-lobed throughout.

Flowers: Male and female flowers separate on same tree; male flowers in catkins in axils of previous year's leaves; female flowers single or in 2- to many-flowered spikes, in axils of current year's leaves; appear in spring soon after leaves.

Fruits: Nuts (acorns) single or in pairs, 2–3 cm long, in bowl-shaped, scaly-fringed cups which cover 1/3 or more of the nut, stalkless or short-stalked; ripen in 1 year.

Habitat: In moist to wet hardwood swamps as an associate of silver maple, green ash, black ash, also on dry, rocky outcrops; occasional in the southern half of the region.

Notes: Acorns were gathered by aboriginal peoples and buried for later use. After boiling or roasting, the flesh of the acorns was eaten as a vegetable, often with grease. The bark was used in medicine. See caution in Introduction. • Bur oak is used for similar wood products as is white oak. Historically, the bark was used for dye.

Food Use By Wildlife: See notes under white oak (pg. 31).

]1 cm

E.T.

BEECH FAMILY (FAGACEAE)

General: Hardwood, deciduous, up to 24 m tall (occasionally over 30 m); trunk straight, up to 90 cm in diameter; crown uneven but generally rounded; several coarse branches; deep spreading root system, may have tap root; bark grey and smooth on young trunks, darker and furrowed when older; twigs stout, reddish-brown, hairless; buds hairless.

Leaves: Alternate, stalked, simple with 7–9 shallow lobes, 10–20 cm long, 7–10 cm wide, dull green, hairless or with small tufts of hair in vein axils beneath; lobes broadest at base, bristle-tipped, sometimes with lateral teeth.

Emma Thurley

Flowers: Tiny, male and female on same tree; male flowers numerous in hanging catkins in axils of previous year's leaves; female flowers single or in small spikes, in leaf axils of current year's leaves; appear before or with leaves in spring.

Fruits: Nuts (acorns), 2–3 cm long, single or in pairs, with a bowl-shaped cup covering about one third of the nut; matures in 2 years.

Habitat: Dry to fresh, sandy to coarse loamy shallow upland sites, occasional on moist sites; an associate of eastern white pine and jack pine on shallow sites, and sugar maple and other hardwoods, including white ash, ironwood, largetooth aspen and white oak, on shallow to deep upland sites; in the southern half of this region, extending into the Algoma Highlands.

Notes: These **leaves and acorns are poisonous to cattle and sheep**.
• Acorns were a very important food source for aboriginal peoples. A variety of methods was used to remove the bitter tannins from the nuts before eating them. Historically, the bark, acorns and ashes were used in medicine. See caution in Introduction.
• The bark was used for tanning and dyeing. The wood is used for furniture, flooring and interior finishing.

1 cm

E.T.

Emma Thurley

General: Hardwood, deciduous, up to 24 m tall; trunk usually straight, forked above into a few large, ascending limbs, up to 90 cm in diameter; crown vase- or umbrella-like, gracefully spreading; roots shallow, wide-spreading; bark dark greyish-brown with wide, deep, intersecting ridges, often scaly; twigs greyish-brown, slightly hairy or hairless, zigzag; buds pale reddish-brown with slightly hairy scales and pointed tips.

Leaves: Alternate, stalked, simple, oval to inversely egg-shaped, pointed at tip, blunt or rounded and unequal at base, 5–15 cm long; upper surface dark green and smooth to rough (sandpapery); underside paler, slightly hairy or hairless; margins coarsely double-toothed.

Flowers: Small, bisexual, in small clusters; stalks slender, undivided, about 1.25 cm long.

Fruits: Winged seeds (samaras), egg-shaped to oval, about 0.5 cm wide, hairless except for fringe around edge; wing deeply notched at tip.

Habitat: Wet organic and moist clayey to sandy sites, also in fresh to dry uplands; an associate of black ash, yellow birch and eastern white cedar in lowland habitats, and sugar maple and other hardwoods in upland sites.

Notes: White elm is often used as an ornamental tree, and its wood is used to make furniture, panelling, barrels, boxes, caskets, crates and boats. • Dutch elm disease resulted in widespread decimation of white elm in the 1950s. Occasionally, large survivors of the disease can be found in this region. Regeneration is often prolific in lowland forests, in damp ditches at the forest edge, and occasionally in upland sites.

Food Use by Wildlife: bud, leaf, seed, sap, twig. • **Mammals:** eastern cottontail, eastern chipmunk, grey and red squirrels, beaver, white-tailed deer. • **Birds:** ring-necked pheasant, wild turkey, yellow-bellied sapsucker, rose-breasted grosbeak, purple finch.

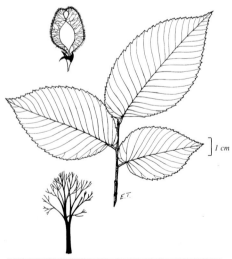

] 1 cm

E.T.

ROSE FAMILY (ROSACEAE)

General: Hardwood, deciduous, up to 21 m tall (occasionally over 25 m); trunk long and straight to short and tapered, up to 60 cm in diameter; crown rounded to elongate; branches reddish-brown; bark smooth and dark reddish-brown to black when young, with horizontal grey marks (lenticels); older bark rough, dark, tending to be flaky, red-brown beneath; twigs reddish-brown, slender; buds blunt, reddish-brown.

Leaves: Alternate, simple, lance-shaped to narrowly oval, tapered to tip and base, 7–13 cm long, 2–4 cm wide, firm; upper side bright green; underside paler, with prominent fringe of white to brown hairs along midrib; margins with fine, incurved teeth.

Flowers: White, 4–6 mm wide, on 3–6 mm long stalks; 20 or more in a 10–15 cm long terminal cluster (raceme); May–June.

Fruits: 1-seeded drupes, fleshy, purplish black, 10 mm in diameter; August–September.

Emma Thurley

Habitat: Dry to moist, sandy to fine loamy upland sites; an associate of sugar maple and other hardwoods, including red maple, basswood, red oak and American beech; in the southern part of this region.

Notes: Wilted leaves are **poisonous to livestock**. • Aboriginal peoples ate these bitter cherries fresh or dried. The dried inner bark was applied to wounds to speed healing. A tea made from the bark was used to treat coughs and colds. See caution in Introduction. • The wood was used to make musket butts, and it is still valuable for furniture.

Food Use by Wildlife: bud, leaf, fruit, bark, twig. • **Mammals:** eastern cottontail, snowshoe hare, eastern and least chipmunks, grey and red squirrels, beaver, black bear, white-tailed deer, moose. • **Birds:** ring-necked pheasant, ruffed grouse, yellow-bellied sapsucker, northern flicker, great-crested flycatcher, eastern kingbird, blue jay, American crow, grey catbird, American robin, cedar waxwing, European starling, northern cardinal, rose-breasted and evening grosbeaks, white-throated sparrow, common grackle, northern oriole.

E.T.

1 cm

AMERICAN BASSWOOD • *Tilia americana*
TILLEUL GLABRE • *Tilleul d'Amérique*

LINDEN FAMILY (TILIACEAE)

Emma Thurley

General: Hardwood, deciduous, up to 25 m tall (occasionally over 30 m); trunk straight, up to 75 cm in diameter; crown rounded; branches slender, arch upwards or spread outwards and turn upwards near tips; root system deep, wide-spreading; bark dark greyish-brown, in long flat ridges; sprouts from the base of old stumps; twigs stout, yellowish-brown, hairless; buds broad, clearly asymmetrical at base, hairless, often reddish.

Leaves: Alternate, stalked, simple, heart-shaped, 11–12 cm wide and slightly longer, hairless to sparsely hairy, with tufts of hair in axils of veins beneath; margins with gland-tipped sharp teeth.

Flowers: Creamy yellow, 1.3 cm wide, very fragrant; 5 petals; in small, flat-topped, long-stalked clusters from the centre of elongate, leafy bracts in leaf axils; June–July.

Fruits: Nut-like, hard, woody, round, 6 mm across, brown-hairy, 1–2 seeded; remain on tree over winter.

Habitat: Dry to moist, sandy to clayey upland sites, as an associate of sugar maple, eastern hemlock, yellow birch and other hardwoods, including red oak, black cherry, white ash and ironwood; occasional in hardwood swamps; in southern half of this region.

Notes: Aboriginal peoples wrapped the bark around weak arms and legs to provide splint-like support. A solution made from boiled bark was applied to burns and sores to ease discomfort. See caution in Introduction. • Aboriginal peoples used troughs made from basswood in maple syrup processing. Inner bark fibre was made into twine, fine string or thread, and was used for mat-weaving, bag-making or with birch bark in the building of wigwams. The soft, light wood is valued for handcarving, modelling, turnery and furniture parts.

Food Use by Wildlife: bud, leaf, seed, bark, twig. • **Mammals:** eastern chipmunk, grey squirrel, deer and white-footed mice, porcupine, white-tailed deer, moose.

1 cm [

E.T.

MAPLE FAMILY (ACERACEAE)

General: Hardwood, deciduous, up to 30 m tall (occasionally over 35 m); trunk usually branch-free for half its length when growing in forest, divides or branches nearly to ground when growing in open, up to 70 cm in diameter (occasionally over 100 cm); crown short, narrow; roots shallow, widely spreading; young bark light-grey, smooth; older bark dark greyish-brown with scaly ridges; twigs red, shiny, hairless, turn greyish with age; buds reddish, shiny, smooth, blunt-tipped.

Leaves: Opposite, reddish-stalked, simple with 3–5 lobes, heart-shaped at base, 5–15 cm wide; lobes sharply pointed, separated by shallow, V-shaped notches (sinuses); upper surface bright green; underside whitish, usually hairless when mature; deep red or scarlet in fall; margins coarsely and irregularly double-toothed.

Brenda Chambers

Flowers: Usually male or female, small, red to yellowish or orange; in dense clusters on slender stalks; appear before the leaves in May.

Fruits: Winged keys (samaras), reddish, about 2 cm long, in pairs; June–July.

Habitat: All moisture regimes and soil textures; in tolerant hardwood stands as an associate of sugar maple, yellow birch and eastern hemlock; a component of pine stands; an associate of yellow birch, eastern white cedar and yellow birch in hardwood swamps.

Notes: Red maple sap, though not as sweet as that of sugar maple (p. 38), was historically used to make syrup and sugar. See caution in Introduction.
• Red maple wood is used for furniture, veneer, plywood, crates and other items. The bark was boiled and used as a dye or to produce a black 'ink.'
• Red maple is named for the reddish colour of its buds, leaf stalks, flowers and fruit and the deep-red colour of its leaves in fall.

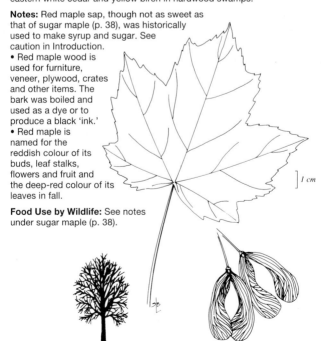

1 cm

Food Use by Wildlife: See notes under sugar maple (p. 38).

Brenda Chambers

General: Hardwood, deciduous, up to 30 m tall (occasionally over 35 m); trunk straight, up to 90 cm in diameter; crown narrow, round; roots wide-spreading, relatively deep; bark dark grey with irregular strips that usually curl outward; twigs reddish-brown, shiny, hairless; buds reddish-brown with sharp-pointed tips and numerous pairs of minutely hairy scales.

Leaves: Opposite, stalked, simple, 3–5 (usually 5) lobed, 9–15 cm wide; top lobe nearly square, usually 3-toothed, slender-pointed, separated from side lobes by wide, rounded notches; upper surface yellowish-green, yellow to brilliant orange and scarlet in fall; underside paler, hairless or occasionally hairy; margins have few, irregular, wavy teeth.

Flowers: Usually male or female, small, yellowish, without petals, in hanging, 5–10 cm long, tassel-like clusters; appear with leaves in April and May.

Fruits: Pairs of plump, winged keys (samaras) with 2 wings nearly parallel, 2.5–4 cm long; stalks slender, usually longer than the keys; mature in fall.

Habitat: Fresh, also moist and dry sandy to clayey upland sites; in pure stands or an associate of yellow birch, red maple, eastern hemlock and other hardwoods, including red oak, basswood, American beech, white ash and ironwood.

Notes: Sugar maple is Canada's national tree. It is the chief source for maple syrup and maple sugar, which contain vitamin B, phosphorus, calcium and enzymes. Aboriginal peoples also fermented sugar maple sap to make vinegar to enhance the flavour of foods such as venison. See caution in Introduction. • Aboriginal peoples used the wood to make a variety of utensils, including large spoons, stirring paddles and ladles to aid in the making of maple syrup. The wood is valued for its strength and used for furniture, flooring, plywood and other lumber products.

]*1 cm*

Food Use of Maples (*Acer* spp.) by Wildlife: bud, leaf, seed, sap, bark, twig. • **Mammals:** eastern cottontail, snowshoe hare, eastern chipmunk, grey, red and flying squirrels, beaver, deer and white-footed mice, porcupine, black bear, raccoon, white-tailed deer, moose. • **Birds:** ruffed grouse, wild turkey, yellow-bellied sapsucker, red-breasted nuthatch, rose-breasted, evening and pine grosbeaks, purple finch, red and white-winged crossbills.

MAPLE FAMILY (ACERACEAE)

General: Hardwood, deciduous, up to 30 m tall; trunk straight, up to 50 cm in diameter (occasionally up to 90 cm); crown broad and rounded; root system usually shallow; bark greyish-brown, flaky and scaly when older; twigs stout, shiny, red to greyish-brown, hairless, with unpleasant odour when broken; buds blunt, reddish, smooth.

Leaves: Opposite, simple, deeply 5-lobed more than halfway to middle of blade, 8–13 cm wide; notches (sinuses) with curved (concave) sides; terminal lobe distinctly narrowed; underside whitened and sometimes hairy; margins sharply toothed or with minor lobes.

Flowers: Small, greenish or reddish, lacking petals, male or female on same or separate trees; in dense, short-stalked, unisexual clusters; April–May.

Fruits: Ribbed, swollen, winged keys (samaras), 3–6 cm long, in pairs with wide-spreading wings, sparsely hairy; ripen early (about the time leaves are fully developed), shed quickly, and germinate soon after; May.

Habitat: In moist to wet hardwood swamps, along river and stream banks, as an associate of red maple, green ash and black ash; occasional in the southern half of the region.

Notes: Historically, the bark was used in medicine and food and the sap was used for food. See caution in Introduction. • The wood was used for bowls. It is used for furniture and products where strength is not required. This tree is planted for ornamental purposes. • The species name *saccharinum* means 'sugary.'

Brenda Chambers, top and bottom

Food Use by Wildlife: See notes under sugar maple (p. 38).

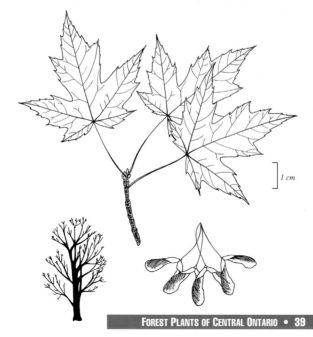

1 cm

BLACK ASH • *Fraxinus nigra*
FRÊNE NOIR

OLIVE FAMILY (OLEACEAE)

Karen Legasy

General: Hardwood, deciduous, up to 20 m tall (occasionally over 25 m); trunk slender, sometimes leaning or bent, extends almost to top of crown, up to 60 cm in diameter; crown open with coarse, upward-growing branches; roots shallow, wide-spreading; bark pale grey, flaky (may also be tight), develops cork-like ridges which are easily rubbed off; twigs light grey, stout, round in cross-section, dull and hairless; buds blackish; side buds small.

Leaves: Opposite, stalked, 30–40 cm long, compound with 7–11 leaflets; leaflets stalkless, oblong to lance-shaped, taper to long, slender point at tip, rounded or slightly tapered at base, 7–10 cm long, 1.5–3 cm wide, dark green, hairless but with rust-coloured tufts of hairs at base when young; margins finely toothed.

Flowers: Small, inconspicuous, with no petals, male or female on same or separate trees, occasionally bisexual and mixed with unisexual flowers on same tree; in dense clusters; appear in early spring before leaves.

Fruits: Single, winged keys (samaras) with a flat, narrowly oblong wing, rounded or blunt at both ends, 2.5–4 cm long, 6–10 mm wide; ripen in fall and remain on tree for most of winter.

Habitat: Moist clayey to sandy uplands to wet organic sites; an associate of balsam poplar, eastern white cedar, white elm, red maple and yellow birch.

Notes: Aboriginal peoples used the inner bark in a remedy for internal ailments. See caution in Introduction. • Aboriginal peoples believed that ash trees had the power to ward off snakes. They wove baskets out of layers of wood stripped from black ash. Its bark was used to cover wigwams.

Food Use of Ashes (*Fraxinus* spp.) by Wildlife: bud, leaf, seed, sap, bark, twig. • **Mammals:** flying squirrels, beaver, white-footed mouse, porcupine, black bear, white-tailed deer, moose. • **Birds:** wild turkey, yellow-bellied sapsucker, cedar waxwing, northern cardinal, evening and pine grosbeaks, purple finch.

1 cm

S|B

Olive Family (Oleaceae)

General: Hardwood, deciduous, up to 21 m tall (occasionally over 25 m); trunk straight, up to 60 cm in diameter; crown narrow, pyramidal to rounded; root system shallow, wide-spreading to deep; bark light grey with distinctive furrowed pattern; twigs stout, light-brown, hairless; buds blunt, brown, 4-sided.

Leaves: Opposite, stalked, 20–58 cm long, compound with 5–9 (usually 7) leaflets; leaflets oval to oblong, gradually taper to tip, rounded at base, 7–13 cm long, 4–8 cm wide, on 6–13 mm long stalks, hairless, dark green above, pale beneath; margins with a few teeth near tip, or untoothed.

Flowers: Small, male and female on separate trees; in dense clusters from axils of previous year's leaves; May–June.

Emma Thurley

Fruits: Flattened, linear-winged keys (samaras), 3–5 cm long, with long wing extending far below seed.

Habitat: Dry to moist, sandy to clayey upland sites; an associate of sugar maple and other hardwoods, including red oak, largetooth aspen, ironwood and basswood; in the southern half of this region.

Notes: Historically, the flowers, seeds, leaves, buds and bark were used in medicine. See caution in Introduction. • Aboriginal peoples used white ash wood to make fish spears, canoes, snowshoes and baskets. The leaves were placed in moccasins for protection against snakes. The strong, tough wood is valuable for manufacturing sporting goods, furniture and agricultural implements.

Food Use by Wildlife: See notes under black ash (p. 40).

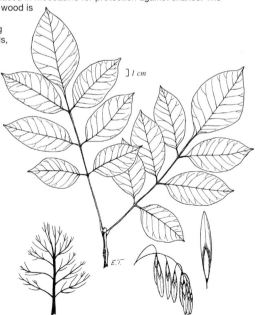

] *1 cm*

E.T.

GREEN ASH • *Fraxinus pennsylvanica*
FRÊNE ROUGE

OLIVE FAMILY (OLEACEAE)

Emma Thurley

General: Hardwood, deciduous, up to 15 m tall (occasionally over 25 m); trunk straight, up to 35 cm in diameter (occasionally up to 45 cm); crown pyramidal; roots deep to shallow, wide-spreading; bark greyish-brown, often tinged red on inner surface, furrowed; twigs grey or brownish, stout, hairy; buds reddish brown, hairy.

Leaves: Opposite, stalked. 25–30 cm long, compound with 5–9 (usually 7) leaflets; leaflets lance-shaped to oblong, gradually tapering to tip and base, 7–12 cm long, 2–4 cm wide, short-stalked, yellowish-green above, paler and hairy below; margins toothless or weakly toothed.

Flowers: Small, lack petals, male and female on separate trees; in dense clusters from axils of previous year's leaves; before the leaf buds enlarge; May.

Fruits: Flattened, linear- to oblong-winged keys (samaras), 2.3–5.5 cm long; seed less than 1/2 of key length.

Habitat: In moist to wet hardwood swamps, along river and stream banks, as an associate of black ash, silver maple, bur oak and red maple; also on upland sites as an associate of sugar maple and other hardwoods; occasional in the southern half of the region.

Notes: This species is also known as 'red ash,' and is variable in leaf and fruit characteristics. • In the past, seeds, leaves, and bark have been used in medicine. See caution in Introduction. • The hard, strong wood is used for agricultural implements, sporting goods and furniture manufacturing.

Food Use by Wildlife: See notes under black ash (p. 40).

] 1 cm

Les arbustes

SHRUBS

The illustrations below show examples of leaf arrangements of shrubs in this section.

simple alternate leaves (pp. 31-84)

simple opposite leaves (pp. 85-102)

palmate opposite leaves (pp. 103-107)

palmate alternate leaves (pp. 108-114)

compound alternate leaves (pp. 115-126)

compound opposite leaves (pp. 127-129)

Brenda Chambers

General: Evergreen, low and spreading or straggling, rarely over 2 m tall; branchlets green, slender and become scaly and brownish with age; branches up to 2 m long, spread from base of plant for about 1/3 of their length, then curve upward and form flat sprays.

Leaves: Needle-like, flat, alternate, spirally arranged but twisted and flattened in 2 rows making branches appear flat, sharp-tipped, narrowed at base, 1–2.5 cm long, 1–3 mm wide; stalks short, twisted; upper surface dark green, underside paler.

Fruiting Structures: Cones in leaf axils; male cones tiny, solitary; female cones tiny, solitary, on short, scaly stalks; May.

Fruits: Bright red, berry-like, with an opening at tip, pulpy, about 7.5 mm long, with 1 brown, bony seed; July.

Habitat: Dry to moist, sandy to fine loamy upland sites to wet organic sites; upland tolerant hardwood stands and hardwood and conifer swamps.

Notes: The **branchlets, leaves and seeds are considered poisonous**. The berry-like fruit, with its reportedly sweet taste but slimy texture, has historically been considered edible. See caution in Introduction. •Ground hemlock may be confused with shrubby growth forms of balsam fir (p. 13), but it can be distinguished by its bright-red fruit and the short, pointed tips of its needles. Balsam fir needles are blunt or notched. • This shrub is also known as 'Canada yew.'

Food Use by Wildlife: twigs • **Mammals:** white-tailed deer, moose.

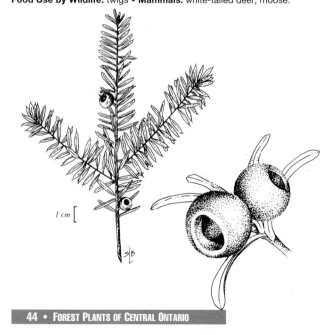

1 cm

General: Evergreen, low, erect or spreading, up to 1.5 m tall (rarely taller); branchlets greenish and smooth, become pale- to dark-brown with ridges and scaly bark; branches curved upward to erect, typically lack a main stem (leader); forms large patches that usually die off from centre.

Leaves: Needle-like, mostly in whorls of 3, awl-shaped, with a tapered, slender, spine-like tip and rounded base, 5–15 mm long, up to 2 mm wide, stiff and erect on younger branches, become more open on older branches, prickly; upper surface with a wide, bluish-white strip along centre.

Fruiting Structures: Cones in leaf axils of previous year's growth; male cones tiny, catkin-like; female cones berry-like; May–June.

Brenda Chambers, both

Fruits: Round, berry-like cones, fleshy, 6–10 mm wide, bluish-black, covered with a fine, bluish-white, waxy powder, contains 1–3 seeds.

Habitat: Rock outcrops in open woods and at the forest edge.

Notes: Creeping juniper (*Juniperus horizontalis, inset photo*), lays flat on the ground with a long trailing stem, many short branches and opposite, scale-like leaves. • The raw berries are said to be **toxic** to livestock. • Common juniper was burned to fumigate a sick person's room. The oil has historically been used to flavour gin. The berries were used to make beer and as a pepper substitute. See caution in Introduction. • Historically, the fresh leaves of creeping juniper were boiled and used cautiously for powerful medicines, because they could cause death.

1 cm

BEBB'S WILLOW • *Salix bebbiana*
SAULE DE BEBB

WILLOW FAMILY (SALICACEAE)

Linda Kershaw

General: Small tree or large coarse shrub, up to 6 m tall; branches and branchlets ascending and wide-spreading, greyish, hairy, reddish-brown when young.

Leaves: Alternate, simple, stalked, inversely egg-shaped to elliptic, pointed at tip, tapered or rounded at base, 3–7 cm long and 1–3 cm wide, thin, usually hairy and greyish on both sides when young, dull green and sometimes hairy above when mature with prominent ridged veins below; margins untoothed, wavy or with irregular teeth.

Flowers: Male and female flowers in catkins on separate plants; female catkins 2–7 cm long, loosely flowered; male catkins 1–3 cm long; scales narrow, tawny, sparsely hairy; appear with leaves in May–June.

Fruits: Slender capsules 5–9 mm long, usually finely hairy; stalk 2–5 mm long; June.

Habitat: Moist to wet thickets, fields, and wet organic sites.

Notes: The leaves of this species have variable shape and margins. In general, willows are variable and identification can be difficult. • Willows contain salicin, which is similar to the active ingredient in aspirin. Historically, the bark and roots of willows were used as a painkiller and anti-fever medicine. Innerbark of Bebb's willow was boiled and the water was then used to stop bleeding of wounds. See caution in Introduction.

Food Use of Willows (*Salix* spp.) by Wildlife: bud, leaf, flower, seed, bark, twig. • **Mammals:** eastern cottontail, snowshoe hare, beaver, porcupine, white-tailed deer, moose. • **Birds:** spruce and ruffed grouse, woodcock, alder flycatcher, Tennessee warbler.

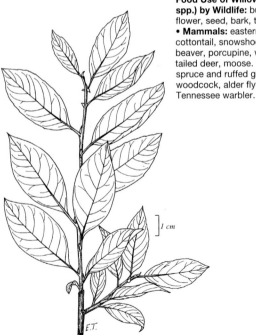

1 cm

E.T.

General: Large, few-stemmed shrub or small tree, 2–6 m tall; branchlets usually hairy, become shiny or have a layer of waxy powder, dark reddish-brown; bark greyish-brown.

Leaves: Alternate, simple, stalked, oblong or elliptic to lance-shaped, pointed or blunt-tipped, tapered at base; blades 3–10 cm long, 1–3 cm wide, firm, bright green and smooth above, whitened and some-times hairy beneath; young leaves have deciduous rusty-coloured hairs; margins irregularly wavy or toothed, mostly above the middle.

Flowers: In stalkless (rarely short-stalked) catkins, male and female on separate plants; female catkins densely flowered, 2–6 cm long, up to 9 cm long in fruit; male catkins 2–4 cm long; scales dark brown to black, long-hairy; fully developed before leaves expand; May–June.

Brenda Chambers

Fruits: Long-beaked, finely hairy capsules, 7–12 mm long, with a short style; June.

Habitat: In damp meadows, along shorelines and on disturbed sites such as flooded ditches.

Notes: There is considerable variation in this species' vegetative characteristics.
• Aboriginal peoples used the bark and catkins in medicines. See caution in Introduction.
• Ornamental 'pussy willows' are the young male catkins.
• The species name *discolor* means 'two-coloured,' referring to the contrast between the dark green upper leaf surface and the pale lower surface. Further notes are included under Bebb's willow (p. 46).

Food Use by Wildlife: see notes under Bebb's willow (p. 46).

1 cm

E.T.

Brenda Chambers

General: Low to mid-sized shrub, 1–3 m tall; branchlets yellow to brown, hairy to woolly, dull; buds have a hood-shaped scale.

Leaves: Alternate, simple, stalked; blades inversely lance- to egg-shaped, short-pointed at tip, tapered at base, 3–10 cm long, 1–3 cm wide, grey-green above, grey-woolly beneath; young leaves hairy, may have rust-coloured hairs; margins smooth or wavy, rolled under; expand after catkins mature.

Flowers: In stalkless catkins, male and female on different plants; female catkins 2–5 cm long; male catkins 1–2.5 cm long; appear early, before leaves; May-June.

Fruits: Slender, long-beaked capsules, 6–9 mm long, finely grey-hairy; June.

Habitat: Dry sandy upland pine, pine-oak and intolerant hardwood mixedwood stands; along lakeshores.

Notes: Historically, the roots were used in medicine. See caution in Introduction. • The species name, *humilis*, means 'low.' Further notes are included in Bebb's willow (p. 46).

Food Use by Wildlife: see notes under Bebb's willow (p. 46).

1 cm

E.T.

WILLOW FAMILY (SALICACEAE)

General: Shrub or small tree, 3–6 m tall; branchlets yellowish to reddish-brown, smooth and very shiny, sometimes hairy when young but soon become hairless and brown.

Leaves: Alternate, simple; blades lance- to egg-shaped, sharp-pointed or gradually tapered to point, tapered to heart-shaped at base, 4–15 cm long, 1.5–4.5 cm wide, green and glossy on both surfaces, paler beneath; young leaves sometimes rusty-coloured and hairy; margins with sharp gland-tipped teeth; leaf-like bracts (stipules) 1–6 mm long, semicircular to kidney-shaped with gland-dotted margins; stalks 5–15 mm long, glandular.

Brenda Chambers

Flowers: In hairy-stalked catkins on short, leafy branches from leaf axils, male or female on separate plants; female catkins 1.5–5 cm long, 1–1.5 cm wide, with green to brownish pistils; male catkins 1.5–4 cm long, densely flowered; scales pale yellow hairy, soon falling; appear as leaves expand; May–June.

Fruits: Slender egg- to cone-shaped, light brown capsules, 4–7 mm long, hairless; June–July.

Habitat: Wet areas, including swamps, marshes, bogs, lakeshores; in roadside ditches.

Notes: Aboriginal peoples used the bark to treat sores and also used it in a smoking mixture. See caution in Introduction.
• The species name *lucida* means 'shining,' referring to the branchlets and leaves. Further notes are included in Bebb's willow (p. 46).

Food Use by Wildlife: see notes under Bebb's willow (p. 46).

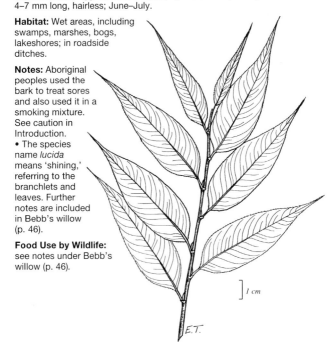

] *1 cm*

E.T.

WILLOW FAMILY (SALICACEAE)

Brenda Chambers

General: Deciduous shrub, low to medium-sized with ascending to erect branches, 1–3 m tall; branchlets yellowish-green to olive brown, slender and slightly hairy; older branches smooth, hairless and dark-brown to almost black; often forms clumps.

Leaves: Alternate, simple, stalked; blades lance-shaped, pointed at tip and base, 2–7 cm long, 0.5–1.5 cm wide; upper surface green, normally smooth and satiny; underside slightly waxy or with thin, silky hairs; slightly silky when young; margins finely toothed; stalks yellowish, 3–10 mm long.

Flowers: In catkins, male and female on separate plants; male catkins round to egg-shaped, 1–2 cm long; female catkins 1.5–2.5 cm long, loosely flowered, up to 4 cm long when in fruit; appear with the leaves in May–June.

Fruits: Egg-shaped to oblong capsules with elongated, slender beaks, slightly hairy, 5–7 mm long.

Habitat: Moist to wet edges of lakes and rivers; in meadows, grassy flats and roadside ditches.

Notes: Slender willow bark and roots were historically used in a remedy to control bleeding, and aboriginal peoples used the bark in a fever and headache remedy. See caution in Introduction.

Food Use by Wildlife: see notes under Bebb's willow (p. 46).

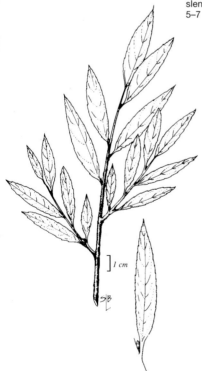

1 cm

General: Deciduous shrub, up to 3 m tall; branchlets brown, hairy and sticky with pale, cork-like spots or lines; pith triangular in cross-section; older branches reddish-brown to grey, hairless; winter buds stalkless with sharp, pointed, curved tips.

Leaves: Alternate, simple, rounded oval to egg-shaped, blunt or pointed at tip, rounded or slightly heart-shaped at base, 4–9 cm long, 2.5–5 cm wide, thin, sticky when young; upper surface bright green; underside paler; margins have numerous regularly spaced, fine, sharp-pointed teeth; stalks 6–12 mm long.

Brenda Chambers

Flowers: In catkins; male catkins long-stalked, slender, scaly, in clusters, about 1 cm long in late summer but lengthen the following spring to about 5–8 cm; female catkins cone-like, 1–2 cm long, long-stalked in small clusters, appear in spring.

Fruits: Winged nutlets shed from woody scales on female cones (catkins) in late summer to fall.

Habitat: Fresh to dry sandy to coarse loamy upland sites, jack pine stands and jack pine mixedwoods; occasional in wet organic black spruce stands.

Notes: You can distinguish green alder from speckled alder (p. 52) by the leaf margins and cones. Speckled alder's leaf margins are double-toothed and more coarsely toothed, whereas green alder's have fine, regularly spaced teeth. Also, speckled alder cones are short-stalked whereas green alder cones have longer stalks. • Aboriginal peoples applied fresh alder leaves to tumours and inflamed areas. The leaves were also wrapped around people suffering from high fever. See caution in Introduction.

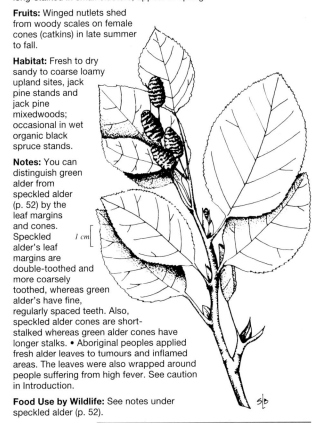

1 cm

Food Use by Wildlife: See notes under speckled alder (p. 52).

SHRUBS

SPECKLED ALDER • *Alnus incana* ssp. *rugosa*
AULNE RUGUEUX

BIRCH FAMILY (BETULACEAE)

Linda Kershaw

General: Deciduous shrub, coarse and spreading, up to 5 m tall; branchlets light reddish-brown and covered with fine, soft hairs; pith triangular in cross-section; buds dark reddish-brown with blunt tips; bark smooth, reddish-brown with cork-like, orange to whitish markings or speckles.

Leaves: Alternate, egg-shaped to oval, usually pointed at tip and rounded at base, 6–10 cm long, 3–6 cm wide; upper surface dark green, smooth, dull and often wrinkled; underside has fine hairs along straight, prominent veins; margins double-toothed, wavy, with sharply and finely toothed edges; stalks 1–2 cm long.

Flowers: In catkins; male catkins stalked, in clusters, 1–2.5 cm long in late summer but lengthen the following spring to become hanging tails 5–8 cm long; female catkins smaller, appear in late summer in tight clusters at branch tips, cone-like with woody scales when mature, about 1 cm long, short-stalked or stalkless.

Fruits: Wingless nutlets shed in fall from female cones (catkins).

Habitat: Wet organic hardwood and conifer swamps, moist clayey to sandy upland intolerant hardwood mixedwood stands.

Notes: Aboriginal peoples used the inner bark to make a yellow to reddish dye. See caution in Introduction. • Alder roots and leaves have nitrogen-fixing nodules, which help to fertilize soil.

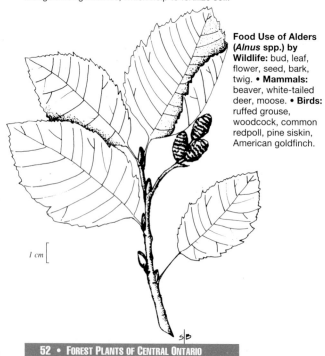

Food Use of Alders (*Alnus* spp.) by Wildlife: bud, leaf, flower, seed, bark, twig. • **Mammals:** beaver, white-tailed deer, moose. • **Birds:** ruffed grouse, woodcock, common redpoll, pine siskin, American goldfinch.

1 cm

BIRCH FAMILY (BETULACEAE)

General: Deciduous shrub, erect, slender, much-branched, 0.5–3 m tall; branchlets usually have fine hairs when young but become hairless, dotted with glands; older stems have smooth, dark-grey to reddish-brown bark with pale, cork-like spots (lenticels).

Leaves: Alternate, simple, circular or kidney- to egg-shaped, widest above middle, blunt to rounded at

Karen Legasy

tip, rounded to tapered at base, 1–4 cm long, 1–2 cm wide, hairless, dotted with yellow glands; upper surface dark green; underside paler to whitish, with 4–5 pairs of prominent veins; margins with coarse, rounded or blunt to pointed teeth; stalks 3–6 mm long.

Flowers: In catkins; male catkins hanging, slender, 12–20 mm long; female catkins erect, cone-like, 12–25 mm long, about 6 mm thick, with numerous overlapping, papery scales; catkins develop in late summer and open the following spring in May–June.

Fruits: Rounded, flattened, winged nutlets.

Habitat: Wet organic conifer swamps, sphagnum bogs.

Notes: Aboriginal peoples used the catkins in an aromatic incense to remedy inflamed mucous membranes. Historically, women drank a tea made from the cones during their menstrual cycle and used it to regain strength after childbirth. See caution in Introduction. • The twigs were used to make basket ribs.

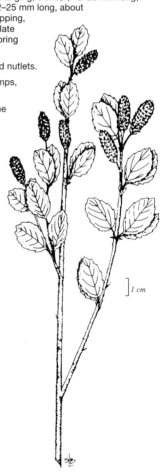

]*1 cm*

BEAKED HAZEL • *Corylus cornuta*
NOISETIER À LONG BEC

BIRCH FAMILY (BETULACEAE)

OMNR

General: Deciduous shrub, 3–4 m tall, often in dense clumps; branchlets hairless to sparsely hairy; bark smooth, pale brown to grey.

Leaves: Alternate, simple, egg-shaped to oval, pointed at tips, rounded or heart-shaped at base, 5–12 cm long, 2.5–7 cm wide; upper surface bright green; underside paler, often hairy; margins irregular, coarsely double-toothed; stalks 8–18 mm long.

Flowers: In catkins; male catkins beige, up to 5 cm long; female catkins tiny, concealed by scales, with crimson, hair-like stigmas protruding; April or May.

Fruits: Round, hard-shelled nuts enclosed in a bristly, beaked husk; beak thin, bristly, tube-like, 3–4 cm long, open and narrowly lobed at the tip; nuts light brown; August–September.

Habitat: All moisture regimes, soil textures and stand types, with the exception of nutrient-poor black spruce organic sites.

Notes: Beaked hazelnuts are rich in protein and oil and low in carbohydrates. Oil from the nuts was historically used as a remedy for toothaches. See caution in Introduction. • Aboriginal peoples made brooms with the twigs and drumsticks with the stems.

Food Use by Wildlife: bud, leaf, nut, bark, twig. • **Mammals:** snowshoe hare, eastern chipmunk, grey and red squirrels, beaver, white-tailed deer, moose. • **Birds:** ruffed grouse, wild turkey, red-bellied woodpecker.

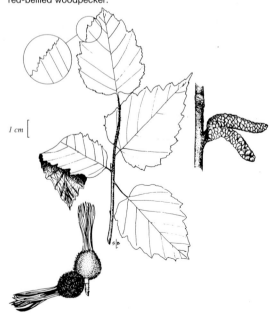

1 cm

BAYBERRY FAMILY (MYRICACEAE)

General: Deciduous shrub, erect, low and much-branched, up to 1 m tall; branchlets hairy and have gland dots; bark reddish-brown to grey or blackish; forms dense patches or thickets from long, spreading underground stems (rhizomes).

Karen Legasy

Leaves: Alternate, simple, long and narrow, 6–12 cm long, 0.5–1 cm wide, tapered at tip and base, lobed, fern-like; upper surface dark green; underside paler, hairy and with resin dots; stalks up to 6 mm long, with pairs of tiny leaf-like bracts (stipules) at base; stipules slightly heart-shaped with long, pointed tips.

Flowers: In catkins, inconspicuous; male catkins slender, dry, scaly, up to 2.5 cm long, clustered near branch tips; female catkins small, at tips of short branches; late April–May.

Fruits: Numerous smooth, rounded nutlets about 4–5 mm long, in round, bur-like, bristly catkins about 2.5 cm across; June–July.

Habitat: Dry to fresh, sandy to coarse loamy pine stands.

Notes: Sweet fern's sweet fragrance is especially noticeable when the leaves and young branches are crushed.
• Aboriginal peoples used sweet fern as a toothache remedy, and also used it to treat diarrhea and skin irritations, such as those caused by poison ivy. They lined blueberry-picking containers with sweet fern, and covered the berries with it so they would not spoil. A 'diet drink' tea was historically made from sweet fern. See caution in Introduction.

Food Use by Wildlife:
bud, leaf, seed, twig.
• **Mammals:** eastern chipmunk, white-tailed deer.
• **Birds:** ruffed grouse.

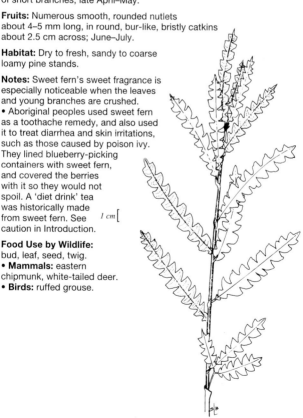
1 cm

Brenda Chambers,
top and bottom

General: Erect shrub, usually 3–4 m tall; branchlets stout, hairless, finely ridged; bark brownish, maturing grey to blackish, with warty spots (lenticels), appears mottled when pale pieces flake off.

Leaves: Alternate, simple, deciduous, highly variable, thick and leathery to very thin; blades egg- to lance-shaped, generally broadest above the middle, pointed, tapered at base, 3–9 cm long, 1–4 cm wide, dull green above, hairless or hairy beneath; fine sharp teeth along margins; stalks 1 cm long, hairy, grooved.

Flowers: Greenish- to yellowish-white, small, bisexual or unisexual; male flowers in crowded clusters of 2–10; female flowers single or 2–3 together; stalks short, in leaf axils; June–July.

Fruits: Round, bright red to yellow, berry-like drupes with 3–5 smooth bony nutlets, about 6 mm in diameter, single or 2–3 together, on short stalks in leaf axils; persist through winter; August–September.

Habitat: Swamps, pond edges, damp thickets.

Notes: This holly has deciduous leaves whereas other species have evergreen leaves. Historically, its bark, berries, buds and wood were used in medicine, and its leaves were steeped for tea. See caution in Introduction.

Food Use by Wildlife: bud, leaf, flower, fruit. • **Mammals:** black bear, white-tailed deer. • **Birds:** ruffed grouse, woodcock, yellow bellied sapsucker, northern flicker, eastern phoebe, blue jay, hermit thrush, American robin, grey catbird, brown thrasher, cedar waxwing, rufous-sided towhee, white-throated sparrow.

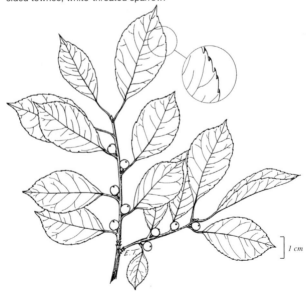

E.T.

1 cm

HOLLY FAMILY (AQUIFOLIACEAE)

General: Deciduous shrub, erect, much-branched, 0.3–3 m tall; terminal branchlets purplish and slender with distantly spaced leaves; lateral branchlets on short spurs, with crowded leaves that often look whorled; bark grey or ashy and mainly smooth with many pale, cork-like markings (lenticels).

Leaves: Alternate, simple, narrowly egg-shaped, usually widest above the middle, short-pointed at tip, tapered to rounded at base, up to 7 cm long and 2.5 cm wide; upper surface bright green, underside paler; margins usually toothless or may have a few scattered teeth; stalks purplish, slender, about 1 cm long.

Karen Legasy

Flowers: Small, either male or female, both sexes on same plant; male flowers in clusters of 2–4; female flowers solitary; stalks thread-like, about 2.5 cm long, from leaf axils; late May.

Fruits: Red, berry-like drupes, about 6 mm in diameter, on slender stalks, with 4–5 slightly ribbed nutlets; August–September.

Habitat: Wet organic sites; occasional in moist to dry, fine loamy to sandy upland sites; in conifer swamps and black spruce-pine upland sites.

Notes: You can recognize mountain holly by its purplish leafstalks.
• Aboriginal peoples made a tonic from the branches and used it for a variety of ailments. **Some sources report the berries to be poisonous** while others say they are edible. They apparently have a strong, bitter taste that makes them unpalatable. See caution in Introduction.

1 cm

5 mm

DOWNY JUNEBERRY • *Amelanchier arborea*
AMÉLANCHIER ARBORESCENT

ROSE FAMILY (ROSACEAE)

Brenda Chambers

General: Perennial shrub or small tree, up to 10 m tall; branchlets purplish when young; older stems light grey and smooth; develops massive, deep root systems.

Leaves: Alternate, simple, oblong to egg-shaped, tapered or sharp-pointed at tip, rounded or heart-shaped at base, up to twice as long as broad, 3–8 cm long, 2–4 cm wide, hairless when mature; young leaves small, folded, hairy beneath, appear at flowering time; margins fine-toothed, with more than twice as many teeth as lateral veins; stalks 1–2.5 cm long, finely hairy.

Flowers: White; 5 petals, 10–15 mm long; in long hanging clusters; appear with leaves in April–May.

Fruits: Berry-like, dark reddish-purple pomes, dry, insipid; June–July.

Habitat: Dry to fresh woods, forest edges and openings, thickets.

Notes: Historically, the fruit and branches of serviceberries were used for food, and the inner bark and roots were used in medicine. See caution in Introduction. • Juneberry wood was used for arrows and pipestems. • The root systems of shrubs in this genus may compete with young conifers.

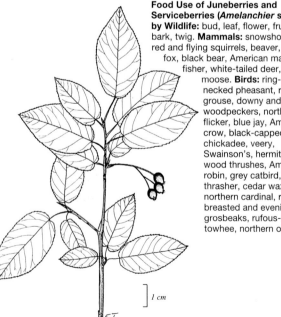

Food Use of Juneberries and Serviceberries (*Amelanchier* spp.) by Wildlife: bud, leaf, flower, fruit, bark, twig. **Mammals:** snowshoe hare, red and flying squirrels, beaver, red fox, black bear, American marten, fisher, white-tailed deer, moose. **Birds:** ring-necked pheasant, ruffed grouse, downy and hairy woodpeckers, northern flicker, blue jay, American crow, black-capped chickadee, veery, Swainson's, hermit and wood thrushes, American robin, grey catbird, brown thrasher, cedar waxwing, northern cardinal, rose-breasted and evening grosbeaks, rufous-sided towhee, northern oriole.

1 cm

E.T.

ROSE FAMILY (ROSACEAE)

General: Deciduous; erect and slender; up to 2 m tall; often growing in clumps; branchlets purplish, mainly hairless.

Leaves: Alternate; on hairless or silky-hairy stalks 2–10 mm long; simple; narrowly egg-shaped to oblong with blunt to pointed tip and tapered base; 3–5 cm long, 1–2.5 cm wide; upper surface green, underside paler; hairless; margins have fine, sharp teeth close together, often tinged purplish-red when unfolding.

Flowers: White; 5 petals 6–10 mm long; individual flowers about 2 cm in diameter; solitary or in small

Karen Legasy

clusters of 2–4 from tops of branches and axils of upper leaves; on hairless stalks; May–June.

Fruits: Round, berry-like pomes; purplish-black; longer than wide, 1–1.5 cm long; on stalks 1–2 cm long; July–August.

Habitat: Dry to moist sites, all upland soil types, conifer and hardwood mixedwood stands; sandy shorelines; bogs and swamps; more frequent on wetter sites than are other serviceberries; openings, clearings and shore thickets.

Notes: You can distinguish mountain juneberry in the field by the fine, close and sharp teeth on its leaves and by its small clusters of 1–4 flowers.
• Aboriginal peoples ate the berries fresh or dried them in the sun and made them into cakes for winter use. See caution in Introduction.

Food Use by Wildlife: See notes under downy juneberry (p. 58).

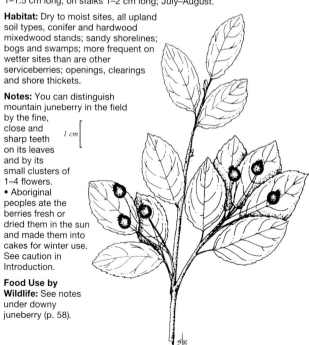

1 cm

ROSE FAMILY (ROSACEAE)

Emma Thurley

General: Perennial shrub or small tree, 5–10 m tall; branchlets purplish and smooth; older stems smooth and grey; develop massive deep root systems.

Leaves: Alternate, simple, stalked; blades egg-shaped to elliptic, long-pointed at tip, rounded to heart-shaped at base, 3–8 cm long, 2–4 cm wide, usually hairless, reddish, folded or partially expanded at flowering time; numerous fine, sharp, pointed teeth on margins, more than twice as many as lateral veins.

Flowers: White; 5 petals; in long hanging clusters; appear with unfolding leaves in April–May.

Fruits: Round, dark reddish-purple to black, berry-like pome, edible, sweet and juicy; stalks 4–5 cm long, July–August.

Habitat: Dry to fresh, sandy to loamy woods; open rocky sites.

Notes: The species name, *laevis*, means 'smooth.' • Aboriginal peoples used the bark for medicinal purposes. See caution in Introduction. See additional notes under downy juneberry (p. 58).

Food Use by Wildlife: See notes under downy juneberry (p. 58).

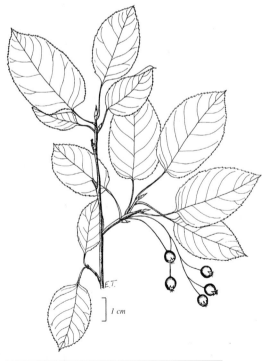

E.T.

] 1 cm

ROSE FAMILY (ROSACEAE)

General: Perennial shrub, straggling or erect, 1–3 m tall, in small loose clumps or solitary; branchlets bright reddish-brown, greyish with age; develop massive deep root systems.

Leaves: Alternate, simple, stalked; blades rounded to bluntly pointed at tip, rounded to heart-shaped at base, 3–7 cm long, 2–4.5 cm wide, folded and woolly when young, hairless after opening; margins finely or coarsely toothed above middle and often nearly to base, with less than twice as many teeth as conspicuous veins.

Flowers: White; 5 petals; numerous, grow in loose hanging clusters; appear while leaves are unfolding and expanding in May–June.

Fruits: Round, dark purple, berry-like pomes, 5–10 mm in diameter, edible, juicy; stalks 1–3 cm long; July–August.

Habitat: Dry to fresh, sandy to loamy woods; open rocky sites.

Notes: The species name, *sanguinea*, means 'blood red.' See additional notes under downy juneberry (p. 58).

Food Use by Wildlife: See notes under downy juneberry (p. 58).

Brenda Chambers

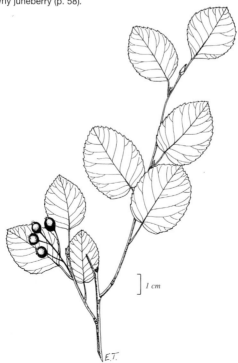

1 cm

E.T.

ROSE FAMILY (ROSACEAE)

General: Perennial shrub, upright, 0.3–2.0 m tall; branchlets purplish-red and covered with silky hairs, soon become smooth and brownish to grey; runners on ground surface spread to form colonies.

Leaves: Alternate, simple, stalked, broadly oval to elliptic, rounded or somewhat pointed at tip, rounded to nearly heart-shaped at base, 2–5 cm long, 2–3.5 cm wide, densely woolly beneath when young but becoming hairless; margins finely and sharply toothed at least on upper 2/3, with more than twice as many teeth as lateral veins; veins irregularly spaced, curved towards tip beyond middle, indistinct near margin.

Flowers: White; 5 petals, 4–8 mm long; in numerous short, dense, erect, 4–10–flowered clusters 1.5–4 cm long; appear before leaves unfold; May–June.

Brenda Chambers

Fruits: Almost round, purplish-black berry-like pomes, about 6 mm in diameter, sweet, juicy; stalks 1–2 cm long; July–August.

> **Habitat:** Open rocky forest.
>
> **Notes:** See notes under downy juneberry (p. 58).
>
> **Food Use by Wildlife:** See notes under downy juneberry (p. 58).

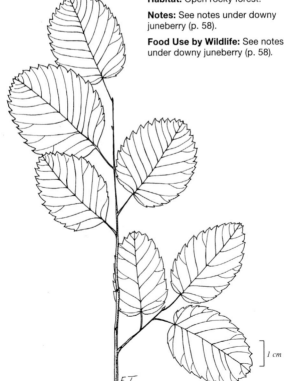

ROSE FAMILY (ROSACEAE)

General: Coarse shrub or small tree up to 10 m tall, with wide-spreading, low, rounded or flat-topped crown, and short, often crooked trunk; branches with rigid, sharp-pointed thorns and crooked twigs; bark loosens into narrow shreds; long taproot develops.

Leaves: Alternate, simple, broadly elliptic to rounded, 2–9 cm long, 1.5–7 cm wide; margins shallowly lobed, sharply toothed; teeth may be gland-tipped.

Flowers: Showy, pink to white, bisexual, similar to apple blossoms; 5 petals; numerous in open flat-topped clusters; May–June.

Fruits: Resemble small apples (pomes), vary with species from orange-red to dark purple-black, 6–12 mm wide, contain 2–5 nutlets; August–October.

Brenda Chambers

Habitat: Disturbed areas, along fencerows and streams; grows best in well-drained areas.

Notes: Hawthorns are easily recognized by their thorns, but hybridization can make species identification difficult. Three species that occur in Central Ontario are golden-fruited hawthorn (*Crataegus chrysocarpa*), black hawthorn (*C. douglasii*) and long-spined hawthorn (*C. succulenta*). • Hawthorns were used by aboriginal peoples for a variety of purposes. A cough medicine was made by soaking the flowers and leaves in boiling water, and back pain was relieved with a tea brewed from the roots. Small cakes were made from the fruits and stored for cooking in winter months. See caution in Introduction. • The hard, heavy wood was used for carving and for making tool handles. The long thorns were used as awls for leather work. • The loggerhead shrike uses the thorns to impale its prey.

Food Use by Wildlife: bud, leaf, fruit, bark, twig. • **Mammals:** eastern cottontail, black bear, raccoon, striped skunk, white-tailed deer. • **Birds:** ring-necked pheasant, ruffed grouse, wild turkey, American robin, cedar waxwing, pine grosbeak.

] *1 cm*

E.T.

Brenda Chambers

Brenda Chambers

General: Perennial shrub, low or mid-sized, up to 2.5 m tall; branchlets grey-brown to purplish, sometimes hairy.

Leaves: Alternate, simple, oval to elliptic, tapering to base, sharp-pointed or tapered at tip, 2–8 cm long, 1–4 cm wide; upper surface dark green, hairless, with a row of dark hair-like glands on midrib; underside paler, sometimes hairy; margins with fine, gland-tipped teeth; stalks 2–10 mm long, with deciduous leaf-like bracts (stipules).

Flowers: White; 5 petals, 4–6 mm long; 5–15 in stalked clusters at leafy branch ends; May–June.

Fruits: Round, purple to black, fleshy, berry-like pomes, 6–10 mm in diameter; in small clusters, sometimes persist in winter; July–September.

Habitat: Wet to moist forest edges, lakeshores, and wet to dry clearings.

Notes: Historically, bark and berries were used in medicine and berries were eaten, although they are very bitter. See caution in Introduction.

Food Use by Wildlife: bud, leaf, fruit. • **Mammals:** black bear, white-tailed deer. • **Birds:** cedar waxwing.

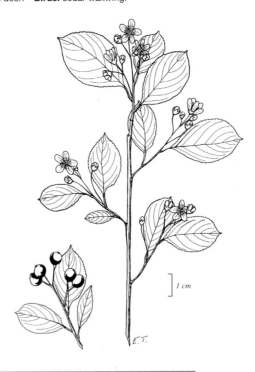

] *1 cm*

E.T.

General: Deciduous shrub or small tree, slender and erect, up to 12 m tall; branchlets smooth, dark reddish-brown, with small, scattered, pale-brown, cork-like markings (lenticels); bark on older stems and branches peels in horizontal papery strips.

Leaves: Alternate, simple, oblong to lance-shaped or narrowly egg-shaped, 4–11 cm long, 1–3 cm wide, short- to long-pointed at tip, blunt to rounded at base; upper surface bright green, shiny; underside paler, hairless; margins have

Karen Legasy

fine, irregular, blunt to rounded, gland-tipped teeth; stalks 1–3 cm long, usually with glands near the blade.

Flowers: White, small, 1.2–1.5 cm wide; 5-petals; stalks 1–2 cm long; in small, flat-topped clusters from leaf axils or crowded at tips of small branches; May–early June.

Fruits: Bright-red, round, juicy cherries, 5–7 mm in diameter, with a single pit or stone in centre; August–September.

Habitat: Dry to moist, sandy to fine loamy upland sites; in pine and intolerant hardwood mixedwood stands; forest openings, edges, roadsides.

Notes: All parts of the cherries, except for the flesh of the fruit, contain a form of cyanide and are poisonous. Raw cherries are considered edible and have been used to make jams and jellies. • Aboriginal peoples used the inner bark to make a tea for coughs and internal ailments, and used the crushed roots in treatments for stomach trouble. See caution in Introduction.

Food Use of Cherries (*Prunus* spp.) by Wildlife: bud, leaf, flower, fruit, bark, twig. • **Mammals:** opossum, eastern cottontail, snowshoe hare, eastern and least chipmunks, grey and red squirrels, white-footed mouse, black bear, raccoon, American marten, fisher, striped skunk, moose. • **Birds:** ring-necked pheasant, ruffed grouse, wild turkey, yellow bellied sapsucker, red-headed and hairy woodpeckers, northern flicker, eastern phoebe, great-crested flycatcher, eastern kingbird, blue jay, American crow, veery, Swainson's, hermit and wood thrushes, American robin, grey catbird, brown thrasher, cedar waxwing, European starling, red-eyed vireo, northern cardinal, rose-breasted and evening grosbeaks, white-throated sparrow, common grackle, orchard and northern orioles.

] 1 cm

CHOKE CHERRY • *Prunus virginiana*
CERISIER DE VIRGINIE

ROSE FAMILY (ROSACEAE)

Brenda Chambers

General: Deciduous shrub, erect, 2–3 m tall (sometimes up to 10 m tall and tree-like), spreads from shoots and often forms thickets; branchlets reddish-brown to purplish-grey, hairless to minutely hairy, have a strong, unpleasant odour when bruised.

Leaves: Alternate, simple, widely oval to egg-shaped, often widest above middle, 4–12 cm long, 2–6 cm wide, usually short-pointed at tip and tapered or rounded at base; upper surface hairless; underside hairless to downy; margins have sharp, fine teeth; stalks 0.5–2 cm long with 1 to several glands at or near base of blade.

Flowers: White, 8–10 mm wide, 5 petals; stalks 4–8 mm long; 10–25 in elongated clusters 5–15 cm long at branch tips; May–June.

Fruits: Deep red or crimson, ripening blackish, round cherries, 8–10 mm in diameter, juicy, with a large stone or pit in centre; August–September.

Habitat: Moist to dry, clayey to sandy upland sites, occasional in wet organic sites; in tolerant hardwood stands and hardwood swamps; streambanks, roadsides.

] 1 cm

Notes: All parts of the choke cherry, except the flesh of the fruit, are poisonous to humans. The bitter, sour-tasting berries are considered edible and have been used to make jelly and wine.
• Aboriginal peoples used the fresh bark in a diarrhea remedy. See caution in Introduction. • You can distinguish choke cherry from pin cherry (p. 65) by the following characteristics: choke cherry has elongated flower clusters whereas pin cherry has more or less flat-topped flower clusters; choke cherry leaves are usually widest above the middle and have a short-pointed tip, whereas pin cherry leaves are usually widest below the middle and often have a tapered point at the tip; choke cherries are deep-red to usually blackish when ripe, whereas ripe pin cherries are bright red.

Food Use by Wildlife: See notes under pin cherry (p. 65).

General: Deciduous shrub, erect, often with numerous branches and twigs, up to 1.5 m tall; branchlets yellowish-brown, hairless to minutely hairy, angled or ridged; older branches purplish-grey with bark peeling off in papery-thin, narrow strips.

Leaves: Alternate, simple, numerous, often crowded, narrowly lance-shaped to oblong, 3–6 cm long, 1–2 cm wide, pointed at tip and base, firm; upper surface dark green, hairless; underside paler, sometimes with fine hairs along veins; margins finely and sharply toothed; stalks 2–6 mm long.

Flowers: White, 5–8 mm wide; 5 rounded petals; 5 small sepals; numerous in dense, narrow, elongated and slightly pyramid-shaped clusters at branch tips; stalks and branches of flower clusters short and finely hairy; June–September.

Fruits: Small, smooth, shiny, papery pods (follicles), splitting open along 1 side, containing few (usually 4) narrow seeds; in clusters of 5–8.

Habitat: Sandy or rocky, usually moist sites; along edges of rivers and lakes, roadside ditches.

Karen Legasy

Notes: The finely hairy branches and stalks of the flower clusters give them a sort of 'fuzzy' appearance. • Aboriginal peoples made what was reportedly one of the best-tasting tea substitutes from the leaves of narrow-leaved meadow-sweet. See caution in Introduction.

Food Use of Meadow-sweets (*Spiraea* spp.) by Wildlife: seed, twig. • **Mammals:** white-tailed deer.

1 cm

General: Perennial shrub, coarse, up to 1.5 m tall; branchlets angled, reddish- to purplish-brown; outer bark of older stems peels off in long papery strips.

Leaves: Alternate, simple, numerous, often overlapping, broadly oval or lance-shaped, sharp-pointed or blunt-tipped, 3–8 cm long, 1–3 cm wide, hairless; margins with coarse sharp teeth; stalks 2–8 mm long.

Flowers: White or pale pink, 4–5 mm in diameter; 5 petals; in numerous open, broadly pyramidal clusters; July–September.

Fruits: Small, linear, few-seeded pods (follicles) opening along 1 side, in clusters of 5–8; often persist over winter.

Habitat: Sandy or rocky, usually moist sites; along edges of rivers and lakes, roadside ditches.

Brenda Chambers

Notes: Some do not recognize this species as distinct from narrow-leaved meadow-sweet (p. 67). • Historically, the leaves have been used as a tea, and the roots, bark and leaves were used in medicine. See caution in Introduction. • The genus name, *Spiraea*, is from the Greek word *speira* meaning 'wreath'; the species name, *latifolia*, means 'broad-leaved.'

Food Use by Wildlife:
See notes under narrow-leaved meadow-sweet (p. 67).

1 cm

E.T.

BUCKTHORN FAMILY (RHAMNACEAE)

OMNR

General: Deciduous shrub, upright to spreading, usually less than 1 m tall, sparsely branched; branchlets green and minutely hairy; mature branches purplish-red to greyish, finely ridged; often forms loose clumps.

Leaves: Alternate, simple, oval to egg-shaped with a short to tapered point at tip, narrowed base, up to 10 cm long and 5 cm wide, largest near branch ends; upper surface green; underside paler; 6–7 pairs of prominent, almost straight veins curve toward tip near margins; margins have rounded to sharply pointed teeth; stalks grooved, 6–12 mm long.

Flowers: Yellowish-green, tiny, about 3 mm wide, with 5 petal-like sepals, short-stalked; in small clusters of 1–5 from axils of lower leaves; late May–early June.

Fruits: Purplish-black, rounded and berry-like, about 6–8 mm in diameter, with 3 flat nutlets, short stalked; August–September.

Habitat: Wet organic to moist upland sites; in conifer swamps, along shorelines and in thickets.

Notes: The berries are poisonous. See caution in Introduction. • Alder-leaved buckthorn can be recognized by its prominent, almost straight veins that curve upward near the margins.

Food Use by Wildlife: fruit.
• **Mammals:** raccoon.
• **Birds:** grey catbird, brown thrasher.

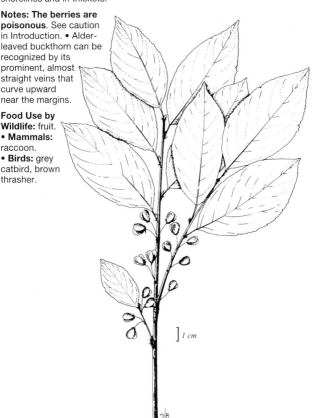

] 1 cm

Brenda Chambers

General: Large shrub or small tree up to 6 m tall; tree form has horizontal tiers of widespreading side branches with upcurved tips; branchlets greenish-red to purple or brownish and glossy, with slender, white pith; **only dogwood with alternately arranged branches**.

Leaves: Alternate, simple, often crowded near branch ends and appearing opposite or whorled; blades thin, egg-shaped or oval, pointed at tip, rounded or tapered at base, 4–13 cm long, 2–7.5 cm wide, dark green above, greyish with fine hairs beneath; veins tend to follow leaf edges to tip; margins untoothed; stalks 1–6 cm long.

Flowers: Creamy-white, small; numerous, in large flat-topped clusters; June.

Fruits: Round berry-like drupes, dark blue-black with waxy powdery coating, 6 mm in diameter, contain one 2-seeded stone; in clusters on red stalks; July–August.

Habitat: All moisture regimes and soil textures; in sugar maple and other hardwood stands and in hardwood swamps.

Notes: Alternate branches and leaves distinguish this species from other dogwoods (pp. 91–92). Alternate-leaved dogwood may hybridize with red osier dogwood (p. 91). • Aboriginal peoples treated sore eyes with a solution made from the bark and roots. A blend for smoking was made by combining other plants with the dried bark from the branches of this species. See caution in Introduction.

Food Use by Wildlife: See notes under red osier dogwood (p. 91).

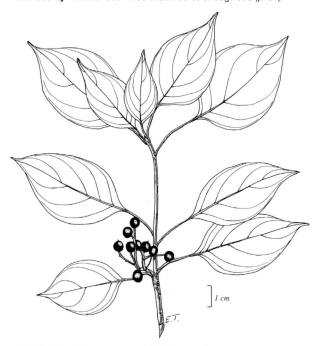

1 cm

E.T.

MEZEREUM FAMILY (THYMELAEACEAE)

General: Deciduous shrub, up to 2 m tall, freely branching like a dwarf tree; branchlets green, turning brown or greyish-brown, jointed; wood soft, brittle; bark very tough and pliable.

Leaves: Alternate, simple; blades egg-shaped to broadly oval or elliptic, broadest at middle, rounded or tapered at either end, 5–8 cm long, 3–5 cm wide, light green, hairless; margins toothless; stalks less than 3 mm long, with dome-shaped, expanded base protecting bud until next season.

Emma Thurley

Flowers: Pale yellow, tubular, 6–9 mm long; 2–5 in hanging clusters with 2–4 hairy bud scales at base; appear as leaves unfold, April–May.

Fruits: Round to elliptic, berry-like (drupes), green at first then purplish-red, 9–12 mm long, contain 1 dark-brown pit; ripen in June-July and fall soon after.

Habitat: Dry to fresh, sandy to loamy, often calcareous upland tolerant hardwood stands; typically in southern part of the region, occasional in northern part.

Notes: The berries, bark and root are considered poisonous. • Aboriginal peoples used the bark to relieve the discomfort of old ulcers and hemorrhoids. See caution in Introduction. • The strong, pliable bark was used for ropes, baskets and in canoe building. • The common name refers to the bark, not to the wood.

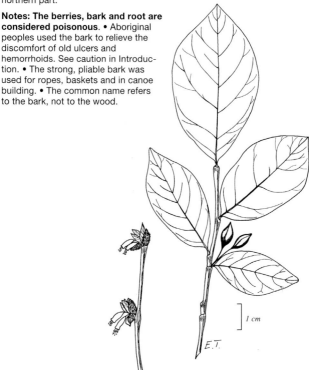

1 cm

E.T.

BEARBERRY • *Arctostaphylos uva-ursi*
ARCTOSTAPHYLE RAISIN-D'OURS

OMNR

General: Evergreen shrub, 5–15 cm tall; stems trailing or spreading on ground with ascending branches, often several metres long; branchlets reddish-brown, finely hairy, sometimes with glands; older branches reddish-brown to greyish-black, hairless or hairy, with papery, peeling bark.

Leaves: Alternate, short-stalked, simple, egg- to spoon-shaped, widest above middle, blunt to rounded at tip, tapered at base, 1–3 cm long, 6–12 mm wide, leathery, firm; upper surface dark green and shiny; underside slightly hairy, paler; margins toothless and flat to slightly rolled downward.

Flowers: White to pinkish or pink-tipped, urn-shaped with 5 very short, rounded lobes, about 5 mm long; in crowded terminal clusters; late May–June.

Fruits: Round, berry-like, red, pulpy, dry, with 1 stone or pit of 5–10 more or less fused nutlets; ripens August.

Habitat: Dry, sandy jack pine stands; dry, sandy or rocky clearings.

Notes: Bearberry leaves were historically used in tobacco mixtures. The berries, considered too dry and tasteless for eating raw, were roasted. A liquid mixture was made from bearberry and eaten to remedy sore or sprained backs. Bearberry was also used in a remedy to ease pain caused by kidney stones. See caution in Introduction.

Food Use by Wildlife: fruit. • **Mammals:** black bear, red-backed and heath voles. • **Birds:** spruce grouse.

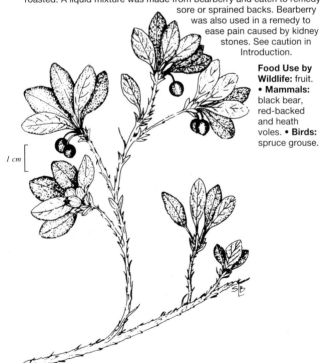

1 cm

HEATH FAMILY (ERICACEAE)

General: Evergreen shrub; stems prostrate and creeping or trailing, sparingly branched, wiry, covered with bristly, brown hairs.

Leaves: Alternate, hairy stalked, simple, oval to broadly egg-shaped, blunt or slightly pointed at tip, rounded or slightly heart-shaped at base, 2.5–7.5 cm long, 1–4 cm wide; upper surface mostly hairless; underside hairy; margins toothless, fringed with brownish hairs.

Brenda Chambers

Flowers: White to pinkish, funnel-shaped with 5 spreading lobes, 1–2 cm long, appear waxy; in clusters from leaf axils or branch tips; early spring.

Fruits: Small, round capsules surrounded by hairy calyxes and containing many dark-brown seeds; late summer.

Habitat: Dry to moist, sandy to fine loamy upland sites; in pine, pine-black spruce and intolerant hardwood mixedwood stands.

Notes: The flowers have historically been eaten in salads. • Aboriginal peoples used the leaves in remedies for urinary tract illnesses. See caution in Introduction. • Trailing arbutus is the provincial flower of Nova Scotia.

Food Use by Wildlife: bud, leaf, seed. • **Mammals:** heath vole. • **Birds:** spruce and ruffed grouse.

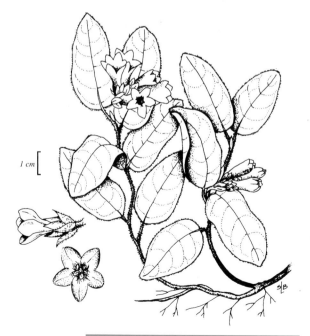

1 cm

OMNR

General: Evergreen shrub, often matted; stems prostrate, trailing or creeping, 20–40 cm long, slightly woody, covered with flat-lying, brownish hairs.

Leaves: Alternate, simple, nearly round to egg-shaped, pointed at tip, tapered at base, small, 2–10 mm long, firm; upper surface dark green to brownish; underside paler with flat-lying, bristly, brown hairs; margins toothless, curled downward; stalks very short.

Flowers: White, bell-shaped with 4 lobes, tiny, 2–3 mm long, hidden among leaves; solitary from leaf axils; early summer.

Fruits: Rounded berries, white, mealy, 5–7 mm in diameter; July–August.

Habitat: Wet organic sites, also in moist to dry clayey to sandy upland sites; in conifer swamps and black spruce-pine stands.

Notes: The white berries have historically been eaten and reportedly have a wintergreen taste. Aboriginal peoples made a tea from the leaves. See caution in Introduction.

Food Use by Wildlife: See notes under wintergreen (p. 75).

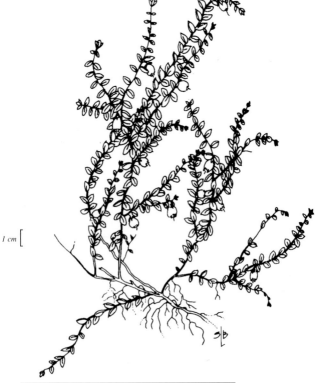

1 cm

HEATH FAMILY (ERICACEAE)

General: Evergreen shrub, 5–15 cm tall; stems slender, creeping on or just below the surface; branches leafy at tips, erect, single or in clumps, mostly hairless.

Leaves: Alternate, crowded near tips of erect branches, short-stalked, simple, oval to inversely egg-shaped, rounded to pointed at tip, narrowed at base, 1–5 cm long, tender when young, firm and leathery when mature; upper surface dark green and shiny; underside paler; margins have obscure, bristle-tipped teeth, rolled slightly downward.

Flowers: White, urn-shaped with 5 small lobes at tip, 5–8 mm long, nodding below leaves on curved or hanging stalks; usually solitary in leaf axils; June.

Fruits: Round, berry-like capsule, red, fleshy, about 10 mm in diameter; ripens in September, often stays on plant throughout winter.

Habitat: Dry to fresh, rocky to clayey upland sites, also wet organic sites; in pine, pine-oak, black spruce-pine, intolerant hardwood mixedwoods and conifer swamps.

Karen Legasy

Notes: The berries have historically been eaten and are reported to taste best after having spent a winter on the plant. The berries and leaves have a distinctive wintergreen flavour. The berries were used to flavour beer. • The leaves were used in remedies for rheumatism, colds, stomach ailments and to restore strength or 'make one feel good.' Wintergreen leaves were wrapped around sore teeth as a remedy for toothache and children chewed the roots to help prevent tooth decay. See caution in Introduction.

Food Use of Wintergreens (*Gaultheria* spp.) by Wildlife: bud, leaf, fruit. • **Mammals:** eastern chipmunk, woodland jumping mouse, white-tailed deer. • **Birds:** spruce and ruffed grouse, wild turkey.

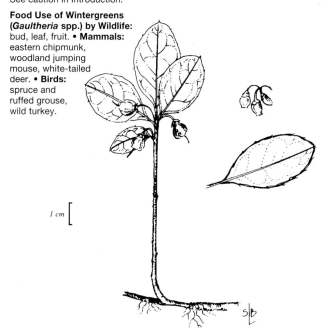

1 cm

BLACK HUCKLEBERRY • *Gaylussacia baccata*
GAYLUSSACCIA À FRUITS BACCIFORMES

HEATH FAMILY (ERICACEAE)

Brenda Chambers

General: Upright, much-branched shrub, up to 1 m tall; branchlets brownish with fine hairs; older stems purplish-grey to black, with peeling outer bark.

Leaves: Alternate, simple, short-stalked, oval to elliptic, tapered at base, rounded or pointed at tip, 2–5.5 cm long, 1–2.5 cm wide, with many tiny golden-yellow resinous glands, dark green above, pale green beneath; margins hairy and toothless.

Flowers: Greenish-white, turning greenish-red, 4–6 mm long, bell-shaped with 5 short lobes; in short, 1-sided lateral clusters; stalks gland-dotted; May–June.

Fruits: Reddish-purple to black berry-like drupes, 6–8 mm across with 10 seed-like nutlets; July–August.

Habitat: Dry to fresh, rocky to sandy pine stands; occasional in wet organic conifer swamps.

Notes: Black huckleberry may be confused with blueberries (pp. 77–78), but the resin dots on the leaves are a reliable distinguishing feature. • Black huckleberries have been used for food and in medicine. Although seedy, they have a sweet, slightly spicy flavour, and can be eaten and prepared like blueberries. See caution in Introduction. • The species name *baccata* means 'berry-bearing.'

Food Use by Wildlife: fruit. • **Mammals:** white-tailed deer. • **Birds:** ruffed grouse, wild turkey, grey catbird, rufous-sided towhee, orchard oriole, pine grosbeak, white-winged crossbill.

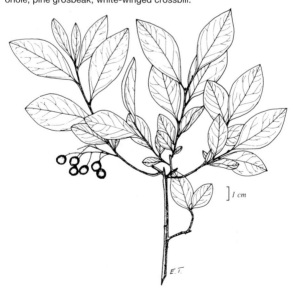

1 cm

E.T.

General: Deciduous shrub, erect with many spreading and ascending branches, up to 60 cm, but usually less than 35 cm tall; branchlets greenish-brown, hairless or finely hairy, with tiny, warty dots; older branches reddish-brown to blackish, hairless, with flaky, ridged bark; forms large patches.

Leaves: Alternate, simple, oval or narrowly lance-shaped, tapered at base and tip, 1–3 cm long, 4–10 mm wide, bright green and hairless on both sides (except for a few hairs on underside along veins); margins with minute, bristle-tipped teeth (use hand lens); stalks short.

Flowers: White to pale pink, bell-shaped with 5 small lobes, less than 6 mm long, nodding in crowded clusters; May–early June.

Fruits: Blueberries, 6–12 mm in diameter; June–August.

Brenda Chambers, top;
Karen Legasy, bottom

Habitat: All moisture regimes and soil textures (except clayey); conifer swamps and upland pine, intolerant hardwood mixedwoods and black spruce-jack pine stands.

Notes: Low sweet blueberry can be distinguished from velvet-leaf blueberry (p. 78) by its leaves. Low sweet blueberry leaves are mainly hairless and have minutely toothed margins, whereas velvet-leaf blueberry leaves are velvety and have toothless margins. • Blueberries are edible raw or cooked. Aboriginal peoples dried the flowers, placed them on hot stones and inhaled the fumes as a remedy for 'craziness.' See caution in Introduction.

Food Use of Blueberries (*Vaccinium* spp.) by Wildlife: bud, leaf, flower, fruit, twig. • **Mammals:** opossum, eastern cottontail, least chipmunk, white-footed mouse, meadow jumping mouse, red-backed, heath and rock voles, red fox, black bear, raccoon, American marten, white-tailed deer. • **Birds:** spruce and ruffed grouse, wild turkey, northern flicker, eastern phoebe, great-crested flycatcher, eastern kingbird, blue jay, black-capped chickadee, veery, hermit and wood thrushes, American robin, grey catbird, brown thrasher, rufous-sided towhee, American tree sparrow, white-throated sparrow, orchard and northern orioles.

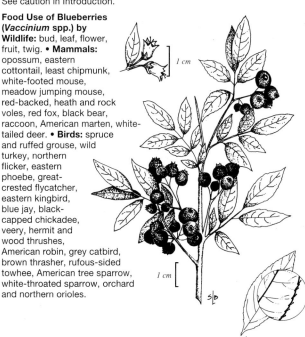

SHRUBS

VELVET-LEAF BLUEBERRY • *Vaccinium myrtilloides*
AIRELLE FAUSSE-MYRTILLE • *Bleuet*

HEATH FAMILY (ERICACEAE)

Brenda Chambers

Karen Legasy

General: Deciduous shrub, low with spreading or ascending branches, up to 50 cm tall; branchlets greenish-brown, densely velvety with whitish hairs; older branches reddish to brown with peeling bark and wart-like dots; spreads to form large patches.

Leaves: Alternate, simple, oval to oblong, pointed at tip and base or somewhat rounded at base, 2.5–5 cm long, 1–2.5 cm wide; upper surface dark green, hairless to downy or velvety; underside paler and downy; margins toothless and with fine hairs; stalks short, hairy.

Flowers: Whitish to pinkish, cylindrical to bell-shaped with 5 small lobes, less than 6 mm long; in crowded clusters at branch tips; May–June.

Fruits: Blueberries, usually with a whitish powder (bloom), 4–7 mm wide; ripen late July–August.

Habitat: All moisture regimes and soil textures (except clayey); conifer swamps and upland pine, intolerant hardwood mixedwoods and black spruce-jack pine stands.

Notes: See notes on low sweet blueberry (p. 77).

Food Use by Wildlife: See notes under low sweet blueberry (p. 77).

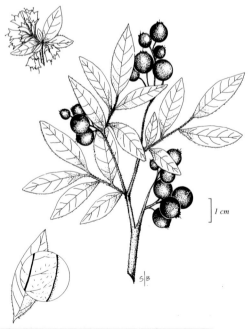

] 1 cm

OVAL-LEAVED BILBERRY • *Vaccinium ovalifolium*
Bleuet • AIRELLE À FEUILLES OVÉES

SHRUBS

HEATH FAMILY (ERICACEAE)

General: Straggling perennial shrub, 0.3–1.2 m tall; branchlets brownish, 4-angled, hairless; older stems purplish-grey to blackish, with peeling bark.

Leaves: Alternate, simple; blades broadly egg-shaped to nearly round, blunt or rounded at tip, 1.5–3 cm long, 0.5–1.8 cm wide, firm, dull above, pale below, hairless; margins usually untoothed; stalks 1-2 mm long.

Flowers: Pinkish, bell-shaped with 5 lobes, 7–10 mm long, solitary, on 1–5 mm long stalks in lower leaf axils of current year's growth; May–June.

Brenda Chambers

Fruits: Round, dark blue berries with a whitish bloom, 6–9 mm in diameter; July–September.

Habitat: Wet organic conifer swamps, rocky mixedwoods and along lakeshores; occasional along the eastern shore of Lake Superior.

Notes: These berries are often poor or disagreeable tasting. See caution in Introduction.

Food Use by Wildlife: See notes under low sweet blueberry (p. 77).

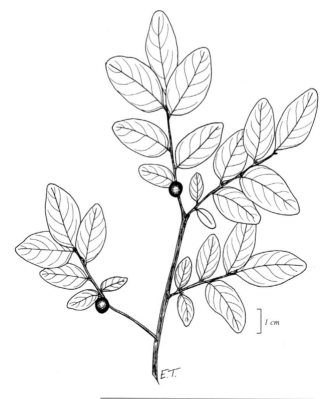

1 cm

E.T.

SHRUBS

SMALL CRANBERRY · *Vaccinium oxycoccos*
AIRELLE CANNEBERGE · Atocas

Karen Legasy

General: Evergreen shrub, delicate, prostrate and trailing with slender, wiry stems and ascending or erect flowering branches up to 20 cm tall; branchlets light- to reddish-brown, minutely hairy; outer bark on older stems peels in pale strips and exposes a smooth, dark inner bark.

Leaves: Alternate, stalkless, simple, narrowly oblong or egg-shaped, pointed at tip, rounded at base, less than 1 cm long, widely spaced; upper surface dark green and shiny; underside whitish; margins toothless, curled downward.

Flowers: Pink, with 4 petals bent back toward base, 5–8 mm long, shooting-star-like, on slender stalks from leaf axils near stem tips; late spring and summer.

Fruits: Round cranberries about 1 cm wide, pale red or pinkish and speckled; ripen August–September.

Habitat: Wet organic to moist upland sites; conifer swamps; open sphagnum bogs.

Notes: Small cranberry may be confused with creeping snowberry (p. 74), but the leaves of that species are rounder and have brown, coarse bristles underneath. Creeping snowberry berries are white and hidden among the leaves whereas those of small cranberry are speckled and reddish and are often more visible. • Small cranberries have been used in preserves, juices, jellies and tarts. See caution in Introduction. • Small cranberry was formerly known as *Oxycoccus microcarpon*.

Food Use by Wildlife: See notes under low sweet blueberry (p. 77).

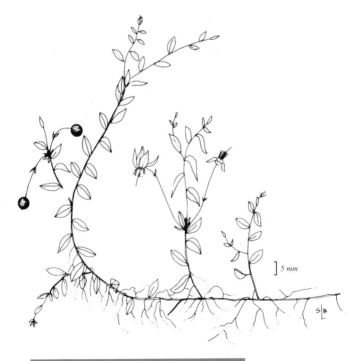

] 5 mm

S|B

HEATH FAMILY (ERICACEAE)

General: Evergreen shrub, prostrate with a trailing, elongated, much-branched, intertwining, slender and cord-like stem often 1 m or more long; branches upright-growing, usually less than 20 cm tall; branchlets light- to reddish-brown and minutely hairy; bark on older branches peels in papery outer layers to reveal a smooth, dark inner layer.

Leaves: Alternate, simple, oblong to elliptic, blunt to rounded at tip, blunt at base, 5–15 mm long, 2–5 mm wide, leathery; upper surface dark green; underside paler; margins toothless, rolled under; stalks very short; July.

Flowers: Light- or pale-pink to whitish, shooting-star-like, with 4 petals or segments spreading to curved backward, about 8–10 mm wide, nodding, on slender, long stalks; solitary or in clusters of 2–6; July.

OMNR

Fruits: Red, oblong to round cranberry, about 1–2 cm in diameter; often remains on plant over winter; ripens late August–September.

Habitat: Wet organic conifer swamps; open sphagnum bogs.

Notes: You can distinguish large cranberry from small cranberry (p. 80) by the leaves. Large cranberry leaves have blunt to rounded tips while small cranberry leaves usually have pointed tips. • The berries have historically been used to make jellies, preserves, tarts and cranberry sauce. See caution in Introduction.

Food Use by Wildlife:
See notes under low sweet blueberry (p. 77).

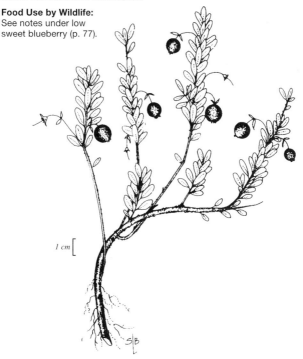

1 cm

BOG ROSEMARY • *Andromeda polifolia* ssp. *glaucophylla*
ANDROMÈDE GLAUQUE

HEATH FAMILY (ERICACEAE)

OMNR

General: Evergreen shrub, erect or trailing, 30–60 cm tall; few branches; branchlets brownish, hairless and round in cross-section; older stems grey to blackish.

Leaves: Alternate, simple, narrowly oblong, rounded or with a tiny point at tip, tapered at base, 2–5.5 cm long, 3–10 mm wide, leathery, firm; upper surface bluish-green to dark green; underside whitened by fine, erect hairs when young; margins toothless, curled under; stalks very short or absent.

Flowers: White to pinkish, bell- or urn-shaped with 5 lobes, less than 6 mm long; in hanging clusters at branch tips; May–June.

Fruits: Small, rounded capsules with long, persistent styles, bluish to brown, about 6 mm in diameter, contain numerous light-brown seeds; July–August.

Habitat: Wet organic conifer swamps, open sphagnum bogs; often forms thickets on boggy margins of lakes.

Notes: Bog rosemary's rounded stems help to distinguish it from bog laurel (p. 85), which has flattened stems. Another distinguishing feature is the leaves. Bog rosemary leaves are alternate and bluish-green with impressed nerves while those of bog laurel are opposite, dark shiny green and smooth. • Aboriginal peoples used the young leaves to brew a tea, but this is not advised. Bog rosemary contains an andromedo-toxin, which, if ingested, can lower blood pressure and cause breathing problems, dizziness, cramps, vomiting and diarrhea. See caution in Introduction.

] *1 cm*

General: Low evergreen shrub, erect and much-branched, up to 1 m tall, often growing in dense patches; branchlets brownish, with minute hairs or small, flaky scales; older stems greyish with shredding outer bark and smooth, reddish inner bark.

Leaves: Alternate, very short-stalked, simple, oval to oblong, short-pointed or rounded at tip, slightly rounded or tapered at base, 1–4.5 cm long, 3–15 mm wide, leathery; upper surface hairless, dull green; underside paler, covered with rusty or white scales; margins toothless or with minute, rounded teeth; smaller toward tip of flowering branches.

Flowers: White, somewhat urn-shaped with 5 lobes, 5–6 mm long; hanging from axils of reduced leaves in 1-sided, elongated terminal clusters on spreading branches; spring–early summer.

OMNR

Fruits: Round capsule, brownish, less than 6 mm in diameter, with persistent slender style, contains numerous minute seeds.

Habitat: Wet organic black spruce bogs and conifer swamps, open bogs and shorelines.

Notes: Leatherleaf can indicate the beginning of bog development by forming floating mats along lakeshores or pond edges.

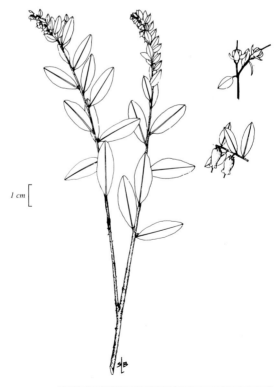

1 cm

General: Evergreen shrub, low, spreading, up to 1 m tall; branchlets densely covered with woolly brown hairs; older stems smooth, hairless, greyish to purplish or reddish-brown.

Leaves: Alternate, simple, narrowly oval to oblong, blunt tipped, rounded or tapered at base, 2–5 cm long, 0.5–2 cm wide, leathery, firm, fragrant when crushed; upper surface dark green, hairless, often wrinkled; underside with brown or rusty woolly hairs; margins toothless, rolled under; stalks short.

Flowers: White, about 1 cm wide; 5 petals less than 6 mm long; on slender stalks in dense, rounded, showy clusters at branch tips; May–June.

Fruits: Small capsules 5–6 mm long, with a slender, persistent style at tip, split open from bottom upwards to release numerous seeds; empty capsules can remain on plant for years; July–August.

Habitat: Wet organic sites, occasional on dry to fresh sandy to loamy upland sites; in conifer swamps and black spruce-pine upland sites.

Notes: Aboriginal peoples made tea from the leaves to drink and to use for medicinal purposes. See caution in Introduction. • They also used the leaves as a tobacco substitute. A brown dye was made from Labrador tea.

Brenda Chambers,
top and bottom

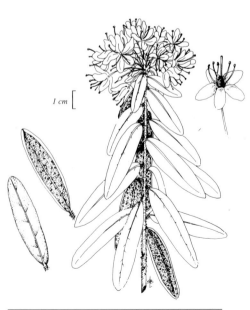

1 cm

General: Evergreen shrub, low and straggling, less than 1 m tall; very few branches; branchlets pale- to dark-brown or blackish and 2-edged.

Leaves: Opposite, simple, narrowly oval or lance-shaped, blunt tipped, rounded to tapered at base, 1–4 cm long, 6–12 mm wide, leathery; upper surface dark green, shiny and hairless; underside whitened with a powdery covering, short hairy; margins toothless, curled under; stalks absent.

Brenda Chambers

Flowers: Pink, saucer-shaped with 5 lobes, 10–15 mm across; on slender, long stalks in terminal clusters; May–June.

Fruits: Round to oval capsules with long, slender, persistent styles, brown, less than 6 mm in diameter, contain numerous seeds; in erect clusters; July–August.

Habitat: Wet organic to moist upland sites; conifer swamps; sphagnum bogs.

Notes: Bog laurel contains a poisonous toxin that may cause severe illness to humans and animals if ingested. See caution in Introduction. See notes on bog rosemary (p. 82).

Food Use by Wildlife: fruit. • **Birds:** ruffed grouse.

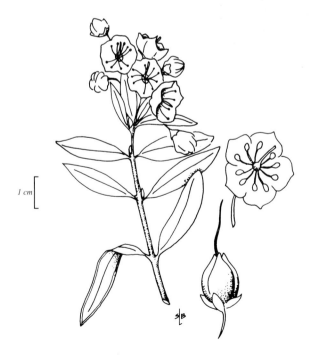

1 cm

SHEEP LAUREL • *Kalmia angustifolia*
KALMIA À FEUILLES ÉTROITES

HEATH FAMILY (ERICACEAE)

General: Evergreen shrub with erect, slender branches, 60–100 cm tall; branchlets brownish, minutely downy, round in cross-section; older branches greyish, hairless.

Leaves: Opposite or in whorls of 3, simple, oblong to oval or elliptic, rounded, blunt or sometimes pointed at tip, narrowed at base, 1.5–5 cm long, 5–20 mm wide, mainly hairless, firm, leathery; upper surface dark green; underside lighter; margins toothless, slightly rolled under; stalks 3–10 mm long.

Flowers: Deep pink to crimson, saucer-shaped with 5 lobes, 0.6–1 cm wide; on long stalks in lateral clusters from leaf axils of previous year's growth (not at branch tips); June–July.

Fruits: Small round capsules up to 6 mm in diameter, with slender, long, persistent styles (almost as long as capsule) at tip; contain many small seeds; often remain on plant for years; late July–August.

Karen Legasy

Habitat: Wet organic sites, also dry to fresh, sandy to coarse loamy upland sites; black spruce-tamarack swamps and black-spruce-pine upland sites.

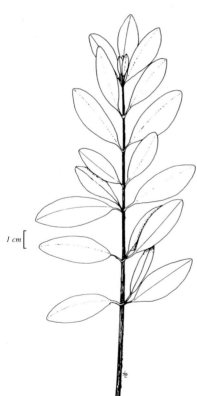

1 cm

Notes: Sheep laurel is considered poisonous and should not be ingested. Aboriginal peoples used sheep laurel to make a remedy for colds and sore backs. It was used in a tonic that was taken in small amounts to remedy bowel ailments. A poultice was made from crushed leaves and applied to the head as a headache remedy. See caution in Introduction.
• The species name '*angustifolia*' means 'with narrow leaves.'

Food Use by Wildlife: fruit.
• **Birds:** ruffed grouse.

WINTERGREEN FAMILY (PYROLACEAE)

General: Evergreen shrub, low, up to 25 cm tall, slightly woody; slender stems (rhizomes) creep at or just below ground level and freely root; flowering stems upright, leafy, single or in groups; branchlets greenish to brownish, hairless, with fine longitudinal ridges.

Leaves: Mainly in whorls, simple, inversely lance-shaped, blunt or pointed at tip, tapered at base, 3–7 cm long, 1–2 cm wide; upper surface dark green, smooth, shiny, with impressed veins, leathery; underside slightly paler, with prominent veins; margins prominently toothed, especially near tip, slightly rolled under; stalks short, grooved.

Brenda Chambers

Flowers: White to rose-pink, saucer-shaped, 10–15 mm wide; 5 petals; on erect or recurved stalks in clusters of 3–10 that extend above leaves; late June–August.

Fruits: Round capsules, 4–8 mm in diameter, usually erect, splitting from tip down, contain numerous, minute seeds; August–September.

Habitat: Dry to fresh, sandy to silty pine and intolerant hardwood mixedwood stands.

Notes: Prince's pine was historically used in remedies for cold in the bladder, consumption, smallpox, kidney stones and stomach troubles. It was also used to purify blood. See caution in Introduction. • Prince's pine is also called 'pipsissewa.'

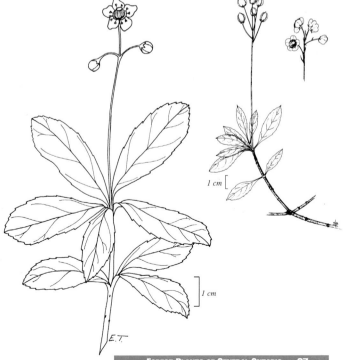

1 cm

PARTRIDGEBERRY • *Mitchella repens*
MITCHELLA RAMPANT

MADDER FAMILY (RUBIACEAE)

General: Small trailing vine, less than 0.5 m long; stems slender, wiry, remain on plant overwinter, root easily; often forms large mats.

Leaves: Opposite, evergreen; blades rounded, blunt at tip, rounded at base, 1–2.5 cm long and as wide, smooth, dark green with a pale midrib, often variegated with white lines above; margins toothless; stalks often as long as blades.

Emma Thurley

Flowers: White to purple-tinged, tubular with coarse hairs inside, usually with 4 (3–8) spreading lobes, 10–15 mm long, fragrant; in pairs; June–July.

Fruits: Bright red double berries (ovaries of 2 flowers united) with indentation and 2 star-shaped marks, contain 8 seeds; persist all winter; ripen in August–September.

Habitat: Dry to moist, sandy to coarse loamy upland sites; in tolerant hardwood, intolerant hardwood mixedwood and pine stands.

Notes: Historically, the whole plant, or more often the vine, was used in medicine, and the berries were used for food and in medicine. The berries are edible, although not very tasty. Aboriginal peoples reportedly used a tea made from this plant to ease childbirth. See caution in Introduction. • Partridgeberry was combined with other plants and used for smoking. • The species name *repens* means 'creeping.'

Food Use by Wildlife: bud, leaf, flower, fruit. • **Birds:** ruffed grouse, wild turkey.

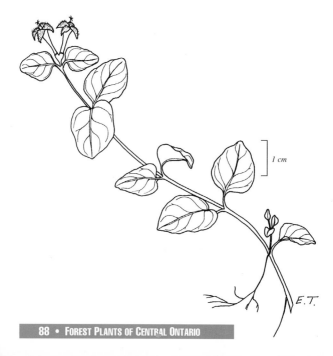

1 cm

E.T.

HONEYSUCKLE FAMILY (CAPRIFOLIACEAE)

General: Small, low evergreen shrub; stems trailing or creeping, 2 m or more long; branches ascending, less than 10 cm tall; branchlets green to reddish-brown, finely hairy, slender and wiry; older stems woody, rarely more than 2 mm in diameter.

Leaves: Opposite, simple, rounded, oval or egg-shaped, wider above middle, blunt-toothed at tip, narrowed at base, 1–2 cm long, with small, bristle-like hairs; margins slightly rolled under; stalks short, hairy.

Flowers: Pinkish-white, bell-shaped with 5 lobes, hairy inside, nodding; usually in pairs, on 3–10 cm long, slender, Y-shaped stalks with 2 tiny bracts at the fork; June–August.

Fruits: Tiny, dry capsules containing a single seed, enclosed by tiny, glandular-hairy bracts; August–September.

Habitat: Wet organic sites to moist to dry clayey to sandy uplands; in many stand types, including conifer and hardwood swamps and upland pine, pine-oak and intolerant hardwood mixedwoods.

Brenda Chambers; top, Linda Kershaw; bottom.

Notes: Aboriginal peoples made a mash from twinflower to remedy inflammation of the limbs. A tea was made from the leaves as a cure for insomnia. See caution in Introduction.

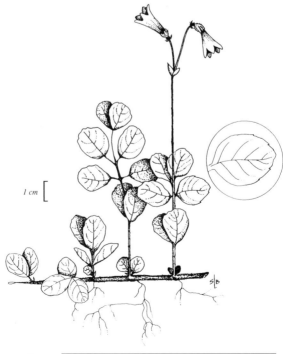

1 cm

Brenda Chambers

General: Deciduous, spreading, low to medium-sized shrub, up to 2 m tall; numerous branches; branchlets and buds rusty-brown, covered with many small, brown spots and fine, white, star-shaped hairs; bark of older branches dark-brownish to greyish, minutely hairy.

Leaves: Opposite, simple, thick, oval to egg-shaped, blunt-tipped, tapered to rounded at base, up to 5 cm long and 3 cm wide; upper surface green to greyish-green with a few star-shaped hairs; underside densely covered with silvery, star-shaped hairs and numerous brown dots or scales; margins toothless; stalks short, grooved, about 1 cm long.

Flowers: Greenish-yellow, 3–5 mm wide, male or female; in dense, elongated clusters from leaf nodes; open before leaves in April–May.

Fruits: Red to yellowish cherry-like drupe, oval, 3–6 mm long, juicy, with a smooth pit, bitter; June–July.

Habitat: Dry to fresh, sandy to loamy, often calcareous upland pine, pine-oak and intolerant hardwood mixedwood stands.

Notes: The juicy berry pulp feels soapy, and the berries have a very bitter taste. Aboriginal peoples used these berries to make 'Indian ice-cream,' which was considered a delicacy. They also used parts of the buffalo berry plant to treat disorders ranging from indigestion to acne. See caution in Introduction.

Food Use by Wildlife: fruit. • **Mammals:** least chipmunk, black bear. • **Birds:** grey catbird, brown thrasher.

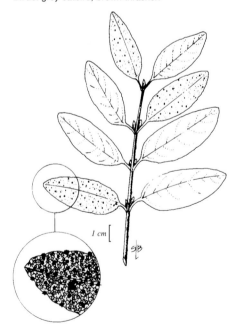

1 cm

DOGWOOD FAMILY (CORNACEAE)

General: Deciduous shrub, erect, ascending or loosely spreading, 1–3 m tall; branchlets greenish and finely hairy; young branchlets bright red to purplish, hairless, with white cork-like oval markings; buds hairy.

Leaves: Opposite, simple, egg- to lance-shaped, tapering to sharp point at tip, narrowed or rounded at base, 5–15 cm long, 2.5–9 cm wide; upper surface dark green, hairless to slightly hairy; underside paler with fine, soft hairs; margins toothless; 5–7 pairs of prominent veins curve toward leaf tip; stalks 0.6–2.5 cm long.

Flowers: Creamy-white, small; in slightly rounded or flat-topped terminal clusters about 4 cm across; June–July.

Fruits: Round, white or bluish-tinged, berry-like drupes, about 5 mm wide, juicy, with an oval to rounded pit; in clusters; August–September.

Habitat: Wet organic hardwood and conifer swamps; moist upland sites.

Karen Legasy

Notes: The berries have historically been considered edible. See caution in Introduction. • Aboriginal peoples smoked the bark in pipes. The bark was also used to make a red dye, and the branches were used to make baskets.

Food Use of Dogwoods (*Cornus* spp.) by Wildlife: bud, leaf, fruit twig.
• **Mammals:** eastern cottontail, eastern chipmunk, grey squirrel, white-footed mouse, beaver, black bear, white-tailed deer, moose.
• **Birds:** ring-necked pheasant, spruce and ruffed grouse, wild turkey, yellow-bellied sapsucker, red-bellied, downy, hairy and pileated woodpeckers, great-crested flycatcher, eastern kingbird, tree swallow, American crow, Swainson's, hermit and wood thrushes, American catbird, brown thrasher, cedar waxwing, European starling, red-eyed vireo, pine warbler, northern cardinal, white-throated sparrow, pine and evening grosbeaks, purple finch.

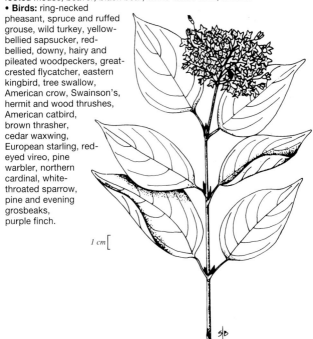

1 cm

Brenda Chambers

General: Coarse erect shrub, often tree-like with one to a few dominant stems, up to 3 m tall; branchlets pink- to yellow-green dotted or streaked with purple or reddish-brown, warty; older stems purplish, with large white pith.

Leaves: Opposite, simple, broadly oval to nearly round, abruptly pointed at tip, broadly rounded at base, 7–15 cm long, 5–12 cm wide, pale to dark green, rough and wrinkled above, greyish and hairy below; 5–8 pairs of veins curve towards the leaf tip; margins untoothed; stalks 12–18 mm long.

Flowers: White to creamy-white, small; in dense flat-topped clusters; June.

Fruits: Round, pale blue to greenish-white, berry-like drupes, 6 mm in diameter, contain a 2-seeded stone; August.

Habitat: Dry to fresh, sandy to loamy, often calcareous upland pine, pine-oak and intolerant hardwood mixedwood stands; also in moist clayey soils in the northeast part of this region (the Little Clay Belt area).

Notes: This species is similar to red osier dogwood (p. 91), which has reddish to purplish stems, narrower leaves with 5–7 pairs of veins and white to bluish fruit. Historically, the bark of round-leaved dogwood has been used for medicinal purposes. See caution in Introduction.

Food Use by Wildlife: See notes under red osier dogwood (p. 91).

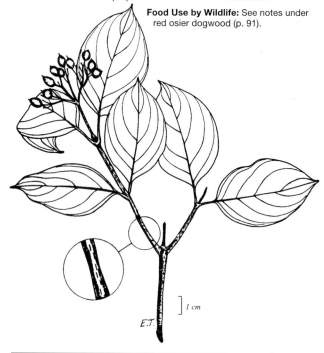

E.T.

] 1 cm

General: Deciduous trailing or climbing and twining vine, often climbing to about 3 m tall, woody; branchlets green to purplish with long, glandular-tipped hairs and purplish-brown spots; older branches brown to grey with shredding bark.

Leaves: Opposite, simple, widely oval to egg-shaped, blunt to pointed at tip, rounded to tapered at base, 5–13 cm long, 2.5–9 cm wide, veiny; upper surface deep green with flattened hairs; underside downy hairy; margins toothless but fringed with shiny, silky hairs; uppermost 1–2 pairs of leaves fused at base to form broad, saucer-like discs around stem; stalks mostly short.

Karen Legasy

Flowers: Orange to yellow, turning reddish, narrowly tubular with 5 slightly spreading lobes, 2–2.5 cm long; in whorled terminal clusters above saucer-like fused leaves; late June–early August.

Fruits: Orangish-red berries, many-seeded; in stalked clusters from centre of terminal, saucer-like fused leaves; ripen August–September.

Habitat: Moist to dry clayey to sandy sites; in hardwood swamps, and upland intolerant hardwood mixedwoods, pine and pine-oak stands.

Notes: You can recognize hairy honeysuckle in the field by its opposite, stalkless or short-stalked, hairy leaves with fringed margins, by its 1–2 pairs of saucer-like fused leaves at the top of its stems, and by its trailing, woody habit.
• The species name *hirsuta* means 'stiffly hairy.'

Food Use by Wildlife: See notes under fly honeysuckle (p. 94).

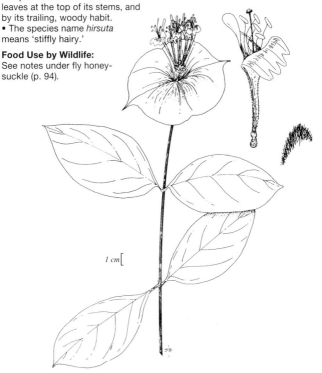

1 cm

FLY HONEYSUCKLE • *Lonicera canadensis*
CHÈVREFEUILLE DU CANADA

HONEYSUCKLE FAMILY (CAPRIFOLIACEAE)

Brenda Chambers

General: Deciduous shrub, erect to straggling or loosely branched, up to 1.5 m tall; branchlets green to purplish, hairless; older branches grey to brownish with bark shredding in thread-like pieces.

Leaves: Opposite, simple, egg-shaped to oblong, blunt to pointed at tip, rounded to slightly heart-shaped at base, 3–9 cm long, 1.5–3 cm wide, thin; upper surface bright green and hairless; underside paler; margins toothless, fringed with hairs; stalks short, fringed with hairs.

Flowers: Pale yellow to yellowish-green, funnel-shaped with 5 short lobes, 12–18 mm long; in pairs, on long, slender stalks from leaf axils; May–June.

Fruits: Red, egg-shaped berries about 6 mm wide, with 3–4 seeds; spreading in long-stalked pairs; ripen late June–July.

Habitat: All moisture regimes and soil textures; in many stand types, including upland tolerant hardwood, intolerant hardwood mixedwoods and pine stands, and hardwood and conifer swamps.

Notes: To help identify fly honeysuckle in the field, look for its mainly hairless mature leaves with their fringed stalks and margins. • Aboriginal peoples used the twigs and bark in a remedy for urinary disorders. See caution in Introduction.

Food Use of Honeysuckles (*Lonicera* spp.) by Wildlife: fruit, twig.
• **Mammals:** snowshoe hare, white-tailed deer, moose. • **Birds:** ruffed grouse, white-throated sparrow, dark-eyed junco, purple finch, American goldfinch.

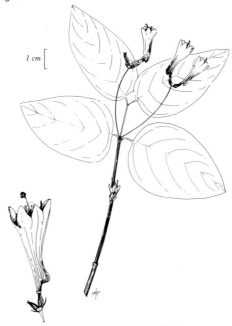

1 cm

HONEYSUCKLE FAMILY (CAPRIFOLIACEAE)

General: Deciduous shrub, usually less than 50 cm tall, erect or ascending with stiff branches; branchlets purplish-red with long, soft, scattered hairs; bark on older branches reddish-brown to grey and peeling to expose a reddish-brown inner layer.

Leaves: Opposite, simple, oval-oblong, widest above middle, blunt or rounded at tip, rounded or tapered at base, 2.5–6 cm long, 1–3 cm wide, firm, crowded near branch ends; upper surface dark green with flat hairs; underside paler and hairy, especially along veins; margins toothless, white hairy, often rolled under; stalks less than 3 mm long.

Derek Johnson

Flowers: Yellowish, funnel-shaped with 5 lobes, about 12 mm long; in pairs on short, hairy stalks; May–June.

Fruits: Blue, round berry, produced from fused ovaries of 2 flowers, short-stalked; July–August.

Habitat: Wet organic conifer swamps and moist clayey to silty intolerant hardwood mixedwood stands; thickets, wet shorelines.

Notes: Also known as *Lonicera caerulea*. • The berries have historically been prepared much like blueberries–eaten raw, cooked or in jellies. See caution in Introduction.

Food Use by Wildlife: See notes under fly honeysuckle (p. 94).

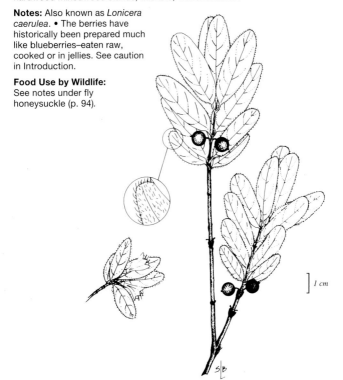

] *1 cm*

SHRUBS

HONEYSUCKLE • *Lonicera dioica*
CHÈVREFEUILLE DIOIQUE

HONEYSUCKLE FAMILY (CAPRIFOLIACEAE)

General: Deciduous, semi-erect, low-growing shrub or vine up to 3 m tall; branches often twining around adjacent plants; branchlets smooth, green or purplish; older stems brown or grey; bark shredding.

Leaves: Opposite, simple, dark green and hairless above, whitened and hairless or hairy beneath; bases of upper 1–4 pairs usually fused around stem and uppermost pair forms a saucer-like disc; lower leaves blunt tipped; toothless margins; stalks short to absent.

Brenda Chambers

Flowers: Greenish-yellow to orange, fading purplish, tubular to funnel-shaped, 12–18 mm long with 2 spreading lips; 2–6 in short-stalked clusters from centre of terminal leaf discs; May–July.

Fruits: Orange-red berries with several seeds; in terminal clusters at centre of terminal leaf discs; July–August.

Habitat: Rocky banks, dry woods and thickets.

Notes: Hairy honeysuckle (p. 93) also grows as a vine and has a pair of fused terminal leaves, but its leaves are hairy. • Historically, the leaves, flowers, roots and berries of glaucous honeysuckle were used in medicine. See caution in Introduction. • Honeysuckles are useful for ornamental plantings.

Food Use by Wildlife: See notes under fly honey-suckle (p. 94).

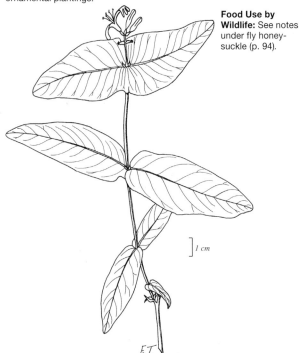

1 cm

General: Deciduous shrub, less than 1 m tall, upright and spreading; branchlets green to reddish and often have tiny hairs in 2 fine lines; older branches brownish to grey.

Leaves: Opposite, simple, egg-shaped to oblong, tapering to a long and sometimes curved point at tip, usually rounded at base, 5–13 cm long, 1.5–6 cm wide; upper surface dark green; underside paler; margins sharply toothed, usually fringed with short hairs; stalks 3–12 mm long.

OMNR

Flowers: Yellow, orange to brownish-red with age, funnel-shaped with 5 lobes, about 2 cm long; in clusters of 2–7 at branch tips or in leaf axils; June to early July.

Fruits: Slender, oblong capsules with long beaks tipped with persistent, thread-like sepals, brown, contain numerous seeds; July–September.

Habitat: Fresh to dry, occasionally moist sites; all soil textures; in pine and intolerant hardwood mixedwood stands; occasional in hardwood swamps.

Notes: The sharply toothed margins of its leaves distinguish bush honeysuckle from members of the closely related honeysuckles (*Lonicera* spp., pp. 93–96). Honeysuckles have leaves with toothless margins. • Aboriginal peoples used the roots in remedies for senility, gonorrhea and urinary disorders. See caution in Introduction.

] *1 cm*

SNOWBERRY • *Symphoricarpos albus*
SYMPHORINE BLANCHE

HONEYSUCKLE FAMILY (CAPRIFOLIACEAE)

Julie Hrapko

General: Deciduous shrub, small, erect to spreading, usually less than 1 m tall, forms low thickets; branchlets light brown when young, purplish to grey and darker with age, hairless or minutely hairy, slender; bark becomes shredding to fibrous; pith brown, small, hollow in centre.

Leaves: Opposite, simple, oval to egg-shaped, blunt to rounded or sometimes minutely pointed at tip, rounded or tapered base, 2–3 cm long, 1–3 cm wide, thin; upper surface dark green, hairless; underside paler, hairless to minutely downy; margins minutely hairy, toothless; leaves on young shoots may be larger with wavy-toothed or lobed margins; stalks short.

Flowers: Pink to white, bell-shaped with short lobes, hairy inside, about 6 mm long; in short clusters of 1–5 at branch tips and in axils of upper leaves; August–October.

Fruits: Round, berry-like drupes, 6–12 mm wide, waxy, white, with tiny dark spot at free end, spongy, contains 2 seeds; solitary or a few together; persist over winter.

Habitat: Sandy or rocky open areas; thickets and open forests.

Notes: These berries are poisonous. See caution in Introduction. You can recognize snowberry by its hollow pith, the dark spot at the free end of berries and its opposite leaves. • Snowberry was probably given its common name for its white berries, which persist over winter.

Food Use by Wildlife: bud, leaf, fruit, bark, twig. • **Mammals:** snowshoe hare. • **Birds:** ring-necked pheasant, ruffed grouse, hermit thrush, pine and evening grosbeaks.

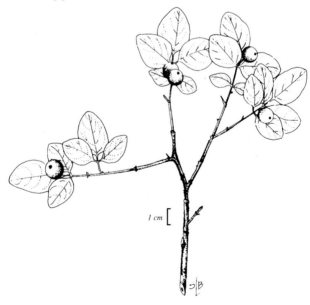

1 cm

HONEYSUCKLE FAMILY (CAPRIFOLIACEAE)

General: Erect, stiffly-branched shrub, up to 5 m tall, spreading at the top; older twigs purplish and ridged; develops suckers from rhizomes.

Leaves: Opposite, simple, oval or oblong, abruptly blunt-tipped, narrowed or rounded at base, 4.5–9.5 cm long, 2.5–5 cm wide, pale green beneath; main vein pale above and brown-hairy beneath; margins toothless or wavy-toothed; stalks grooved, 0.5–2 cm long.

Flowers: Creamy white, ill-scented, small; in short-stalked, flat-topped terminal clusters 5-10 cm across; June.

Fruits: Nearly round or elliptic drupes, 6–9 mm long, whitish-yellow, turn pinkish, then bright blue, and finally blue-black with a waxy powdery coating; July–September.

Brenda Chambers, top and bottom

Habitat: Moist to dry, clayey to sandy upland pine and intolerant hardwood mixedwood stands, and wet organic conifer swamps.

Notes: Historically, the fruit and leaves of wild raisin were used for food, and the fruit and bark were used in medicine. The berries are still eaten. See caution in Introduction • This shrub is planted for ornamental purposes.

Food Use of *Viburnum* spp. by Wildlife: bud, leaf, fruit, twig.
• **Mammals:** snowshoe hare, eastern chipmunk, red squirrel, beaver, white-tailed deer, moose. • **Birds:** ring-necked pheasant, ruffed grouse, wild turkey, pileated woodpecker, great-crested flycatcher, Swainson's and hermit thrushes, American robin, brown thrasher, cedar waxwing, European starling, northern cardinal.

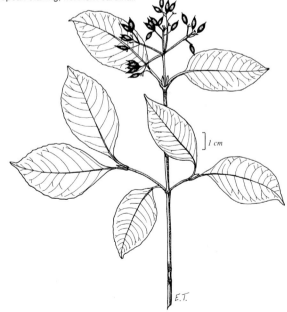

] *1 cm*

E.T.

General: Large shrub, up to 6 m tall, sometimes tree-like with spreading top, forms thickets occasionally; branchlets slender and brownish, later become purplish-brown to grey and sometimes ridged; winter buds grey, long and narrow.

Leaves: Opposite, simple; blades elliptic to lance- or egg-shaped, abruptly and sharply pointed at tip, rounded or tapered at base, hairless; margins finely toothed; stalks 0.5–2.5 cm long, grooved, with winged margins.

Flowers: Creamy-white, sweet-scented; in terminal clusters 5–10 cm across; May–June.

Brenda Chambers

Fruits: Nearly round to ellipsoid drupes, blue-black with a waxy powdery coating, up to 12 mm long, contain 1 large flat stone; August–October.

Habitat: Wet to moist forest edges, streambanks and roadsides; occasional in the southern part of this region.

Notes: Historically, nannyberries were used for food, and both fruit and bark were used in medicine. See caution in Introduction. • This shrub is planted for ornamental purposes.

Food Use by Wildlife: See notes under northern wild raisin (p. 99).

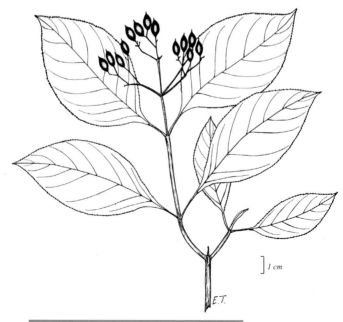

] 1 cm

E.T.

HONEYSUCKLE FAMILY (CAPRIFOLIACEAE)

General: Small, erect or spreading shrub, up to 1.5 m tall; branchlets light brown or grey, hairless or slightly hairy.

Leaves: Opposite, simple, egg-shaped or round to broadly lance-shaped, pointed at tip, rounded or heart-shaped at base, 3.5–9 cm long, 1.5–5.5 cm wide, hairless; margins coarsely toothed; stalks grooved, less than 1 cm long or absent, with a pair of bristle-like bracts (stipules) at the base which are usually longer than the stalk.

Flowers: Creamy white, small; in terminal stalked clusters 2.5–7.5 cm across; May–June.

Fruits: Elliptical dark purple-black drupes, 6-9 mm long, contain 1 flat stone; in open clusters; August–September.

Habitat: Dry to fresh, sandy to loamy, often calcareous upland pine, pine-oak and intolerant hardwood mixedwood stands; uncommon in the southern part of this region.

Brenda Chambers

Notes: Historically, downy arrow-wood bark was used in medicine. See caution in Introduction. • This shrub is planted for ornamental purposes.

Food Use by Wildlife: See notes under northern wild raisin (p. 99).

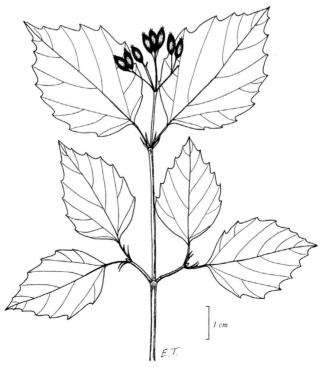

1 cm

E.T.

Brenda Chambers,
top and bottom

General: Low, straggly shrub, up to 2 m tall, with prostrate branches rooting at nodes and tips; branchlets and buds rusty-hairy; older stems purplish-brown and sometimes slightly ridged; develops suckers from rhizomes.

Leaves: Opposite, simple, broadly oval to almost round, abruptly pointed at tip, round or heart-shaped at base, 10–20 cm long, 7–18 cm wide, covered when young with light brown hair which persists on veins beneath; margins finely toothed; stalks hairy, 1–6 cm long, bearing a pair of bristle-like bracts (stipules) at base.

Flowers: White, of 2 kinds; outer flowers large, showy, sterile; inner flowers numerous, less conspicuous, fertile; in short-stalked, saucer-shaped clusters up to 13 cm across; May–June.

Fruits: Round to oval berry-like drupes, about 9 mm long, green, turning crimson to purple-black as they mature; July–September.

Habitat: Moist to fresh, silty to sandy upland tolerant hardwood stands, in association with sugar maple, yellow birch and eastern hemlock; in the southern half of this region.

Notes: Also known as *Viburnum lantanoides.* • Historically, hobblebush bark and leaves were used in medicine. See caution in Introduction. • This shrub is planted for ornamental purposes. • The name, *alnifolium*, means 'alder-leaved.'

Food Use by Wildlife: See notes under northern wild raisin (p. 99).

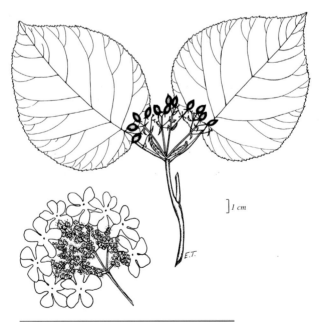

]1 cm

E.T.

HONEYSUCKLE FAMILY (CAPRIFOLIACEAE)

General: Deciduous shrub, upright and coarse, 1–4 m tall; branchlets grey to brownish-grey.

Leaves: Opposite, simple, somewhat maple-leaf-shaped, with 3 pointed, deeply cut, spreading lobes, rounded or slightly heart-shaped at base, 5–11 cm long and wide; upper surface dark green, smooth; underside paler, hairless or with a few hairs; margins with coarse and rather wavy teeth; stalks grooved, 1–4 cm long.

Flowers: White, of 2 kinds; outer flowers showy, sterile, 15–25 mm wide; inner flowers much smaller, fertile; in showy, flat-topped clusters up to 15 cm wide, on stalks at branch tips from between uppermost pair of leaves; June–July.

Fruits: Orange to red, rounded, 8–12 mm long, juicy and cherry-like drupes with 1 flat pit; in loose clusters; August–September.

Brenda Chambers

Habitat: Moist to wet areas; low, cool sites; swamps and bogs; in thickets along shores, in forest openings and in wet ditches at forest edges.

Notes: Also called *Viburnum opulus*. • The fruit is rich in vitamin C, but it is apparently very sour when raw. High-bush cranberry has been eaten as a cooked fruit and its juices have been used to make cold beverages and jelly. See caution in Introduction. • The species name *trilobum* means 'three-lobed' and refers to the leaves. • Also see notes on mooseberry (p. 105)

Food Use by Wildlife: See notes under northern wild raisin (p. 99).

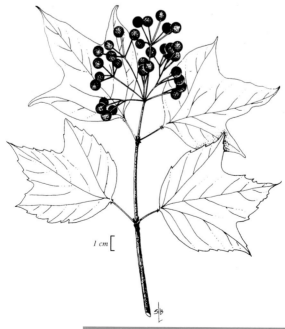

1 cm

Brenda Chambers

General: Low shrub up to 2 m tall, with slender ascending branches; branchlets green and hairless or minutely hairy; older stems reddish or purplish-grey; develops suckers from rhizomes.

Leaves: Opposite, simple, maple-leaf-like, 3-lobed, sharp-pointed at tips, round or heart-shaped at base, 6–12 cm long, sparsely hairy above, downy with small yellow and blackish resinous dots beneath; margins coarsely toothed; leaves at ends of some branches may have poorly developed lobes; stalks hairy, 1–3 cm long, with a pair of bristle-tipped bracts (stipules).

Flowers: Creamy white, small; numerous, in long-stalked terminal clusters 2.5–9 cm across; June.

Fruits: Round to oval, berry-like drupes, approximately 1 cm in diameter, green at first, then red turning dark blue or purple-black, with one hard, flattened stone; July–October.

Habitat: Dry to fresh, sandy to loamy upland oak-pine and intolerant hardwood mixedwood stands; in the southern part of this region.

Notes: Aboriginal peoples made a tea from the inner bark to relieve cramps and treat colic. See caution in Introduction. • This shrub is planted for ornamental purposes. • The species name *acerifolium* means 'maple-leaved.'

Food Use by Wildlife: See notes under northern wild raisin (p. 99).

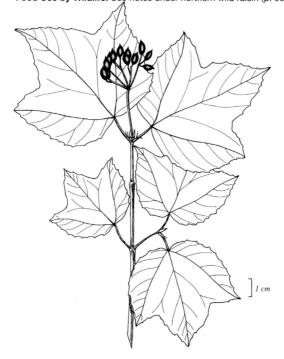

1 cm

HONEYSUCKLE FAMILY (CAPRIFOLIACEAE)

General: Deciduous shrub, erect to spreading or straggling; usually less than 2 m tall; branchlets purplish-brown or reddish, hairless and often angled or ridged; older branches grey to brownish; numerous branches.

Leaves: Opposite, simple, usually somewhat maple-leaf-shaped, with 3 shallow, sharply pointed lobes, tapered to heart-shaped at base, 4–12 cm long, 2.5–12 cm wide; upper surface dark green, hairless; underside paler, with hairs on veins; leaves near branch tips may not be lobed; margins coarsely toothed; stalks smooth, 0.5–4 cm long.

Flowers: Whitish, 4–7 mm across, with 5 spreading petals joined at base in a short tube; in small, rounded clusters about 2.5 cm wide; usually at ends of short, side branches that have 1 pair of leaves; June–July.

Fruits: Yellow to orange or red, rounded, 6–12 mm long, cherry-like drupe, juicy, with a large, flat, egg-shaped pit, strong-scented; July–August.

Habitat: Wet organic sites; moist, clayey to sandy upland sites; conifer and hardwood mixedwood stands; common in the boreal forest to the north; rare in this region.

Brenda Chambers, top;
Linda Kershaw, bottom

Notes: Mooseberry can be distinguished from high-bush cranberry (p. 103) by its leaves. High-bush cranberry leaves are deeply lobed, whereas mooseberry leaves are shallowly lobed. Also, high-bush cranberry is often more than twice the size of mooseberry and its flower clusters are much larger and showier and grow at the branch tips rather than on short side branches. • The species name *edule* means 'edible' and refers to the berries, which have been eaten raw or used to make jelly. See caution in Introduction. See notes on high-bush cranberry (p. 103).

Food Use by Wildlife: See notes under northern wild raisin (p. 99).

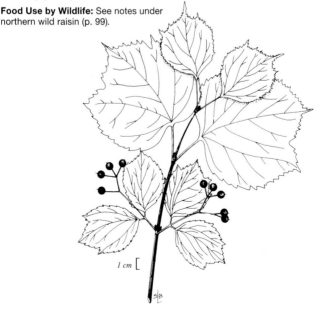

1 cm [

General: Deciduous tall shrub or small bushy tree, 3–5 m tall; trunk short, often crooked and divided into a few fairly straight, slender branches growing up from the ground; branchlets yellowish-green to purplish-grey, covered with short, grey hairs, appear dull and velvety; bark greenish-grey to greyish-brown, thin and flaky; buds grey-hairy, with a pair of visible scales.

Leaves: Opposite, simple, maple leaf with 3 prominent, pointed lobes and usually 2 smaller lobes near base, 5–12 cm long, 5–10 cm wide; upper surface dark green, hairless; underside paler with fine hairs; margins coarsely toothed; stalks reddish-tinged, as long as or longer than blade.

Flowers: Greenish-yellow, about 5 mm wide, male, female, or bisexual, with both sexes in same cluster; in dense, branched, 6–10 cm long, erect terminal clusters on long, slender stalks; appear after leaves in late May to early July.

Brenda Chambers, top and bottom

Fruits: Slender-stalked pairs of reddish-tinged winged keys (samaras) about 2 cm long (including wing); in clusters; July and August.

Habitat: All moisture regimes, all soil textures; in all stand types except dry tolerant hardwoods, and some very dry pine stands and nutrient poor boreal conifer wetlands.

Notes: Aboriginal peoples treated sore eyes with a lotion made from the pith of mountain maple twigs. See caution in Introduction.

Food Use of Maples (*Acer* spp.) by Wildlife: bud, leaf, seed, twig.
• **Mammals:** snowshoe hare, eastern chipmunk, red squirrel, beaver, white-tailed deer, moose.
• **Birds:** ruffed grouse.

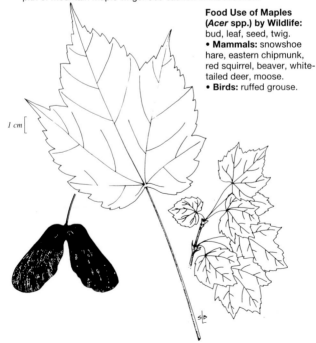

1 cm

MAPLE FAMILY (ACERACEAE)

General: Large coarse shrub or small tree up to 10 m tall; branchlets greenish to reddish-brown and hairless; bark of older stems and branches has distinct pale vertical stripes; shallow, wide-spreading root system.

Leaves: Opposite, simple, maple leaf with 3 long-tapering sharp-pointed lobes (may be unlobed on fast growing shoots or sometimes has smaller lobes near base), 10–18 cm long and nearly as wide, rounded or heart-shaped at base; upper surface hairless and brighter green than lower; margins finely and sharply toothed; stalks 2.5–8 cm long.

Emma Thurley

Flowers: Greenish-yellow, typically male or female on separate shrubs, sometimes bisexual, 3–6 mm in diameter; 5 petals; on slender stalks in loosely arching clusters 7–14 cm long; May–June.

Fruits: Stalked pairs of widely divergent winged keys (samaras), each 2.5–3 cm long (including wing); July–August.

Habitat: Dry to moist upland sites, all soil textures; in sugar maple, eastern hemlock, yellow birch and other hardwood stands; less common in northern part of this region.

Notes: Aboriginal peoples eased the swelling of limbs with a paste made from soaking the bark of striped maple. Cough and cold symptoms were treated with a tea brewed from the bark. Vomiting was induced by scraping the inner bark onto a cloth, boiling it and then drinking liquid squeezed from the cloth. See caution in Introduction. • This species is planted as an ornamental shrub. • The common name, striped maple, refers to the pale vertical stripes on the bark of this shrub.

Food Use by Wildlife: See notes under mountain maple (p. 106).

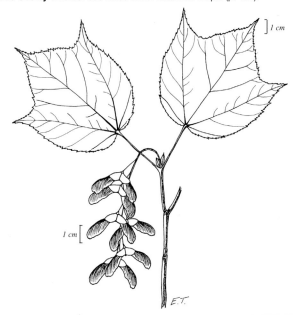

] *1 cm*

1 cm [

E.T.

SWAMP BLACK CURRANT • *Ribes lacustre*
GADELLIER LACUSTRE

GOOSEBERRY FAMILY (GROSSULARIACEAE)

Karen Legasy

General: Deciduous shrub, low, with spreading and ascending branches, up to 1 m tall; branchlets pale brown, minutely hairy, ridged and covered with slender, sharp prickles and longer thorns or spines at nodes; older branches have greyish, peeling bark and exposed, often blackish inner layer.

Leaves: Alternate, simple, 4–8 cm long and wide, 3–5 lobed; lobes pointed, deeply cut, again irregularly lobed; base heart-shaped to squared; upper surface dark green; underside paler, usually hairless, sometimes with scattered hairs; margins have rounded, coarse teeth; stalks slightly hairy.

Flowers: Greenish-purple, tiny, saucer-shaped with 5 petals; petals about 1.3 mm long; in hanging, slender, elongated clusters; May–June.

Fruits: Purple-black, round berries, bristly with gland-tipped hairs, 8 mm in diameter; ripen late July–August.

Habitat: Moist, medium loamy to fine loamy upland sites, occasional in sandy to coarse loamy fresh sites, wet organic sites; in intolerant hardwood mixedwoods and hardwood and conifer swamps.

Notes: Aboriginal peoples ate the berries fresh or cooked, but they are not very palatable. See caution in Introduction. • Be careful when touching this shrub; the sharp prickles can give a painful stab.

Food Use of Currents and Gooseberries (*Ribes* spp.) by Wildlife: fruit. **Mammals:** least chipmunk, raccoon, American marten, fisher, moose. • **Birds:** American robin, grey catbird, brown thrasher, cedar waxwing.

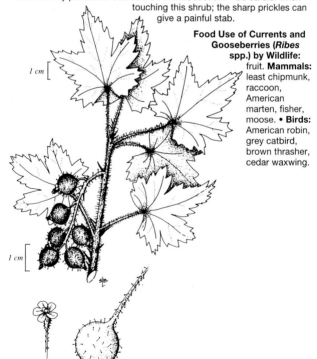

1 cm

1 cm

GOOSEBERRY FAMILY (GROSSULARIACEAE)

General: Deciduous shrub, low, spreading, straggling or reclining on ground with stems often rooting, up to about 1 m tall; stems lacking prickles; branchlets grey to brownish, ridged and minutely hairy; bark peels off older branches to reveal an often reddish-purple to blackish inner layer.

Leaves: Alternate, simple, maple-leaf-like, 4–10 cm long, 5–10 cm wide, 3–5-lobed; lobes wide, cut less than halfway to leaf base, pointed to rounded at tips; leaf base shallowly heart-shaped to squared; 2 sides of leaf are almost parallel; upper surface dark green, almost hairless; underside paler, usually hairy; margins have rounded to abruptly pointed teeth; stalks sparsely hairy, 2.5–6 cm long.

Karen Legasy

Flowers: Greenish-purple, saucer-shaped, less than 6 mm wide; 5 petals; in elongated, hanging clusters from scaly buds in leaf axils; stalks slender, with scattered, gland-tipped hairs; June.

Fruits: Bright-red berries, smooth, 6–9 mm in diameter; in hanging clusters; ripens July–August.

Habitat: Wet organic sites, moist to fresh, clayey to sandy upland sites; in hardwood and conifer swamps, intolerant hardwood mixedwoods and tolerant hardwood stands.

Notes: Historically, the berries have been used to make jams, jellies and pies, but they are apparently very sour when raw. See caution in Introduction. • In addition to the different berry colours, you can distinguish red currant from northern wild black currant (*Ribes hudsonianum*) by its stems and leaves. Red currant has spreading to reclining stems that often root and its maple-like leaves have no yellow resin dots on the underside, whereas northern wild black currant is ascending to erect and its leaves have yellow resin dots on the underside.

Food Use by Wildlife: See notes under swamp black currant (p. 108).

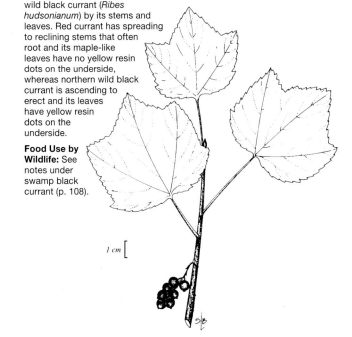

1 cm

GOOSEBERRY FAMILY (GROSSULARIACEAE)

Karen Legasy

General: Deciduous shrub, low with ascending to erect branches, up to 90 cm tall; branchlets greyish, often with scattered prickles and 1–3 slender, sharp, 3–8 mm long spines where leafstalks join branches; greyish outer bark on older stems eventually peels away to reveal a smooth, reddish-brown to blackish layer.

Leaves: Alternate, simple, 2.5–6 cm long and wide, 3–5-lobed; lobes with pointed tips; base wedge- to heart-shaped; upper surface dark green, hairless to slightly hairy; underside paler and hairy; margins hairy and coarsely toothed; stalks 1–3 cm long.

Flowers: Greenish-yellow to purplish, narrowly bell-shaped, 6–9 mm long; 5 petals; loose clusters of 2-3 from leaf axils; stalks short, usually hairless; bracts at base of flower tiny, leaf-like, fringed with long hairs; June.

Fruits: Smooth, round, bluish-black berries, 8–12 mm in diameter; ripen July–August.

Habitat: Moist to dry, clayey to sandy, upland intolerant hardwood mixedwoods and lowland hardwood swamps.

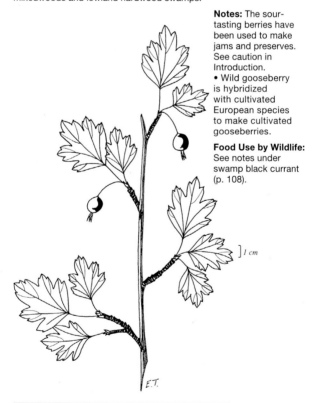

Notes: The sour-tasting berries have been used to make jams and preserves. See caution in Introduction.
• Wild gooseberry is hybridized with cultivated European species to make cultivated gooseberries.

Food Use by Wildlife: See notes under swamp black currant (p. 108).

] *1 cm*

E.T.

GOOSEBERRY FAMILY (GROSSULARIACEAE)

General: Small shrub, 0.9–1.2 m tall; branches upright, greyish and fine hairy when young; outer bark smooth, reddish to black with age, peeling off to expose reddish inner layer.

Leaves: Alternate, simple, rounded, 3–7.5 cm long, 3.5–10 cm wide, with 3–5 pointed lobes, heart-shaped or blunt at base; upper surface dark green, sparsely hairy to hairless; underside paler, with fine hairs; golden resinous dots on both surfaces; margins coarsely double toothed; stalks 3–5 cm long, resin-dotted, hairy.

Bill Crins

Flowers: Creamy-white to yellowish, bell-shaped, about 9 mm long, with a persistent bract longer than the stalk; 5 petals, 2.5–3 mm long; in many-flowered hanging clusters 2.5–7.5 cm long, from leaf axils; May–June.

Fruits: Round, black berries, 6–9 mm in diameter, in hanging clusters; July–August.

Habitat: Damp soil along streams, on wooded slopes, in low wet woods, open meadows and on rocky ground.

Notes: The cultivated black currant (*Ribes nigrum*) is related to this species, and is distinguished by its greenish-white to purplish flowers and shorter flower bracts. It may be found along roadsides as an escape from cultivation. • The fruit of wild black currant is eaten fresh and in preserves and beverages. Aboriginal peoples ate the currants raw and also dried them for winter use, often cooking them with sweet corn. Historically, all plant parts were used in medicines. See caution in Introduction.

Food Use by Wildlife: See notes under swamp black currant (p. 108).

] *1 cm*

E.T.

GOOSEBERRY FAMILY (GROSSULARIACEAE)

General: Deciduous shrub with low and trailing to ascending stems, less than 1 m tall, with a distinctive, skunk-like odour when crushed, lacks prickles; branchlets brown to purplish-grey, minutely hairy to hairless and somewhat ridged or angled; older branches often blackish and smooth as outer bark peels off.

Leaves: Alternate, simple, 4–8 cm wide, wider than long, maple-leaf-like, 3–5-lobed (sometimes 7); lobes egg-shaped, pointed; base deeply heart-shaped; upper surface dark green, hairless; underside paler with fine hairs; margins doubly toothed; leafstalks finely hairy, 3–5.5 cm long.

Flowers: Yellow-green to purplish saucer-shaped, less than 6 mm wide; 5 petals; in loose, elongated, upright clusters 2.5–6 cm long; flowerstalks and tiny leaf-like bracts at base of flower have gland-tipped hairs; June.

Brenda Chambers

Fruits: Red berries bristly with glandular hairs, about 6 mm in diameter; July–August.

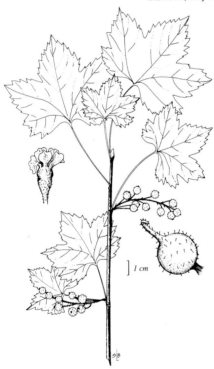

1 cm

Habitat: Dry to moist, sandy to clayey upland sites and wet organic sites; in upland tolerant hardwood and intolerant hardwood mixedwood stands and lowland hardwood and conifer swamps.

Notes: You can distinguish skunk currant in the field by crushing some of its leaves and smelling its distinctive, skunk-like odour. The berries have a disagreeable taste.
• Aboriginal peoples used the roots in a remedy for back pain. See caution in Introduction.

Food Use by Wildlife: See notes under swamp black currant (p. 108).

General: Low shrub, less than 1 m tall; branches spreading or upright, brownish-grey, fine hairy, with scattered weak prickles or gland-tipped hairs and 1–3 sharp spines at nodes; spines on older stems firm, 1 cm long; outer bark peels revealing inner brownish-purple to black bark.

Emma Thurley

Leaves: Alternate, simple, 3–7.5 cm long and wide, 3–5-lobed, blunt or rounded at tip, shallow to deeply heart-shaped at base; upper surface dark green, soft-hairy; underside glandular-hairy along veins; margins coarsely toothed; stalks 2.5–4 cm long, finely hairy, with scattered gland-tipped hairs.

Flowers: Greenish-yellow, bell-shaped, 6–9 mm long, 2–3 together on slender, glandular-hairy stalks from short branches of older growth; May–June.

Fruits: Large, edible berries, reddish-purple to black when ripe, 8–12 mm across, many-seeded, covered with stiff pale brown prickles; July–August.

Habitat: Dry to moist, sandy to loamy, tolerant hardwood stands.

Notes: Aboriginal peoples removed the bristles of gooseberries by placing the gooseberries in baskets over hot coals until the bristles were singed off. The berries were then made into winter preserves. Historically, the bark was used in medicine. See caution in Introduction.

Food Use by Wildlife: See notes under swamp black currant (p. 108).

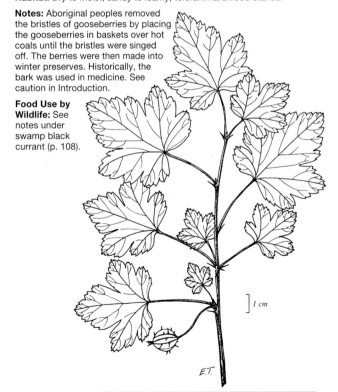

] *1 cm*

E.T.

Emma Thurley

General: Perennial shrub, straggling, 1–2 m tall, widely branched; stems lacking prickles but with glandular, reddish, bristly hairs; bark freely shedding.

Leaves: Alternate, simple, long-stalked, glandular-hairy, maple-like, 10–20 cm wide and long, 3–5-lobed to 1/2 or a 1/3 of length, generally soft hairy on both surfaces; margins sharply and irregularly toothed.

Flowers: Rose-purple, bisexual, 3.5–5 cm in diameter; 5 petals; several in a loose cluster at branch ends; June–July.

Fruits: Dome-shaped, somewhat flattened, dull red raspberries, 1 cm in diameter; dry, insipid; July–August.

Habitat: In moist, shaded forest edges and roadsides.

Notes: Thimbleberry (*Rubus parviflorus*) is somewhat similar to purple flowering raspberry in growth habit and foliage, but it is easily distinguished by its white (rather than purplish) flowers, its typically 3-lobed (rather than 3–5-lobed) leaves and its more palatable fruit. • Historically, the *coureurs de bois* used purple flowering raspberry leaves in their leather shoes to protect their feet.

Food Use by Wildlife: See notes under wild red raspberry (p. 117).

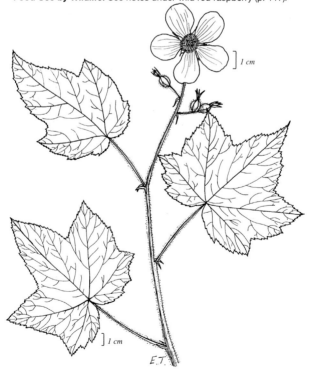

] *1 cm*

] *1 cm*

E.T.

ROSE FAMILY (ROSACEAE)

General: Deciduous shrub, low, trailing or creeping, slender; stems soft-hairy with erect, leafy flowering branches; long, trailing shoots taper to slender, whip-like rooting tips.

Leaves: Alternate, compound; leaflets 3 (rarely 5), egg-shaped to vaguely diamond-shaped, taper to a point at tip, 2–10 cm long, 1–4.5 cm wide, usually hairless; central leaflet tapers to point at base; lateral leaflets rounded to tapered at base; margins sharply toothed.

Karen Legasy

Flowers: White to pale pink, about 1 cm wide; 5 petals; in loose terminal clusters or 1–2 flowers in leaf axils; May–June.

Fruits: Dark-red to purple raspberries, juicy, not easily picked or separated from their cores (receptacles); July–September.

Habitat: Wet organic sites to moist clayey to sandy upland sites, occasional on fresh to dry upland sites; in hardwood and conifer swamps, intolerant hardwood mixedwoods and tolerant hardwood stands.

Notes: Dwarf raspberry is quite easy to distinguish from wild red raspberry (p. 117). Dwarf raspberry is a low, trailing plant with whip-like runners, whereas wild red raspberry is a bristly, erect shrub up to 2 m tall. Dwarf raspberry usually has 3 leaflets per compound leaf while wild red raspberry has 3–7 (often 5). Also, wild red raspberries are easier to pick than dwarf raspberries. See notes on wild red raspberry (p. 117).

Food Use by Wildlife: See notes under wild red raspberry (p. 117).

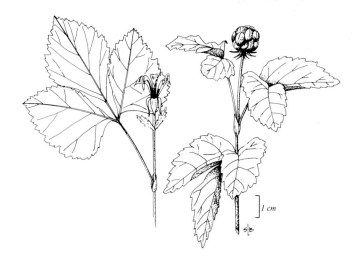

1 cm

ROSE FAMILY (ROSACEAE)

General: Perennial trailing shrub, up to 30 cm tall; whip-like, branches 2–4 m long root at tips; stems brown to purplish red with scattered hooked prickles.

Leaves: Alternate, compound with 3–5 leaflets, 6–18 cm wide and long, green, thin, mostly hairless; terminal leaflet egg-shaped to nearly elliptic, tapered to point, often with small lobes above middle, rounded at base; lateral leaflets asymmetrical and/or deeply lobed; leaves of flowering canes usually smaller, with 3 leaflets; margins sharply toothed; stalks with fine hairs and linear, persistent leaf-like bracts (stipules).

Flowers: White, 10–15 mm long; 5 petals; solitary or in terminal clusters of 2–5 on nearly erect stems; June.

Brenda Chambers

Fruits: Rounded red clusters of large juicy drupelets firmly attached to their cores (receptacles), raspberry-like; July–August.

Habitat: Dry, open, sandy or rocky woods and thickets.

Notes: The species name *flagellaris* means 'whiplike,' referring to trailing stems.

Food Use by Wildlife: See notes under wild red raspberry (p. 117).

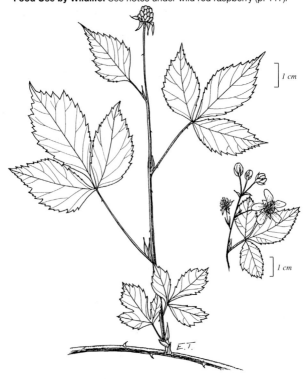

1 cm

1 cm

ROSE FAMILY (ROSACEAE)

General: Deciduous shrub; stems erect, arching, spreading, woody and prickly or bristly, up to 2 m tall; young branches sparsely to densely bristly and usually with slender, gland-tipped hairs; older stems brownish, smoother, shedding their papery bark.

Leaves: Alternate, compound with 3–5 (rarely 7) leaflets; leaflets egg-shaped to oblong with sharp, pointed at tip, rounded to tapered at base, 5–10 cm long; upper surface dark-green, hairless to slightly hairy; underside greyish- or whitish-hairy; margins toothed; when there are 5 leaflets, the middle 2 are closer to the top leaflet than to the bottom 2; stalks bristly hairy.

Karen Legasy

Flowers: White to greenish-white; 5 petals; in small clusters of 2–5 at branch tips; June–July.

Fruits: Red or amber raspberries, usually drop from their cores (receptacles) intact; July–August.

Habitat: Wet organic sites, moist to dry, clayey to sandy upland sites; in hardwood and conifer swamps, intolerant hardwood mixedwood, and tolerant hardwood stands; abundant on disturbed, open ground.

Notes: The berries were an important food source for aboriginal peoples, who ate them fresh, dried them for winter use or made them into jelly. These peoples made an eyewash with the bark of roots. See caution in Introduction.

Food Use of Raspberries (*Rubus* spp.) by Wildlife: bud, leaf, fruit.
• **Mammals:** opossum, eastern cottontail, snowshoe hare, eastern and least chipmunks, grey squirrel, deer mouse, meadow and woodland jumping mice, red fox, black bear, raccoon, American marten, fisher, white-tailed deer, moose. • **Birds:** ring-necked pheasant, ruffed grouse, woodcock, wild turkey, northern flicker, eastern phoebe, great-crested flycatcher, eastern kingbird, blue jay, American crow, veery, wood thrush, American robin, grey catbird, brown thrasher, cedar waxwing, red-eyed vireo, northern cardinal, rose-breasted, evening and pine grosbeaks, indigo bunting, rufous-sided towhee, white-throated sparrow, rusty blackbird, common grackle, orchard and northern orioles.

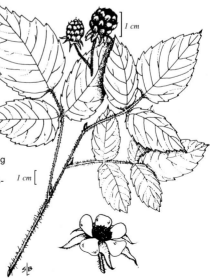

1 cm

1 cm

Brenda Chambers

General: Perennial shrub, about 2 m tall, with erect or high-arching stem (canes); older canes angled or ridged, reddish-purple to brown, sometimes with scattered weak prickles.

Leaves: Alternate, palmately compound with 5 (on first year canes) or 1–3 (on second year cane) leaflets, thin, hairless; leaflets broadly egg- to lance-shaped, long-pointed; leaves on first year canes have a long-stalked terminal leaflet 10–20 cm, a shorter-stalked central pair of leaflets and a stalkless basal pair; margins sharply toothed.

Flowers: White, 1–2 cm long, 5–8 mm wide; 5 petals; stalks glandless, 2–4 cm long, with 1–2 cm long leaf-like toothed bracts (stipules) at base; up to 25 in elongate leafy clusters on prominent stalks; June–July.

Fruits: Round to thimble-shaped blackberries, 12 mm long, dryish or juicy; small drupelets are hard to separate from their receptacles; July–September.

Habitat: Open rocky areas in woods; along roadsides, streambanks and lakeshores.

Notes: Another common name is 'Canada blackberry.'

Food Use by Wildlife: See notes under wild red raspberry (p. 117).

] *1 cm*

E.T.

ROSE FAMILY (ROSACEAE)

General: Perennial shrub, up to 2 m tall; stems (canes) mostly erect, high-arching, with gland-tipped hairs and straight stiff bristles; flowering canes (usually second year stems) brown to purplish-red, ridged, with broad-based prickles and some gland-tipped hairs.

Leaves: Alternate, palmately compound with 5 (on first year canes) or 1–3 (on second year cane) leaflets; leaves on first year stems have a terminal leaflet 5–20 cm long, 3–10 cm wide, oval to broadly lance-shaped, sharp-pointed at tip, rounded or heart-shaped at base, stalked (as are lateral pair) and a basal pair of unstalked leaflets; all leaflets hairy; margins have sharp double teeth; stalks have gland-tipped hairs and scattered prickles.

Emma Thurley

Flowers: White; 5 petals, 1–2 cm long, 5–8 mm wide; up to 20 or more in a cluster, on a long glandular-hairy stalk, with scattered prickles; June–July.

Fruits: Long, thimble-shaped blackberries, up to 2.5 cm long, a cluster of fleshy single-seeded drupes on fleshy receptacle; July–September.

Habitat: Dry to moist forest openings, roadsides.

Notes: Natural hybrids occur (mainly with bristly blackberry, *Rubus setosus*) and identification is difficult. • Historically, the fruit was used for food, drink, purple dye, and in medicine, green twigs for black dye, branches in drinks, and bark and roots in medicine. See caution in Introduction.

Food Use by Wildlife:
See notes under wild red raspberry (p. 117).

] *1 cm*

] *1 cm*

E.T.

ROSE FAMILY (ROSACEAE)

Brenda Chambers

General: Deciduous shrub, bushy, up to about 1 m tall (usually less); branchlets reddish, covered with many straight, slender prickles 3–4 mm long; thorns persist on older branches, often present to base of stem.

Leaves: Alternate, compound with 5–7 leaflets; leaflets egg-shaped to oval, blunt to pointed at tip, rounded to slightly heart-shaped base, 2–5 cm long; upper surface dull green, hairless; underside paler, minutely hairy; margins sharply toothed; stalks with a pair of small leaf-like bracts (stipules) at base.

Flowers: Pink, saucer-shaped; 5 petals, 2–3 cm long; usually solitary, sometimes a few near branch tips; June–July.

Fruits: Bright-red, fleshy rosehips, rounded to egg-shaped, about 2 cm long, smooth, contain many stiff-hairy achenes; August–September.

Habitat: Moist to dry, loamy to sandy upland sites and wet organic sites; in intolerant hardwood mixedwood stands and conifer swamps.

Notes: Rosehips are high in vitamin C and have been used to make jelly and tea. The petals have historically been used in salads and the dried leaves were used to make a tea substitute. Rose plants were also used in diarrhea remedies. See caution in Introduction. • See notes on smooth wild rose (p. 121) for distinguishing features.

Food Use of Roses (*Rosa* spp.) by Wildlife: bud, leaf, fruit.
• **Mammals:** snowshoe hare, black bear.
• **Birds:** ring-necked pheasant, ruffed grouse, Swainson's thrush.

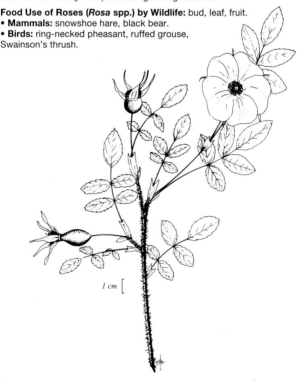

1 cm

General: Deciduous shrub, erect, up to 1.5 m tall; stems and branchlets lack prickles or have a few straight, slender prickles, especially near base, often densely prickly near base on vigorous shoots; branchlets reddish-purple.

Leaves: Alternate, stalked, compound, with 5–7 (sometimes 9) leaflets and a pair of small leaf-like bracts (stipules) at base of stalk; leaflets oval to egg-shaped, widest above middle, pointed to rounded at tip,

Karen Legasy

rounded to wedge-shaped at base, 1–4.5 cm long; margins sharply toothed to below middle.

Flowers: Pink, saucer-shaped; 5 petals, 2–3 cm long; solitary or a few in terminal clusters; May–early July.

Fruits: Red rosehips, smooth, round to egg- or pear-shaped, 1–1.5 cm in diameter, contain many stiff-hairy achenes; August–early October.

Habitat: Moist to dry, loamy to sandy upland sites; open forests, fields, clearings, roadsides and shorelines.

Notes: Smooth wild rose can be distinguished from prickly wild rose (p. 120) by its stems, which lack prickles or have a few scattered prickles. The stems of prickly wild rose are densely covered with prickles. Also, smooth wild rose's egg-shaped leaflets are widest above the middle, whereas those of prickly wild rose are usually widest below the middle.

Food Use by Wildlife: See notes under prickly wild rose (p. 120).

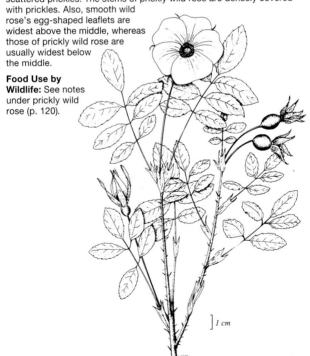

1 cm

Brenda Chambers

General: Deciduous small tree or shrub, up to 10 m tall; branchlets greenish-brown to reddish with pale, cork-like, elongated spots (lenticels), usually hairless; older branches reddish-brown; bark scaly; winter buds sticky.

Leaves: Alternate, stalked, compound with 11–17 leaflets; leaflets lance-shaped to narrowly oblong, taper to a pointed tip, 5–10 cm long, 1–2.5 cm wide, 3–5 times as long as wide; upper surface yellowish-green, hairless; underside paler; margins finely and sharply toothed.

Flowers: White, saucer-shaped, about 7 mm in diameter; 5 petals; in dense, round clusters 5–15 cm in diameter; June–July.

Fruits: Bright-red, round berries about 7 mm in diameter; in clusters; ripen August–September.

Habitat: Dry to moist, sandy to fine loamy upland sites, wet organic sites; black spruce-pine, intolerant hardwood mixedwood and tolerant hardwood upland stands and conifer swamps; wet roadsides and shorelines.

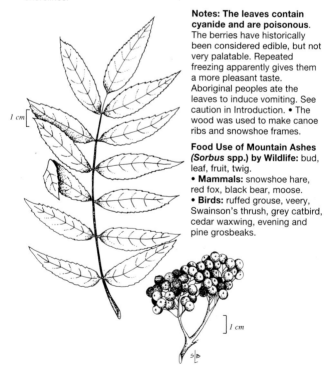

Notes: The leaves contain cyanide and are poisonous. The berries have historically been considered edible, but not very palatable. Repeated freezing apparently gives them a more pleasant taste. Aboriginal peoples ate the leaves to induce vomiting. See caution in Introduction. • The wood was used to make canoe ribs and snowshoe frames.

Food Use of Mountain Ashes (*Sorbus* spp.) by Wildlife: bud, leaf, fruit, twig.
• **Mammals:** snowshoe hare, red fox, black bear, moose.
• **Birds:** ruffed grouse, veery, Swainson's thrush, grey catbird, cedar waxwing, evening and pine grosbeaks.

1 cm

1 cm

s/b

ROSE FAMILY (ROSACEAE)

General: Deciduous small tree or shrub, up to 10 m tall; branchlets greenish-brown to reddish with pale, cork-like, elongated spots (lenticels), normally hairless; older branches reddish-brown; bark scaly; winter buds sticky.

Leaves: Alternate, stalked, compound with 11–17 leaflets; leaflets narrowly oval to oblong, blunt to rounded at tip with a short, abrupt point, 3.5–8 cm long, 1.5–3 cm wide, 2–3 times as long as wide; upper surface bluish-green, underside paler to whitish and slightly hairy; margins finely to sharply toothed.

Flowers: White, saucer-shaped; 5 petals, 4–5 mm long; in dense, round clusters 6–16 cm in diameter; June–July.

Karen Legasy

Fruits: Bright-red or scarlet, round berries 8–10 mm in diameter; in dense clusters; ripen August–September.

Habitat: Wet organic sites, moist to dry, fine loamy to rocky upland sites; in hardwood and conifer swamps, and upland intolerant hardwood mixedwood, tolerant hardwood and black spruce-jack pine stands.

Notes: One way to distinguish showy mountain ash from American mountain ash (p. 122) is by its leaflets. Showy mountain ash leaflets are 2–3 times longer than wide, with a relatively round, short point, whereas American mountain ash leaflets are 3–5 times longer than wide with a longer, more slender point. Also, the flowers and fruit of showy mountain ash are slightly larger and showier than those of American mountain ash.

Food Use by Wildlife: See notes under American mountain ash (p. 122).

] 1 cm

] 1 cm

Brenda Chambers

General: Small evergreen shrub, tufted, erect, 10–20 cm tall, from slender, woody, extensively creeping underground stems; stems reddish-brown, from a woody base.

Leaves: Alternate, compound with 3 leaflets; leaflets triangular to oblong, widest above middle, usually 3-toothed at tip and tapered at base, 10–25 mm long, leathery; upper surface dark green, shiny, usually hairless; underside paler, hairless or with short, yellowish hairs; margins smooth except for the 3 teeth at tip; stalks up to 3 cm long.

Flowers: White; 5 petals, 8–10 mm wide; 1–6 in loose, usually flat-topped clusters; June–July.

Fruits: Small, densely hairy, dry, hard, single-seeded (achenes); in small, hairy clusters; July–August.

Habitat: Sand, gravel or rock; dry, exposed areas; along ridges.

Notes: Cinquefoil plants were historically used in remedies for fever, sore throat and piles. They were also used in mouthwash and lotion. See caution in Introduction.

1 cm

CASHEW FAMILY (ANACARDIACEAE)

General: Deciduous shrub, erect, bushy, less than 1 m tall; branches erect to ascending; branchlets minutely hairy, later become hairless, ridged, brownish-grey, with wart-like markings; usually in patches from creeping underground stems.

Leaves: Alternate, compound with 3 leaflets; leaflets egg-shaped, short to tapered at tip, narrowed to rounded at base, 5–15 cm long and 2.5–6 cm wide; terminal leaflet distinctly stalked; 2 side leaflets short-

Karen Legasy

stalked or almost stalkless; upper surface, hairless, dark green, dull to shiny; underside paler, minutely hairy along veins; margins toothless to irregularly toothed or lobed; stalks 5–25 cm long.

Flowers: Greenish-white to yellowish, tiny, 2–3 mm long; 5 petals; in dense clusters from leaf axils; June–July.

Fruits: Whitish, round, berry-like, 5–6 mm wide, dry, contain 1 white seed; often stay on plant through winter; late July–August.

Habitat: Dry to moist, rocky to clayey, often calcareous upland sites.

Notes: This species is also known as *Rhus radicans* ssp. *negundo*.
• **Eating poison ivy leaves or fruit can be fatal.** Poison ivy contains a toxic chemical that can cause a **severe allergic reaction** which takes the form of an itchy rash, swelling or even blisters. If you come into contact with poison ivy, wash the affected area with a strong soap. Consult a physician if a severe reaction results. • Aboriginal peoples dried poison ivy leaves and fruit and added them to fire pits which were lit when an enemy was approaching. Burning poison ivy releases its poisonous toxin as tiny droplets which are carried through the air on ashes and dust particles.
• A vine-like form of poison ivy is found in parts of southern Ontario.

Food Use by Wildlife: fruit.
• **Birds:** ring-necked pheasant, ruffed grouse, wild turkey, downy, hairy and pileated woodpeckers, northern flicker, eastern phoebe, American crow, black-capped chickadee, grey catbird, brown thrasher, cedar waxwing, European starling, yellow-rumped warbler, white-throated sparrow, dark-eyed junco, purple finch.

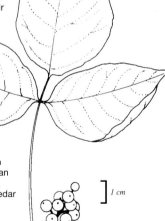

1 cm

GRAPE-WOODBINE • *Parthenocissus inserta*
vigne vierge (P. quinquefolia)

GRAPE FAMILY (VITACEAE)

Emma Thurley

General: Woody, high-climbing vine; tendrils sparsely branched, usually lacking adhesive discs.

Leaves: Alternate, long-stalked, palmately compound with 5 leaflets; leaflets elliptic to egg-shaped, long-tapered at tips, wedge-shaped at base, 5–12 cm long, dark green and shiny above, paler beneath, short-stalked; margins coarsely and sharply toothed above middle.

Flowers: Greenish, small, about 5 mm across; 25–200 or more in widely forked, branching clusters lacking a main central axis; branches forking in pairs; June–July.

Fruits: Round, bright blue berries, 8–10 mm in diameter, with a thin layer of flesh around 3–4 seeds; stalks often bright red; August–September.

Habitat: Moist soil in woods and thickets; open ground along roadsides; uncommon in the southern part of this region, occasional in northern part.

Notes: This species is also known as *Parthenocissus vitacea* • Grape-woodbine is closely related to five-leaved Virginia-creeper (*P. quinquefolia*) which has a high-climbing habit and closely grouped clusters of flowers from a central axis and often has adhesive discs at the tips of its tendrils. Five-leaved Virginia-creeper generally occurs to the south of this region. • Historically, bark and twigs of grape-woodbine were used in medicine or food. See caution in Introduction.

] 1 cm

E.T.

BUTTERCUP FAMILY (RANUNCULACEAE)

General: Slender woody perennial vine, trails on ground and over trees, shrubs and fences, climbs up to 5 m or more; stems round, smooth, light brown to reddish-purple.

Leaves: Opposite, hairless, compound in 3s; leaflets long-stalked, broadly egg-shaped or slightly heart-shaped, sharp-pointed, 5–9 cm long and 2.5–6 cm wide; margins toothless to irregularly toothed, often deeply cut or lobed; stalks 5–9 cm long, twining around nearby supports.

Flowers: Usually white or cream-coloured, male or female on separate plants, sometimes bisexual; no petals; 4–5 petal-like sepals; in large open, leafy clusters 1–8 cm long from leaf axils; July–August.

Emma Thurley

Fruits: Hairy, brown achenes with feathery persistent styles 2.5–3.8 cm long, forming fluffy heads in large clusters; September–October.

Habitat: Forest edges, thickets, along roadsides.

Notes: All plant parts are poisonous and can cause skin irritation, severe abdominal pain and perhaps death. • This species may be confused with purple virgin's bower (*Clematis verticillaris*), which has a similar climbing habit, but generally less teeth on its leaf margins. Flowers are required to differentiate these 2 species. The sepals of virgin's bower are white to cream-coloured and relatively small (6–10 mm long), whereas those of purple virgin's bower are purple to bluish purple and much larger (30–50 mm long).

COMMON ELDERBERRY • *Sambucus canadensis*
SUREAU BLANC

HONEYSUCKLE FAMILY (CAPRIFOLIACEAE)

Brenda Chambers

General: Deciduous shrub, erect, up to 3 m tall; stems soft or barely woody, essentially hairless, with a large, white pith; branchlets yellowish-grey; older stems greyish-brown, thick and warty.

Leaves: Opposite, compound with 5–11 (usually 7) leaflets; leaflets egg-shaped to oval, pointed at tip, tapered to rounded at base, 5–15 cm long, 2.5–5.5 cm wide, stalkless or short-stalked; upper surface bright green, hairless; underside paler, hairless or hairy along veins; margins sharply toothed; stalks 2.5–5 cm long.

Flowers: White, about 3 mm wide, very fragrant; 5 petals; numerous in long-stalked, compound, flat-topped to slightly rounded terminal clusters 10–18 cm wide; July.

Fruits: Purplish-black, round, berry-like drupes with 3–5 roughened pits, juicy, up to 6 mm wide; in compound clusters; August–September.

Habitat: Wet to moist areas; swamps; roadsides and forest edges.

Notes: People have often been poisoned by eating unripe or uncooked elderberries or by using the hollowed-out stems as peashooters. Aboriginal peoples used the inner bark of common elderberry in a toothache remedy. See caution in Introduction. • See notes on red-berried elder (p. 129) for distinguishing features.

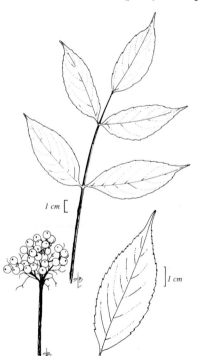

Wildlife Use of Elderberries (*Sambucus* spp.) by Wildlife: bud, leaf, fruit, twig. • **Mammals:** snowshoe hare, white-tailed deer, moose. • **Birds:** ring-necked pheasant, ruffed grouse, yellow-bellied sapsucker, red-headed and red-bellied woodpeckers, eastern phoebe, eastern kingbird, red-breasted and white-breasted nut-hatches, veery, Swainson's, hermit and wood thrushes, American robin, grey catbird, brown thrasher, cedar waxwing, European starling, red-eyed vireo, northern cardinal, rose-breasted and pine grosbeaks, indigo bunting, swamp and white-throated sparrows, rusty blackbird.

1 cm

1 cm

HONEYSUCKLE FAMILY (CAPRIFOLIACEAE)

General: Deciduous shrub, erect, up to 4 m tall; branches and stems soft to barely woody; branchlets yellowish-brown and hairy; older branches greyish-brown, thick, warty and with a large, orange to brown pith.

Leaves: Opposite, compound with 5 (sometimes 7) leaflets; leaflets egg- to lance-shaped, pointed at tip, narrowed to rounded and often unequal at base, 5–13 cm long, 2.5–5.5 cm wide, stalked;

Brenda Chambers

upper surface green; underside paler and soft-hairy to hairless; margins sharply toothed; stalks 2.5–5 cm long.

Flowers: Whitish, 3–6 mm wide; 5 petals; numerous in elongated, rounded or pyramid-shaped clusters 5–13 cm long; May–June.

Fruits: Purple-black, rounded, berry-like, 5–6 mm wide; in pyramidal clusters 4–10 cm long; July–August.

Habitat: Wet organic sites and moist to dry, clayey to sandy upland sites; in hardwood swamps and tolerant hardwood stands.

Notes: The berries, bark, leaves and roots are considered poisonous. See caution in Introduction. The flowers, crushed leaves and branchlets have a very disagreeable odour. • Red-berried elder can be distinguished from common elderberry (p. 128) by its leaflets, which are often uneven at the base, and by its flower clusters, which are rounded to pyramidal rather than flat-topped like those of common elderberry.

Food Use by Wildlife: See notes under common elderberry (p. 128).

1 cm

HERBS

To assist in the identification of herbs, a chart which is based on both floral and leaf characteristics is provided. Examples of flower types are illustrated first. Further detail on leaf characteristics is included in the glossary (p. 435). Once key identifying features of a herb are noted, the user can work through the chart to determine the group or groups of species to which the plant may belong. Page numbers are provided in the chart for quick reference. It is important to note that the same plant species may be found in different parts of the chart, because of variations in features such as flower colour and leaf type. Following is an explanation of how to use the herb chart.

1. Look at the plant for leaf features such as opposite/alternate or toothed/toothless and make a note of these features. If flowers are lacking, go to the chart and examine the possibilities.

2. Look for flower features such as colour, shape and number of parts (petals, sepals). Flower shapes are divided into several types– tubular, aster-like, tiny and crowded, petals alike and petals not alike.

3. Examine the possibilities in the chart and refer to the pages listed.

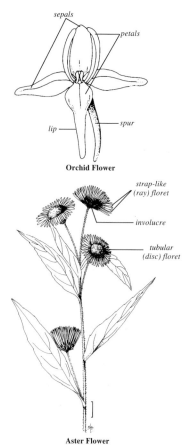

sepals

petals

lip

spur

Orchid Flower

strap-like (ray) floret

involucre

tubular (disc) floret

Aster Flower

Les plantes herbacées

**Flowers regular,
(petals alike);
leaves dissected**
shown: Tall buttercup
Ranunculus acris

**Flowers irregular
(petals not alike)**
shown: Blunt-leaf orchid
Platanthera obtusata

**Flowers tiny, crowded;
leaves compound with >3 leaflets**
shown: Bristly sarsaparilla
Aralia hispida

**Flowers in dense,
aster-like heads**
shown: Bog aster
Aster nemoralis

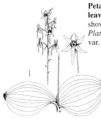

**Petals not alike;
leaves basal**
shown: Large round-leaved orchid
Platanthera orbiculata
var. *macrophylla*

**Petals alike,
joined in a tube**
shown: Harebell
Campanula rotundifolia

**Petals alike, 3 parts;
leaves opposite**
shown: White trillium
Trillium grandiflorum

**Petals alike,
4 parts; leaves
toothed, alternate**
shown: Evening primrose
Oenothera biennis

**Petals alike, 5 parts; leaves
compound with 3 leaflets**
shown: Common wood-sorrel
Oxalis acetosella spp. *montana*

Petals alike, 6 parts
shown: Blue bead lily
Clintonia borealis

FLOWERS	LEAVES			
	Opposite		Alternate	
White	teeth	no teeth	teeth or dissected	no teeth
tubular		Melampryum (p. 263)	Campanula (pp. 259-60)	Campanula (pp. 259-60)
Dense heads or aster like			Aster (pp. 280-84) Chrysanthemum (p. 275) Lactuca (p. 272)	Aster (pp. 280-87) Anaphalis (p. 277) Antennaria (p. 278)
Tiny and crowded	Mentha (p. 253) Lycopus (p.252)		Achillea (p. 276) Prenanthes (p. 269)	Maianthemum (pp. 157-59) Polygonum (p. 174)
petals not alike		Melampyrum (p. 263)	Viola (pp. 214-18) Dicentra (p. 223)	Polygala (p. 220)
petals alike -3 parts				
petals alike -4 parts	Circaea (p. 231)	Galium (pp. 264-6) Cornus (p. 241)	Cardamine (p.225)	Epilobium (p. 230) Maianthemum (pp. 157, 159)
petals alike -5 parts	Mitella (pp.193-4)	Cerastium (p.180) Stellaria (p. 181) Claytonia (p.176)	Campanula (pp. 259-60) Anemone (pp. 184-5) Orthilia (p. 244)	Campanula (pp. 259-60) Parnassia (p. 196) Monotropa (pp. 250-51)
petals alike -6 parts		Trientalis (p. 242) (7 petals)		Maianthemum (p. 158) Allium (p. 154)
Yellow				
tubular		Melampyrum (p.263)	Pedicularis (p. 262)	
Dense heads or aster like			Solidago (pp. 288-9, 291) Lactuca (p. 272)	Solidago (pp. 290-1)
Tiny and crowded				
petals not alike		Melampyrum (p.263)	Viola (p. 219) Pedicularis (p.262)	
petals alike -3 parts				
petals alike -4 parts			Oenothera (p. 228)	
petals alike -5 parts		Hypericum (p. 226)	Caltha (p. 189) Ranunculus (pp. 187-8)	Monotropa (pp. 250-51)
petals alike -6 parts		Medeola (p. 168)		Uvularia (p.160)

LEAVES

Basal	Compound	
	3 leaflets	>3 leaflets
	Trifolium (p. 210)	
Petasites (p. 279) *Chrysanthemum (p. 275)*	*Trifolium (p. 210)*	
Antennaria (p. 278)	*Panax (p. 236)* *Heracleum (p. 239)*	*Actaea (p. 190)* *Sanicula (p. 238)* *Aralia (pp. 233-35)* *Thalictrum (p. 191)*
Plantanthera (p. 142) *Viola (pp. 214-8)* *Goodyera (pp. 146-7)* *Spiranthes (p. 145)* *Sanguinaria (p. 222)*		*Lathyrus (p. 206)* *Dicentra (p. 223)* *Sanicula (p. 238)*
	Trillium (pp. 164, 166-7)	
	Cardamine (p. 225)	
Mitella (pp. 193-4) *Parnassia (p. 196)* *Pyrola (pp. 245-49)* *Drosera (p. 173)* *Dalibarda (p. 201)* *Tiarella (p. 195)* *Moneses (p. 243)*	*Menyanthes (p. 240)* *Coptis (p. 182)* *Fragaria (pp. 197-8)* *Panax (p. 236)* *Oxalis (p. 211)* *Heracleum (p. 239)*	*Aralia (pp. 233-35)* *Osmorhiza (p. 237)*
Hepatica (p. 186) *Allium (p. 154)*		
Hieracium (p. 271) *Taraxacum (p. 268)*	*Trifolium (p. 207)*	
	Trifolium (p. 207)	
Plantanthera (p. 143) *Viola (p. 219)* *Corallorhiza (p. 148)* *Cypripedium (p. 139)*	*Trifolium (p.207)*	*Corydalis (p. 224)* *Lathyrus (p. 206)*
	Aquilegia (p. 183) *Waldsteinia (p. 199)*	*Geum (p. 202)* *Agrimonia (p. 203)* *Hypericum (p. 226)*
Clintonia (p. 156) *Erythronium (p. 155)*		*Caulophyllum (p. 192)*

FLOWERS	LEAVES				
Orange	**Opposite**		**Alternate**		
	teeth	no teeth	teeth or dissected	no teeth	
tubular					
Dense heads or aster like					
Tiny and crowded					
petals not alike			*Impatiens* (p. 221)		
petals alike -3 parts					
petals alike -4 parts					
petals alike -5 parts					
petals alike -6 parts					
Pink or Red					
tubular		*Apocynum* (p. 257)		*Polygala* (p. 220)	
Dense heads or aster like			*Cirsium* (p. 274) *Lactuca* (p. 272)		
Tiny and crowded	*Veronica* (p. 261) *Eupatorium* (p. 273)		*Achillea* (p. 276)	*Rumex* (p. 175)	
petals not alike	*Veronica* (p. 261)			*Polygala* (p. 220)	
petals alike -3 parts					
petals alike -4 parts			*Epilobium* (pp. 229-30)	*Epilobium* (p. 229)	
petals alike -5 parts	*Geranium* (p. 227)	*Apocynum* (p. 257) *Claytonia* (p. 176)	*Anemone* (pp. 184-5)	*Monotropa* (pp. 250-51)	
petals alike -6 parts				*Streptopus* (p. 161)	

LEAVES

Basal	Compound	
	3 leaflets	>3 leaflets
Hieracium (p. 270)		
	Trifolium (pp. 208-10)	
Rumex (p. 175)	Trifolium (pp. 208-10)	
Plantanthera (p. 144)	Trifolium (pp. 208-10)	Corydalis (p. 224)
Corallorhiza (pp. 148-50)	Arisaema (p. 153)	
Cypripedium (p. 138)		
Asarum (p. 179)	Trillium (p. 165)	
Pyrola (p. 247)	Aquilegia (p. 183)	Potentilla (p. 200)
Parnassia (p. 196)		
Sarracenia (p. 172)		
Drosera (p. 173)		
Hepatica (p. 186)		

FLOWERS	LEAVES				
Violet/Purple or Blue	**Opposite**		**Alternate**		
	teeth	no teeth	teeth or dissected	no teeth	
tubular	*Scutellaria (p. 254)*	*Clinopodium (p. 255)*	*Campanula (pp. 259-60)*	*Campanula (pp. 259-60) Mertensia (p. 258)*	
Dense heads or aster like			*Aster (pp. 280-87) Lactuca (p. 272) Cirsium (p. 274)*	*Aster (pp. 280-87)*	
Tiny and crowded	*Eupatorium (p. 273) Mentha (p. 253)*	*Clinopodium (p. 255)*			
petals not alike	*Scutellaria (p. 254)*	*Listera (p. 140) Prunella (p. 256) Veronica (p. 261)*	*Viola (p. 212)*	*Polygala (p. 220)*	
petals alike -3 parts					
petals alike -4 parts	*Epilobium (pp. 229-30)*			*Epilobium (pp. 229-30)*	
petals alike -5 parts	*Geranium (p. 227)*		*Campanula (pp. 259-60)*	*Campanula (pp.259-60) Mertensia (p. 258) Geocaulon (p. 178)*	
petals alike -6 parts				*Streptopus (p. 162)*	
Green or Brown					
tubular					
Dense heads or aster like					
Tiny and crowded	*Urtica (p. 177)*	*Rumex (p. 175)*			
petals not alike				*Epipactis (p. 152)*	
petals alike -3 parts		*Galium (pp. 264-66)*			
petals alike -4 parts		*Galium (pp. 264-66)*			
petals alike -5 parts				*Geocaulon (p. 178)*	
petals alike -6 parts		*Medeola (p. 168)*		*Streptopus (p. 162) Polygonatum (p. 163)*	

LEAVES

	Basal	Compound	
		3 leaflets	>3 leaflets
	Campanula (p. 259) *Mertensia (p. 258)*		
	Viola (pp. 212-3, 217) *Corallorhiza (pp. 148-9)*		*Lathyrus (p. 206)* *Vicia (pp. 204-5)*
	Iris (p. 170)		
			Thalictrum (p. 191)
	Monotropa (pp. 250-51)		*Geum (p. 202)* *Potentilla (p.200)*
	Sisyrinchium (p. 169) *Hepatica (p. 186)*		*Caulophyllum (p. 192)*
	Aralia (pp. 233-35) *Typha (p. 171)* *Plantago (p. 267)*	*Arisaema (p. 153)*	*Aralia (pp. 233-35)* *Sanicula (p. 238)*
	Plantanthera (pp. 141, 143) *Corallorhiza (p. 149)* *Epifagus (p. 151)*		
	Asarum (p. 179)		*Sanicula (p. 238)*
	Pyrola (pp. 245-49) *Aralia (pp. 233-35)* *Orthilia (p. 244)*		*Aralia (pp. 233-35)*
			Caulophyllum (p. 192)

General: Perennial herb; flowering stems 10–55 cm tall, erect, hairy, leafless, with 1 (rarely 2) flower at top.

Leaves: Basal, stalkless, simple, 2, oblong, taper to point at tip, thick, up to 20 cm long and 7.5 cm wide, with prominent, parallel veins, sparsely hairy; upper surface dark green; underside silvery.

Flowers: Pink with reddish veins, occasionally white; lower (lip) petal pouch-like, inflated, egg-shaped, narrowest base, about 6.3 cm long; upper part of interior surface has long, white hairs; lateral petals narrower and longer than sepals; sepals petal-like, greenish-purple, lance-shaped, 3.5–5 cm long; May–July.

Fruits: Capsules, brown, 4.5 cm long, erect.

Habitat: Dry upland pine, pine-black spruce and intolerant hardwood mixedwood stands, and wet black spruce organic sites.

Notes: Pink lady's-slipper is also commonly called 'moccasin flower.' • A close relative is showy lady's-slipper (*Cypripedium reginae*) (inset photo), the largest of our native orchids (up to 80 cm in height). It is occasional in the Central Region, occuring in open to semi-shaded, moderately acidic to neutral wetland habitats. Showy lady's-slipper flowers are predominantly white to cream-coloured, with a rose-purple streaked, pouch-shaped lip. • Aboriginal peoples used the roots in a sedative, and in an epilepsy medicine. See caution in Introduction. • If you are allergic to pink lady's-slipper, contact with it may severely irritate your skin, and ingesting it may cause internal irritation. • Pink lady's-slipper takes 10 years from germination to reach the flowering stage and should not be picked. • The genus name *Cypripedium* is Latin for 'Venus's slipper.'

*Brenda Chambers, top;
Bill Crins, bottom*

**Food Use of Lady's-slippers
(*Cypripedium* spp.) by Wildlife:**
leaves, flowers. • **Mammals:**
white-tailed deer.

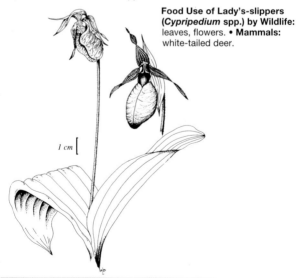

1 cm

ORCHID FAMILY (ORCHIDACEAE)

General: Perennial herb; flowering stems unbranched, 10–70 cm tall, with leaves.

Leaves: Alternate, clasping stem, simple, egg- to lance-shaped pointed at tip, up to 20 cm long and 5–10 cm wide, bright green with parallel veins, slightly downy.

Flowers: Yellow, 1–2, terminal; lower (lip) petal inflated, sac-like, widest below middle, usually purple-veined, about 5 cm long; lateral petals purple-brown, wide-spreading, lance-shaped to linear, flat or spirally twisted, slightly longer than lip; sepals 3, petal-like, greenish-yellow, lance-shaped; June–August.

Fruits: Capsules, oval, ridged, contain numerous dust-like seeds.

OMNR

Habitat: Moist to fresh calcareous sites; tolerant hardwood and mixedwood stands; uncommon.

Notes: Remedies for headache, insomnia and pain have been made from the roots of yellow lady's-slipper. See caution in Introduction.

Food Use by Wildlife: See notes under pink lady's-slipper (p. 138).

1 cm

ORCHID FAMILY (ORCHIDACEAE)

Frank Boas

General: Perennial herb; flowering stems single, very slender, 6–15 cm tall, hairless or slightly hairy above leaves.

Leaves: Opposite, single pair near middle of stem, stalkless, clasping stem, simple, widely heart- to egg-shaped, 1–3 cm long, 0.8–4 cm wide.

Flowers: Purplish-brown to pale green; lower (lip) petal narrow, twice as long as lateral petals, divided into 2 slender lobes; 5–16 small flowers in elongated terminal cluster; June–July.

Fruits: Capsules, brown, egg-shaped, small; July–August.

Habitat: Wet organic conifer swamps.

Notes: The species name *cordata* means 'heart-shaped' and refers to the leaves.

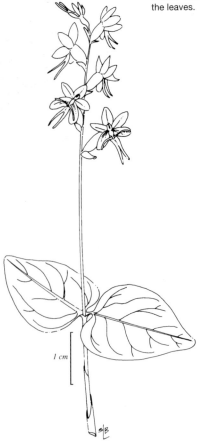

1 cm

ORCHID FAMILY (ORCHIDACEAE)

General: Perennial herb; flowering stems leafless (scapes), 5–40 cm tall, much taller than leaf, hairless; from elongated, fleshy roots.

Leaves: Basal, solitary, inversely egg-shaped, rounded or blunt at tip, tapering at base, stalked, 2.5–12 cm long, 1–5.5 cm wide.

Flowers: Green to greenish-white; lower (lip) petal lance-shaped, blunt tipped, 6–10 mm long, 2 mm wide, fleshy and downward-pointing with a tapered, 4–8 mm long spur; lateral petals narrow, arching and ascending; upper sepal petal-like, rounded, arching as a hood; June–July.

Fruits: Capsules, erect, 8 mm long.

Habitat: Wet organic conifer swamps.

Notes: This plant is also known as *Lysiella obtusata* and *Habenaria obtusata*. • Mosquitoes and small moths pollinate blunt-leaf orchid.

*OMNR, top;
Bill Crins, bottom*

1 cm

Linda Kershaw

General: Perennial herb; flowering stems single, 20–40 cm tall, with 1–5 scale-like, narrow, lance-shaped bracts, hairless; from fleshy roots.

Leaves: Basal, usually 2, opposite, stalkless, clasping stem, simple, rounded, blunt-tipped, 5–25 cm long and wide, usually flattened on ground; upper surface glossy; underside silvery.

Flowers: Greenish-white, fragrant; lower (lip) petal linear-oblong, blunt-tipped, somewhat curved under, 10–24 mm long, up to 5 mm wide with a cylindrical spur thickened toward tip; stalks up to 1 cm long; 5–25 in loose, elongated terminal clusters; July–August.

Fruits: Capsules, brown, erect, 1.5–2 cm long, 4–6 mm thick.

Habitat: Dry to moist tolerant hardwood and intolerant hardwood mixedwood stands; uncommon.

Notes: This plant is also known as *Habenaria orbiculata*. • Hooker's orchid (*Platanthera hookeri*) also has a pair of rounded leaves lying flat on the ground, but its leaves are not typically as shiny as those of large round-leaved orchid. The flowering stem of Hooker's orchid is usually naked, or occasionally has a single bract. Its flowers are yellowish-green, and in cross-section, are shaped like ice tongs. • The species name *orbiculata* means 'round' and refers to the leaves.

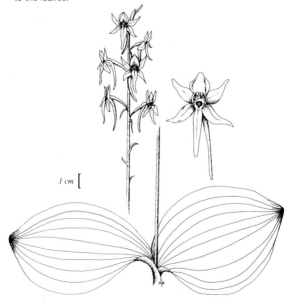

1 cm

General: Perennial herb; flowering stems hairless, thick, 10–40 cm tall; from fleshy, thickened underground stems (rhizomes).

Leaves: Alternate, sheathing, simple, narrowly oblong to lance-shaped, usually pointed at tips, 2–10 cm long, 0.5–3 cm wide, soft, flat.

Flowers: Green to yellowish-green; lower (lip) petal lance-shaped, blunt-tipped, 3–9 mm long, usually 1–2 mm wide below middle, stiff, fleshy with a somewhat club-shaped spur 2–8 mm long (almost equal to or shorter than lip); sepals and lateral petals egg-shaped; in narrow, cylindrical spikes 1.5–9 cm long; June–August.

Fruits: Capsules, erect, with many minute seeds.

Habitat: Moist forest edges and ditches, wet organic sites.

Notes: This plant is also known as *Limnorchis hyperborea* or *Habenaria hyperborea*.

*Linda Kershaw, top;
Derek Johnson, bottom*

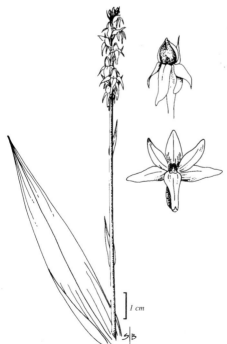

1 cm

Brenda Chambers

General: Perennial herb; flowering stems slender, 20–90 cm tall, leafy.

Leaves: Oval, elliptic or lance-shaped, 5–20 cm long, 2–7 cm wide, reduced upwards on stem.

Flowers: Lilac, rarely white, fragrant; lower (lip) petal 8–13 mm long and wide (or slightly wider), with 3 fringed, fan-shaped parts and a slender, club-shaped spur; lateral petals oblong or narrowly spoon-shaped, toothed or toothless near tip; June–August.

Fruits: Capsules, erect, with many minute seeds.

Habitat: Wet forest edges, stream edges; uncommon.

Notes: Small purple-fringed orchid is also known as *Blephariglottis psycodes* and *Habenaria psycodes*. • The name *psycodes* means 'butterfly-like' and refers to the appearance of the flower.

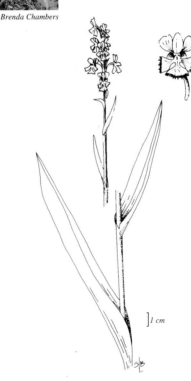

] *1 cm*

ORCHID FAMILY (ORCHIDACEAE)

General: Perennial herb; flowering stems 3–50 cm tall; from thick, fleshy, short roots.

Leaves: Basal, simple, linear to lance-shaped, pointed at tips, up to 20 cm long, 5–10 mm wide, gradually reduced upwards on stem.

Flowers: Creamy white, 6–12 mm long; 2 lateral petals and 3 petal-like sepals overlap to form an arching hood over the downturned lower (lip) petal; spirally arranged in dense terminal spike; August–September.

Fruits: Capsules, dry, many-seeded.

Habitat: Damp or dry roadsides, grassy flats on sandy soil.

Notes: Several *Spiranthes* species are found in this region, including shiny ladies'-tresses (*S. lucida*), nodding ladies'-tresses (*S. cernua*) and northern ladies'-tresses (*S. gracilis*). All 4 species have spirally arranged flowers, but those of northern ladies'-tresses are in a single row. The flowers of shiny ladies'-tresses and nodding ladies'-tresses are arranged similarly to those of hooded ladies'-tresses, but, the lip of shiny ladies'-tresses has a prominent yellow stripe, and the lip of nodding ladies'-tresses lacks a constriction. • The flowers of hooded ladies'-tresses have a strong, almond-like fragrance.

Brenda Chambers

] *1 cm*

D.A. Sutherland

General: Perennial herb, glandular-hairy; leafless flowering stems (scapes) 10–25 cm tall, with several small scales; from slightly creeping underground stems (rhizomes).

Leaves: In a basal rosette, egg-shaped, 1–3 cm long, 0.5–2 cm wide, 5-nerved, with scattered, horizontal dark veins and white blotches, taper to a sheathing stalk.

Flowers: Greenish-white; lower (lip) petal inflated, sac-like, tip curved under; petals and sepals 4 mm long, with upper sepal and 2 lateral petals fused into a hood; in 1-sided, elongated or spike-like terminal clusters 2.5–6 cm long; July–August.

Fruits: Capsules.

Habitat: All moisture regimes; sandy to coarse loamy soils, intolerant hardwood mixedwoods, black spruce-jack pine stands, wet organic conifer swamps.

Notes: You can recognize dwarf rattlesnake-plantain by its usually 1-sided flower cluster and by its dark green, net-patterned leaves, which are typically smaller than those of other rattlesnake-plantains.

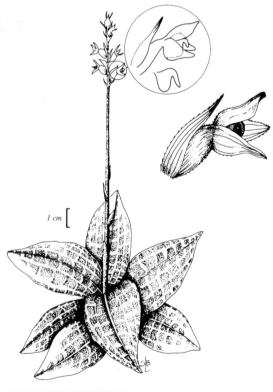

1 cm

ORCHID FAMILY (ORCHIDACEAE)

General: Perennial; flowering stems erect, 10–35 cm tall; from a short underground stems (rhizomes) with thick fleshy roots.

Leaves: In a basal rosette, egg- to lance-shaped, 3–5 cm long, 1–2.5 cm wide, 5- to 9-veined, dark green, usually mottled with dark and light green and with a network of interlacing veins bordered with white.

Flowers: Whitish to greenish-white; lower (lip) petal about 5–7 mm long, 3–4 mm wide, upturned along sides, strongly sac-like at base, narrowed at tip into a blunt beak about as long as body; lateral petals 5–6 mm long, 2–3 mm wide, joined with upper sepal to form a hood; lateral sepals petal-like, 5–6 mm long, 2.5–3.5 mm wide; spirally arranged in loose terminal clusters 4–10 cm long; July–August.

Fruits: Capsules, 3-valved, with many seeds.

Habitat: Dry to fresh upland conifer stands; wet organic conifer swamps.

Notes: The largest of the rattlesnake-plantains is green-leaved rattlesnake-plantain (*Goodyera oblongifolia*) (inset photo), which has a distinctive, prominent white midrib and lacks the checkered leaf pattern of the other species.
• The species name *tesselata* means 'like a mosaic' and refers to the pattern on the leaves.

Brenda Chambers, both

] 1 cm

E.T.

Emma Thurley

General: Perennial herb; flowering stems slender, 8–30 cm tall, yellowish; from a whitish, intricately branched, coral-like underground stem (rhizome).

Leaves: Alternate, 2–5, thin, semi-transparent, scale-like, sheathing the stem.

Flowers: Greenish-yellow to slightly brown-tinged; lower (lip) petal almost as long as lateral petals, notched at tip, whitish, rarely spotted with red or purplish dots; lateral petals similar to sepals but often red-dotted; sepals petal-like, 4.5–6 mm long, narrowly oblong to tongue-shaped, bluntish, greenish-yellow; stalks very short, grow upward at first but eventually bend downward; 2–10 in elongated terminal clusters up to 8 cm long; May–July.

Fruits: Capsules, greenish, about 1 cm long, nodding; July–August.

Habitat: Wet organic eastern white cedar swamps and upland intolerant hardwood mixedwood stands.

Notes: Early coral-root, at 8–30 cm tall, is just over half the size of spotted coral-root (p. 149), which is 15–50 cm tall. The stems of early coral-root have leaf sheaths near their bases only, whereas spotted coral-root has leaf sheaths to above the middle of its stems. Spotted coral-root produces more flowers (10–30) than early coral-root (2–10). • See notes on spotted coral-root (p. 149).

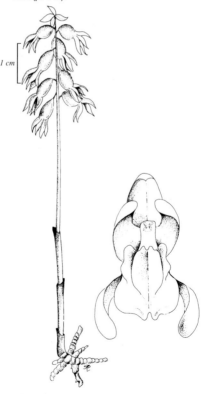

1 cm

ORCHID FAMILY (ORCHIDACEAE)

General: Perennial herb; flowering stems slender, 15–50 cm tall, purplish or pale yellow to pale green; from a branched, coral-like, underground stem (rhizome).

Leaves: Alternate, several, scale-like, with thin, semi-transparent tubular sheaths about 7.5 cm long.

Flowers: Brownish-purple lateral petals and petal-like sepals, with petals smaller than sepals; lower (lip) petal white, spotted with crimson, about 5 mm long, with prominent spur; short-stalked, in elongated terminal clusters of 10–30 flowers; June–August.

Fruits: Capsules, egg-shaped, up to 2.5 cm long, nodding.

Habitat: Fresh to moist pine and intolerant hardwood mixedwood stands.

Notes: This orchid is also known as 'large coral-root.' • Spotted coral-root is saprophytic, meaning that it derives its nutrients from decaying organic matter. • The genus *Corallorhiza* was named for its coral-like roots. See notes on early coral-root (p. 148).

Brenda Chambers

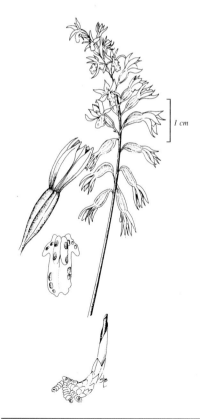

1 cm

General: Perennial herb; flowering stems stout, 15–50 cm tall, purplish; from a branched, coral-like underground stem (rhizome).

Leaves: Alternate, scale-like, forming thin, semi-transparent, tubular sheaths near stem base.

Flowers: Purplish-red, lance- to egg-shaped, 8–20 mm long, nodding; lower (lip) petal tongue-shaped, reddish-purple, striped, about as long as lateral petals; lateral petals and sepals translucent, purple-veined, form a broad, arching hood above lip; 7–25 in loose terminal clusters; May–June.

Fruits: Capsules, elliptic, strongly bent downward, 1.5–2 cm long.

Habitat: Fresh to moist tolerant hardwood and intolerant hardwood mixedwood stands.

Notes: This orchid is saprophytic. • Striped coral-root was named for its purple-striped flowers and coral-like roots.

Brenda Chambers

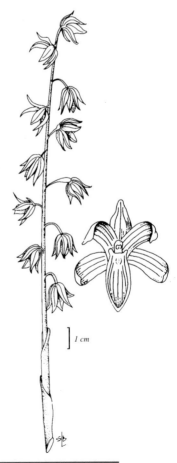

1 cm

General: Perennial herb; flowering stems slender, 15–45 cm tall, freely branched, pale brown, usually marked with fine, brown-purple stripes, lack green pigment.

Leaves: Alternate, scattered, scale-like, triangular to egg-shaped, 2–4 mm long.

Flowers: In 2 forms, numerous, scattered along branches; upper flowers sterile, showy, white, tubular, about 1 cm long, often with 2 brown-purple stripes; lower flowers abundantly fertile, not opening, about 5 mm long; August–October.

Fruits: Capsules, about 5 mm long, split open at tip; seeds tiny, numerous.

Habitat: Dry to fresh, upland tolerant hardwood stands with American beech; uncommon.

Notes: The genus name *Epifagus* is derived from the Greek, *epi* 'upon' and the Latin *fagus* 'beech,' because this herb is a parasitic plant that grows under and feeds on the roots of American beech trees.

Emma Thurley

1 cm

E.T.

HELLEBORINE • *Epipactus helleborine*
ÉPIPACTIS À FEUILLES LARGES

ORCHID FAMILY (ORCHIDACEAE)

General: Perennial; flowering stems erect, leafy, up to 80 cm tall; often forms extensive woodland colonies.

Leaves: Alternate, simple, stalkless, numerous, egg- to lance-shaped, pointed, clasping stem; lower leaves up to 10 cm long; upper leaves progressively shorter and narrower.

Flowers: Lower (lip) petal constricted near middle, fleshy, 1–1.5 cm. long and 4–8 mm wide; lateral petals and sepals similar, dull green to whitish or pinkish, purple-veined, egg- to lance-shaped, 10–14 mm long; in many-flowered, loose terminal clusters 5–25 cm long with narrow, lance-shaped bracts; lower bracts longer than flowers; July–August.

Fruits: Capsules, egg-shaped to ellipsoid, 1 cm long; lowermost mature as uppermost flower buds open.

Habitat: Dry to moist, sandy to clayey, upland tolerant hardwood stands with sugar maple, red oak and other hardwoods; forest edges.

Emma Thurley

Notes: Helleborine originated in Europe, where it was widely used in medicine. It has successfully invaded a wide variety of woodland habitats in central Ontario. The leaves resemble those of a lady's slipper orchid (*Cypripedium* sp.).

] *1 cm*

E.T.

ARUM FAMILY (ARACEAE)

General: Perennial; flowering stems single, 10–60 cm tall, with several papery sheaths at base; root short, thick, vertical underground stem (corm).

Leaves: Usually 2, compound, with 3 leaflets; leaflets elliptic to egg-shaped; lateral leaflets sharp-angled at base and asymmetrical, green on both sides; lateral veins join into 1 vein along margin; stalks 25–60 cm long, with bases sheathing flowering stems.

Flowers: Tiny, yellow, at base of a flowering tube ('the preacher in the pulpit') surrounded by a large tubular bract (spathe) that curls over in a pointed hood above; spathe green, brown, purple or striped; May–June.

Brenda Chambers

Fruits: Bright-red, rounded berries, 1- to 3-seeded, about 1 cm in diameter; in an egg-shaped cluster.

Habitat: Moist tolerant hardwood stands, often with yellow birch and eastern hemlock.

Notes: The size and colour of the spathe and the size and shape of the leaflets are variable. • Historically, the dried root (corm) was used in food and medicine. The roots are considered **dangerous** when fresh. Aboriginal peoples eliminated this by pounding and drying the nutritious roots before using them for flour. The plant was combined with black snakeroot (p. 238) and the bark of cherry (pp. 65–66) to make a medicine for coughs and fevers. **The berries are poisonous**, and the whole plant, particularly the root, is extremely acrid, even caustic to the tongue. See caution in Introduction.

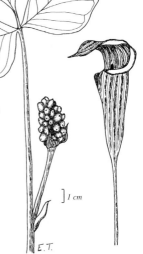

Food Use by Wildlife: berries. • **Birds:** ring-necked pheasant, wild turkey, wood thrush.

1 cm

E.T.

Emma Thurley

General: Perennial or biennial; flowering stems erect, firm, 15–60 cm tall; from white, coated, egg-shaped to conical 2–6 cm long bulbs clustered in a crown on short underground stems (rhizomes).

Leaves: Usually basal, 2 or 3, flat, simple, lance-shaped to elliptic, 10–30 cm long (including slender stalk), 2–6 cm wide, appear in early spring, shrivel before flowers appear.

Flowers: White; 6 'petals' (3 petals and 3 petal-like sepals), 5–7 mm long; rounded terminal clusters; June–July.

Fruits: Capsules, short, egg-shaped to rounded, depressed at tip with 3 deep lobes; 1–2 black seeds in each lobe.

Habitat: Fresh to moist, loamy to clayey tolerant hardwood stands; weakly to moderately calcareous sites.

Notes: This plant has a distinct onion-like odour. Aboriginal peoples ate the roots and nutritious bulbs either fresh or cooked. Historically, the bulb was also used in medicine. See caution in Introduction. • Wild leek bulbs can taint cows' milk.

] 1 cm

E.T.

General: Low perennial herb; flowering stem 10–20 cm tall; from a deep, solid, scaly underground organ (corm); offshoots from corm base may form colonies of many 1-leaved, sterile shoots and a few 2-leaved, fertile plants.

Leaves: Appear basal, but near middle of half-buried stem, 1 pair, flat lance-shaped to elliptic, 10–20 cm long, 2–4 cm wide, taper at both ends, usually mottled with purplish-brown.

Flowers: Yellow (rarely tinged purple), often spotted near base inside or darker-coloured outside, single, lily-like, nodding; 6 'petals' (3 petals and 3 petal-like sepals), lance-shaped, spreading, 18–40 mm long; April–May.

Fruits: Capsules, egg-shaped, narrowest at base.

Brenda Chambers

Habitat: Dry to moist, sandy to clayey tolerant hardwood stands; occasional in hardwood swamps.

Notes: This plant has distinctive, purplish-brown, mottled leaves and usually grows in patches. It was given its common name for its markings, which were said to resemble those of a brook trout. Another common name, 'adder's tongue lily,' refers to the purplish points the young leaves have as they emerge in spring. The leaves gradually disappear after the forest canopy develops.

1 cm

E.T.

BLUE BEAD LILY · *Clintonia borealis*
CLINTONIE BORÉALE

LILY FAMILY (LILIACEAE)

General: Perennial herb; flowering stem up to 40 cm tall; usually in patches.

Leaves: Basal, 2–4 (usually 3), simple, oblong or tongue-shaped, short-pointed at tip, clasping at base, up to 30 cm long, 4–9 cm wide, dark green, thick, leathery, upright-growing.

Flowers: Greenish-yellow, bell-shaped; 6 'petals' (3 petals and 3 petal-like sepals); in loose clusters of 2–8 flowers on tip of single, leafless stalk (scape); May–July.

Fruits: Bead-like berries, dark blue, shiny, oval to rounded, with several seeds; in clusters; ripen August.

Habitat: All moisture regimes and soil textures; all stand types except tolerant hardwood stands on dry calcareous sites.

Notes: Blue bead lily may be confused with three-leaved smilacina (p. 158), but blue bead lily leaves are basal whereas three-leaved smilacina leaves grow alternately along the stem. Also, when the leaves of three-leaved smilacina are held up to the light, they are slightly transparent and the parallel veins are clearly visible. • **These berries are poisonous.** The young leaves were considered edible in salads or as boiled greens and reportedly taste like cucumber. Aboriginal peoples used the leaves of blue bead lily as a plaster for bruises and sores. See caution in Introduction. • Blue bead lily is also commonly called 'yellow clintonia.'

*Brenda Chambers, top;
OMNR, bottom*

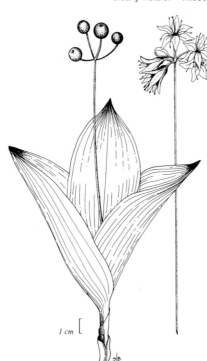

1 cm

LILY FAMILY (LILIACEAE)

General: Perennial herb; flowering stems curved upward, up to 1 m tall, slightly zigzag, reddish at nodes; from fleshy, brownish, jointed, underground stems (rhizomes) often 1 cm or more thick.

Leaves: Alternate, 5–12 or more along stem, arranged in 2 vertical rows, spreading, stalkless, simple, oblong to elliptic, pointed, 7–15 cm long, 2–7 cm wide, hairy along margins and on underside; veins parallel.

Flowers: Creamy-white, 3–5 mm wide; numerous, in showy terminal clusters 3–15 cm long; May–July.

Fruits: Berries, rounded, greenish, speckled with red or brown, mature translucent ruby-red speckled with purple, 1- to 2-seeded.

Brenda Chambers, both

Habitat: Dry to moist, sandy to clayey upland sites, tolerant hardwood and intolerant hardwood mixedwoods and pine-oak stands, forest edges.

Notes: Also known as *Smilacina racemosa*. • The leaves and young shoots are edible raw or cooked. In the past, the roots were used for medicine and food, the leaves were used in medicine and the berries were used for food. See caution in Introduction.

Food Use by Wildlife: See notes under Canada mayflower (p. 159).

1 cm

E.T.

LILY FAMILY (LILIACEAE)

Brenda Chambers

General: Perennial herb; flowering stems up to 20 cm tall, slender, erect; grows in patches.

Leaves: Alternate, 2–4 (usually 3), stalkless, simple, lance-shaped to oval, tapering to pointed tip, sheathing or clasping stem at base, 5–12.5 cm long, 1–5 cm wide, hairless, ascending.

Flowers: White; 6 'petals' (3 petals and 3 petal-like sepals), 8 mm wide; along slender stem above leaves; June–July.

Fruits: Dark-red, rounded berries; ripen in July.

Habitat: Wet organic conifer swamps; sphagnum bogs.

Notes: Also known as *Smilacina trifolia* • Three-leaved smilacina may be confused with Canada mayflower (p. 159). The leaves of Canada mayflower are shorter and have a heart-shaped, stalkless base that does not clasp the stem, whereas the leaves of three-leaved smilacina clasp the stem. The flowers of Canada mayflower have 2 petals and 2 petal-like sepals while the flowers of three-leaved smilacina have 3 petals and 3 petal-like sepals. See notes on blue bead lily (p. 156). • The berries have historically been considered edible, but they may cause diarrhea. See caution in Introduction.

Food Use by Wildlife: See notes under Canada mayflower (p. 159).

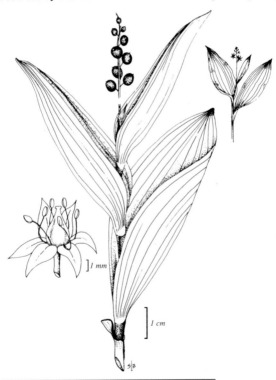

LILY FAMILY (LILIACEAE)

General: Perennial herb; flowering stems erect, 5–25 cm tall, often bent or zigzagged; spreads by slender, branching underground stems (rhizomes) and often forms extensive patches of single leaves.

Leaves: Alternate, 1–3 (usually 2) on flowering stems, stalkless or occasionally short-stalked, simple, egg-shaped, sharp- to blunt-tipped, heart-shaped at base, 2–10 cm long, 1.5–5.5 cm wide, dark green, hairless or sometimes finely hairy on underside.

Flowers: White, about 5 mm wide; 4 'petals' (2 petals and 2 petal-like sepals); in elongated, 1.5–5 cm long, terminal clusters with stalks usually longer than clusters; May–June.

Fruits: Speckled, pale-red, rounded berries, hard and green at first; ripen in July.

Habitat: All moisture regimes, soil textures and stand types.

Notes: Canada mayflower is also commonly known as 'wild lily-of-the-valley.' • This species may be confused with three-leaved smilacina (p. 158). Canada mayflower has 2 petals and 2 petal-like sepals, while three-leaved smilacina has 3 petals, 3 petal-like sepals and more elongated leaves that often clasp the stem. • Aboriginal peoples used the root in a sore-throat medicine. The berries are reported to be bitter-tasting and may cause diarrhea. See caution in Introduction.

Food Use of *Maianthemum* spp. by Wildlife: seed. • **Mammals:** snowshoe hare, eastern chipmunk, white-footed mouse. • **Birds:** ruffed grouse.

Karen Legasy, top; Brenda Chambers, bottom

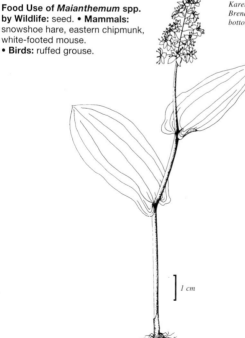

1 cm

LILY FAMILY (LILIACEAE)

General: Perennial herb, hairless, 20–50 cm tall at flowering time (to 1 m at maturity); stem appears to pierce leaves, forked above middle, with a few bladeless sheaths below; from slender, short underground stems (rhizomes) with fleshy roots.

Leaves: Alternate, simple, stalkless, with lower edges surrounding stem (appear pierced by stem), 0–2 below fork, 4–8 above fork, oval or oblong to lance-shaped, 4.5–13 cm long, 1.5–4 cm wide, usually whitish-hairy beneath.

Flowers: Yellow, nodding, lily-like; 6 distinct 'petals' (3 petals and 3 petal-like sepals), 2.5–5 cm long, taper to a slender point; April–June.

Fruits: Capsules, rounded or blunt at tip, 3-lobed in cross-section, with few seeds in each lobe.

Habitat: Dry to moist tolerant hardwood stands, often with basswood, red oak and ironwood, in southern part of our region.

Notes: Another common name is 'bellwort.'
• The entire plant has been used in food and medicine. Aboriginal peoples used the root to treat a variety of ailments, including stomach ailments. See caution in Introduction.

Brenda Chambers, both

• The species name *grandiflora* means 'large-flowered.'

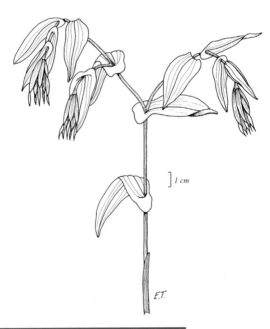

] *1 cm*

E.T.

LILY FAMILY (LILIACEAE)

General: Perennial herb; stems simple or forked, less than 30 cm tall, sparingly hairy (especially at nodes).

Leaves: Alternate, stalkless, simple, lance-shaped to oval, tapering to a long, pointed tip, rounded at base, 5–9 cm long, 2–3.5 cm wide, slightly clasping to not clasping stem; veins parallel with fine hairs beneath; margins toothless to minutely toothed, fringed with fine hairs.

Flowers: Rose-coloured to whitish, bell-shaped, about 1 cm long; 6 'petals' (3 petals

Karen Legasy

and 3 petal-like sepals), spreading near tips; stalks with a bend or twist in middle, 1–3 cm long; solitary, at nodes and hanging below leaves; May–June.

Fruits: Red berries with numerous seeds, rounded to oblong, about 1 cm long; ripen in July.

Habitat: Dry to fresh, sandy to clayey upland sites, occasional in wet organic sites; in most stand types except very dry pine and black spruce-pine stands.

Notes: White mandarin (p. 162) is a similar plant, but it has 1 (sometimes 2) greenish-white flowers per distinctly kinked stalk, its leaves have a heart-shaped, strongly clasping base and their margins don't have a fringe of fine hairs. • The roots of rose twisted-stalk were historically used in a cough remedy. The berries may cause diarrhea if eaten in quantity. See caution in Introduction.

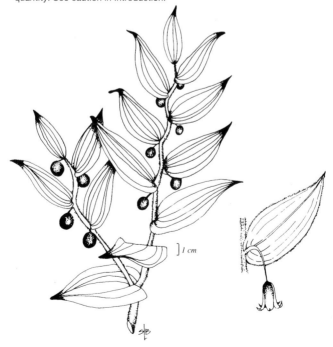

] *1 cm*

LILY FAMILY (LILIACEAE)

Brenda Chambers

General: Perennial herb; flowering stems stout, 40–100 cm tall, usually forked, hairless at nodes; from short, thick underground stems (rhizomes).

Leaves: Alternate, stalkless, clasping stem, simple, egg-shaped to oblong, abruptly pointed at tip, heart-shaped at base, 6–12 cm long, 2–5.5 cm wide, hairless; margins toothless or very fine- toothed.

Flowers: Greenish-white to purplish; 6 'petals' (3 petals and 3 petal-like sepals) spreading from near middle, 15 mm long; stalks 3–5 cm long, jointed and sharply bent above middle; single or in pairs, hanging below leaves; June–July.

Fruits: Elliptic, red berries, about 15 mm long, with numerous seeds.

Habitat: Moist to fresh sites, warm river valleys in the northern part of our region.

Notes: Another common name is 'twisted-stalk.'
• A close relative, rose twisted-stalk (p. 161), has finely hairy nodes and leaf undersides; its leaves are also stalkless, but lack a heart-shaped, clasping base.
• Aboriginal peoples used the berries and stems of this species for a variety of ailments. See caution in Introduction. • The genus name *Streptopus* is from the Greek *streptos*, 'twisted' and *pous*, 'foot or stalk,' referring to the sharply bent stalks of the flowers and fruits.

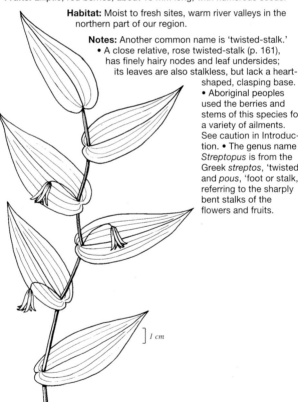

1 cm

E.T.

General: Slender perennial herb; flowering stems erect or arching, up to 1 m tall; from whitish, 1–1.8 cm thick underground stems (rhizomes) with many joints.

Leaves: Alternate, stalkless, simple, lance-shaped to broadly oval, 5–15 cm long, 1.5–7.5 cm wide, pointed; underside paler, finely hairy along prominent veins.

Brenda Chambers

Flowers: White to greenish, bell-shaped, 10–13 mm long, with 6 short lobes (3 petals and 3 petal-like sepals); usually in pairs, on long stalks, hanging from leaf axils; May–July.

Fruits: Dark-blue to black berries, several-seeded; usually in pairs.

Habitat: Dry to moist, rocky to clayey, upland tolerant hardwood and eastern white pine stands, hardwood swamps.

Notes: Aboriginal peoples ate the roots raw or cooked and made flour from them. A tea, also made from the roots, was used to treat coughs. Headaches were treated by inhaling the fumes from a root solution poured over hot stones. See caution in Introduction. • The species name *pubescens* means 'hairy,' and refers to the hairiness on the veins of the undersides of the leaves.

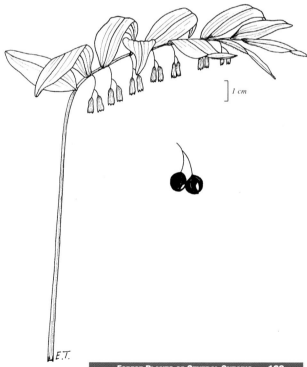

] *1 cm*

E.T.

WHITE TRILLIUM • *Trillium grandiflorum*
TRILLE GRANDIFLORE

LILY FAMILY (LILIACEAE)

Brenda Chambers

General: Low perennial herb; flowering stems 20–40 cm tall; from short, stout underground stems (rhizomes).

Leaves: In a whorl of 3 at top of stem, stalkless, simple, broad, egg-shaped, gradually narrowed below middle to base and pointed at tip, 8–12 cm long, net-veined.

Flowers: White; 3 petals, oblong to lance-shaped, pointed, wavy-edged, 4–6 cm long; sepals lance-shaped, shorter than petals, 3–5 cm long; stalk erect, 5–8 cm long; single, terminal; May.

Fruits: Rounded, 6-angled, single, black berries with many seeds.

Habitat: Moist to dry, sandy to coarse loamy tolerant hardwood stands; uncommon at high elevations of Algonquin Park and in northern part of our region.

Notes: The petals turn from white to pinkish before they wither. There are many colour variations in trillium petals caused by infections from micro-organisms; for example, green petals or white petals with green stripes. Aberrant flower and leaf forms are also found. • Historically, white trillium roots were used in medicine. Aboriginal peoples used them to treat a variety of female maladies. See caution in Introduction. • White trillium is the floral emblem of Ontario. It is the most widely distributed trillium in this province. • The species name *grandiflorum* means 'large-flowered.'

Food Use by Wildlife: leaves, flowers. • **Mammals:** white-tailed deer.

] *1 cm*

E.T.

LILY FAMILY (LILIACEAE)

General: Low perennial herb; flowering stems 15–40 cm tall; from short, stout, brown underground stem (rhizome) up to 3 cm thick.

Leaves: In a whorl of 3 at top of stem, stalkless, simple, broad, 4-sided, narrowing to base and tip, 4–19 cm long and broad, net-veined.

Flowers: Maroon-red; 3 petals lance-shaped to oval, pointed at tip, 2–4 cm long, spreading; sepals green, about same length as petals; stalk erect, 3–10 cm long; single, terminal; May.

Fruits: Flattened, egg-shaped, 6-angled, single, dark red berries with many seeds.

Habitat: Moist to dry, sandy to coarse loamy upland tolerant hardwood stands; hardwood swamps.

Notes: Another common name, 'wake-robin,' refers to the flowering time of this plant, which coincides with the return of robins in spring. • A pale yellow variant of red trillium (inset photo) has been found. • Historically, red trillium roots were used in medicine. See caution in Introduction. • Pollinating carrion flies are attracted to the flower by its colour and carrion odor.

Food Use by Wildlife: leaves, flowers.
• **Mammals:** white-tailed deer.

Brenda Chambers, both

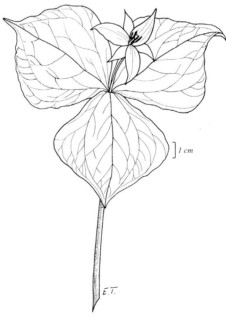

] *1 cm*

E.T.

LILY FAMILY (LILIACEAE)

General: Low perennial herb; flowering stems 20–40 cm tall; from short, thick, brown underground stems (rhizomes).

Leaves: In a whorl of 3 at top of stem, stalked, thin, simple, egg-shaped, rounded at base, pointed at tip, 5–10 cm long when flowers expand and larger at maturity, net-veined.

Flowers: White; 3 petals, streaked with purple at base, egg- to lance-shaped, 2–4 cm long, with wavy margins; sepals shorter than petals; stalk erect, 2–5 cm long; single, terminal; May–June.

OMNR

Fruits: Egg-shaped to elliptic, single, bright, shiny red berries with many seeds, 1.5–2 cm long, 3-angled, erect.

Habitat: Dry to fresh tolerant hardwood, conifer and intolerant mixedwood stands; less common in the northern part of our region.

Notes: The young leaves have a distinctive bronzy tint. Unlike those of white trillium (p. 164) or red trillium (p. 165), the leaves of painted trillium are stalked. • Historically, trillium roots were used in medicine. See caution in Introduction. • The species name *undulatum* means 'wavy,' referring to the wavy edges of the petals.

Food Use by Wildlife: leaves, flowers. • **Mammals:** white-tailed deer.

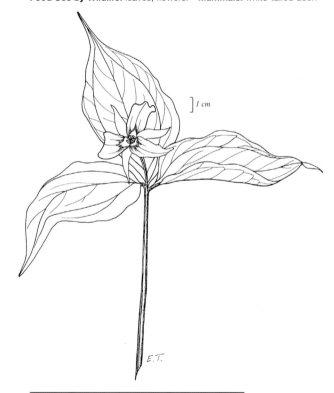

] *1 cm*

E.T.

LILY FAMILY (LILIACEAE)

General: Perennial herb; flowering stems 15–60 cm tall; 1 or several slender, unbranched stems; from brown underground stems (rhizomes) up to 3 cm thick.

Leaves: In a whorl of 3 at top of stem, short stalked, simple, diamond-shaped to rounded, pointed at tip, narrowed at base, 4–15 cm long and wide, net-veined.

Flowers: White, rarely pink, about 3.8 cm wide, nodding; 3 petals with tips bent slightly backward; 3 sepals; stalk curved, 0.5–2.5 cm long; solitary, terminal; May–June.

Fruits: Red to purplish berries, egg-shaped; late June–July.

Habitat: Moist to fresh, sandy to clayey intolerant hardwood mixedwood and eastern white cedar-yellow birch stands; hardwood swamps.

Notes: The species name *cernuum* comes from the Latin word *cernuus*, which means 'drooping' or 'nodding,' and refers to the flowers.

Karen Legasy, top;
Brenda Chambers, bottom

Food Use by Wildlife: leaves, flowers. • **Mammals:** white-tailed deer.

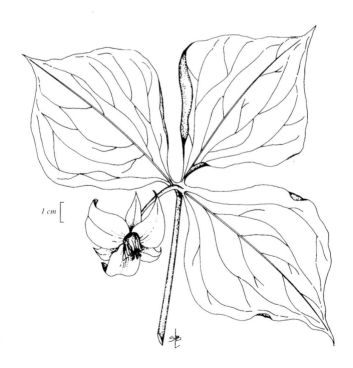

1 cm

LILY FAMILY (LILIACEAE)

General: Perennial herb; flowering stems slender, erect, 15–50 cm tall, woolly; from horizontal white tuber 3–8 cm long.

Leaves: Simple, stalkless, in 2 whorls; lower whorl near middle of stem, with 6–10 oval to lance-shaped leaves 6–12 cm long and 2–3 cm wide; upper whorl at top of stem with 3, pointed, oval leaves up to 7 cm long.

Flowers: Yellowish-green, nodding; 6 'petals' (3 petals and 3 petal-like sepals) 8–10 mm long; stalks 15–25 mm long; in clusters of 3–9 nodding flowers from centre of upper leaf whorl; May–June.

Fruits: Rounded, dark-purple to blue berries, with few seeds.

Habitat: Dry to moist, sandy to clayey tolerant hardwood stands.

Notes: This herb was named for its edible root, which tastes like starchy cucumber. The roots were used in medicine and for food. See caution in Introduction.

Emma Thurley

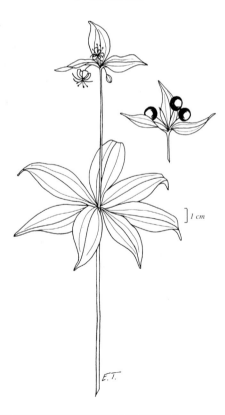

] *1 cm*

E.T.

IRIS FAMILY (IRIDACEAE)

General: Perennial herb, 10–60 cm tall, erect, tufted, usually pale green; flowering stems unbranched (rarely slightly forked), pale, distinctly flattened and winged along edges.

Leaves: Basal, grass-like, pointed at tips, 10–50 cm long, over 6 mm wide, half the flowering stem height or longer; margins finely toothed.

Flowers: Violet-blue, with yellow centre; 6 'petals' (3 petals and 3 petal-like sepals), slender point at tips; stalks short, at stem tips, usually exceeded by a pointed bract; June–July.

Fruits: Capsules, widely oval to rounded, whitish-green to straw-coloured or pale brown, 3–6 mm long.

Habitat: In ditches and wet clearings.

Notes: This plant gets its common name from the way the flowers appear like 'blue eyes' on grass-like stems.

Linda Kershaw, top;
Derek Johnson, bottom

1 cm

IRIS FAMILY (IRIDACEAE)

General: Perennial herb; flowering stems erect, 60–90 cm tall, unbranched or with 1–2 branches above, taller than leaves; from thick, fleshy, horizontal underground stems (rhizomes) covered with fibrous roots and either deeply buried or near surface.

Leaves: Basal leaves stalkless, simple, sword-like, 20–80 cm long, 0.5–3 cm wide, pale green to greyish, often purplish at base when fresh, firm, erect; stem leaves alternate, prolonged but shorter than basal leaves.

Flowers: Blue-violet with darker veins and yellow, green and white colouring toward base or middle of flower, 6.3–10 cm wide; 3 petal-like sepals spread horizontally with 3 petal-like styles arching over them; 3 petals, narrower than sepals, erect; May–July.

Fruits: Capsules, oblong, 3-sided, 1.5–6 cm long, thick, firm; inner surface looks varnished; seeds D-shaped, brown, 5–8 mm long; open late and often persist over winter.

Habitat: Wet areas, swamps, marshes, wet fields, ditches, shorelines.

Brenda Chambers, both

Notes: The rootstock is extremely poisonous. • Aboriginal peoples used the roots in poultices for sores, inflammation, burns and any other type of pain. See caution in Introduction.
• Aboriginal peoples also used blue flag leaves to make various shades of green dye and to weave mats and baskets.
• Some people believed snakes would shun the scent of blue flag, so they carried a piece of the root and handled it from time to time to protect themselves from poisonous snakes.

1 cm

General: Perennial herb, 1–2.7 m tall; stems filled with pith; from coarse, creeping underground stems (rhizomes).

Leaves: Alternate, sheathing, simple, long, flat and narrow (grass-like), up to 2.5 cm wide, taller than stem, greyish-green, somewhat spongy.

Flowers: Tiny, numerous, male and female in adjacent terminal spikes each up to 15 cm long; female spike lowermost, 1.2–3.5 cm thick, greenish, dark brown at maturity; male spike above female, cream-coloured, bright yellow at maturity, deteriorates after pollen release leaving a bare stalk; May–July.

Fruits: Dry, hard, single-seeded (achenes), elliptic, about 1 mm long, designed to float with numerous long, slender hairs at base.

Karen Legasy

Habitat: Marshes, occasionally colonizes wet ditches and clearings.

Notes: The roots contain carbohydrates, and have historically been eaten raw or baked. See caution in Introduction.
• The leaves can be used to weave baskets, rugs and other items.

Food Use by Wildlife: plant.
• **Mammals:** muskrat (primary food source).

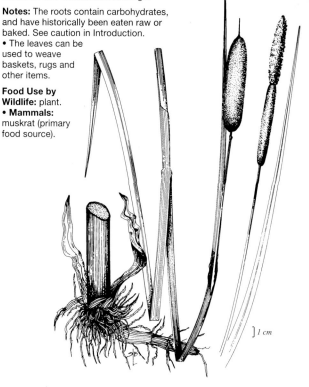

] *1 cm*

PURPLE PITCHER PLANT • *Sarracenia purpurea*
SARRACÉNIE POURPRE

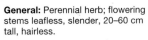

Karen Legasy, top;
Linda Kershaw, bottom

General: Perennial herb; flowering stems leafless, slender, 20–60 cm tall, hairless.

Leaves: In a basal rosette, simple, pitcher-shaped with an erect, broad, flap-like hood, 10–30 cm long, green or yellowish, purple-veined or sometimes veinless, spreading to ascending, curved, winged; inner surface of hood covered in dense, stiff, downward-pointing hairs; base narrows to a stalk; .

Flowers: Purplish-red or occasionally yellowish, rounded, 5–7 cm wide; 5 petals arch over yellowish, umbrella-like style; solitary, nodding; May–August.

Fruits: Capsules, granular on surface, 5-sectioned, contain numerous seeds.

Habitat: Open sphagnum bogs.

Notes: Aboriginal peoples and used the roots to make an infusion for curing smallpox. See caution in Introduction. • The hollow, inflated leaves were used as drinking cups by people out in the forest. • Pitcher plants are carnivorous. They attract and trap insects in their leaves with the stiff hairs in their hoods. The hairs force the insects down through the smooth-surfaced, narrow neck and into the cavity below. The plants then release secretions to prevent the insect from escaping and to dissolve it.

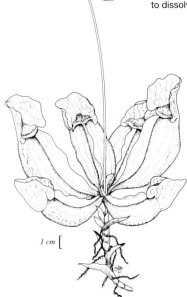

1 cm

General: Perennial herb, small; flowering stems erect, 5–25 cm tall, slender, hairless, rises from rosette of leaves.

Leaves: In a basal rosette; blades at least as wide as long, 4–10 mm long, abruptly narrow to a flat, hairy stalk 1.3–5 cm long; upper surface and margins covered with reddish, glandular hairs that release sticky fluid.

Flowers: White to red or pinkish, about 6 mm wide; 5 petals; 3–15 flowers along 1 side of stem; June–August.

Fruits: Capsules; seeds elongated and tapered at both ends.

Habitat: Wet organic black spruce stands; nutrient-poor soils, wet sand, sphagnum bogs and wet fields; silty and boggy shores.

OMNR

Notes: Two related species are slender-leaved sundew (*Drosera linearis*), with narrow leaves, and spatulate-leaved sundew (*D. intermedia*), with spatula-shaped leaves. • Sundew plants eat insects, which are attracted to the glistening, sticky fluid on the glandular leaf hairs. When an insect lands on 1 hair, the other hairs sense it and bend over the insect, making it adhere to their sticky fluid. The hair glands then release an acid and enzymes that dissolve all the soft parts of the insect, and other leaf glands absorb the released nutrients. The same insects used for food often pollinate these plants.

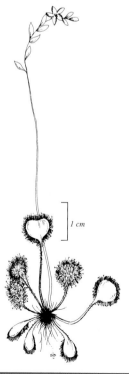

1 cm

BUCKWHEAT FAMILY (POLYGONACEAE)

Karen Legasy

General: Perennial; stems 0.3–3 m long, reddish, twining, laying on ground or nearly erect, freely branching and climbing, sparingly hairy.

Leaves: Alternate, stalked, simple, widely triangular-egg-shaped or sometimes arrowhead-shaped, pointed at tip, deeply heart-shaped at base, 5–12 cm long, 2.5–10 cm wide, finely hairy; leaf-like bracts (stipules) abruptly bent and bristly at base; margins wavy.

Flowers: White or pink-tinged, small; in branched clusters, from leaf-axils or terminal; June–September.

Fruits: Dry, hard, single-seeded (achenes), 3-angled, oblongish and pyramid-shaped or inversely egg-shaped, 3–4 mm long, hairless, shiny, black.

Habitat: Moist to dry, fine loamy to sandy upland pine and tolerant hardwood stands; dry roadsides, in cutovers and other disturbed sites; occasional in hardwood swamps.

Notes: Fringed bindweed forms vigorous mats in cutovers.

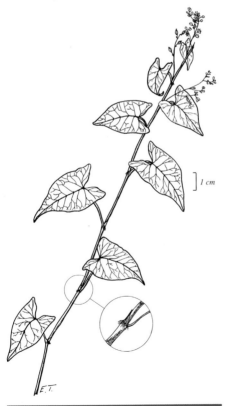

1 cm

E.T.

BUCKWHEAT FAMILY (POLYGONACEAE)

General: Perennial or annual herb; flowering stems slender, erect or almost so, 15–50 cm tall, unbranched or branched; from woody, horizontal or creeping underground stems (rhizomes).

Leaves: Alternate, simple, narrowly arrow-shaped, blunt or pointed at tip, 2-lobed at base, 2.5–10 cm long, covered with minute, nipple-shaped projections (papillae); basal leaves long-stalked; stem leaves short-stalked or stalkless.

Flowers: Reddish to yellowish, small; 3 petals and 3 petal-like sepals; on stalks in erect, loose, narrow clusters; June–September.

Fruits: Dry, hard, single-seeded (achenes), nut-like, shiny, yellowish-brown, 3-angled.

Habitat: Disturbed areas, roadsides, fields.

Notes: The crushed leaves and stems may irritate skin and cause an allergic reaction, and the pollen may trigger an **allergic reaction** which takes the form of respiratory problems. • The leaves are rich in vitamin C and apparently have a sour taste like that of rhubarb. The leaves have been found to contain **oxalic acid**, which can interfere with calcium metabolism. See caution in Introduction.

Ray Demey

Food Use by Wildlife: seed.
• **Mammals:** eastern cottontail, white-footed mouse. • **Birds:** ring-necked pheasant, ruffed grouse, woodcock, wild turkey, American tree sparrow.

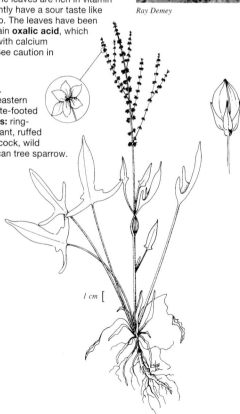

1 cm

CAROLINA SPRING BEAUTY • *Claytonia caroliniana*
CLAYTONIE DE CAROLINE

PURSLANE FAMILY (PORTULACACEAE)

Brenda Chambers

General: Perennial herb; stems delicate, 3–30 cm tall; from deep rounded tubers 0.5–2.7 cm thick; plants disappear by early summer.

Leaves: Opposite, 1 pair on stem below flowers, stalked, simple, broadly lance-shaped to narrowly oval, 3–6 cm long (including stalk), 1–2 cm wide.

Flowers: White or pink with deeper pink veins, showy; 5 petals, 0.5–1.5 cm long; 2–11 in loose terminal clusters; April–May.

Fruits: Capsules, 3–6 seeded, open by inrolling sections.

Habitat: All moisture regimes and soil textures, tolerant hardwood stands.

Notes: Virginia spring beauty (*Claytonia virginica*) is a similar species, but it has a thicker corm (1–5 cm) and narrower (2–10 mm wide) leaves without distinct stalks. • The leaves are high in vitamins A and C, and the small, starchy roots can be cooked and eaten. Historically, the tuber was used for food. See caution in Introduction. • As the common name suggests, this plant is one of the first to appear in the spring. Its flowers close at night and during storms, when pollinating insects are not active. The stems and leaves wither and disappear soon after the plant flowers.

Food Use by Wildlife: seeds. • **Mammals:** white-footed mice.

1 cm

E.T.

General: Perennial herb; flowering stems 30–200 cm tall, with few, 0.75–2 mm long stinging bristles on leaves and upper stem; from spreading underground stems (rhizomes).

Leaves: Opposite, stalked, simple, lance-shaped, 5–15 cm long, rounded at base; margins coarsely toothed.

Flowers: Greenish, 1–2 mm long, no petals, male or female on same plants; leaf-like bracts (stipules) at base of flower clusters linear to lance-shaped, 5–15 mm long, sparsely hairy; in many-flowered, branched clusters from axils of upper leaves and longer than leaf stalks; June–September.

Fruits: Egg-shaped, flattened, erect, dry, hard, single-seeded (achenes), 1.5 mm long, enclosed by 2 sepals.

Habitat: Moist, coarse loamy to clayey, hardwood stands, with balsam poplar and black ash; dry calcareous tolerant hardwood stands.

Bill Crins

Notes: The introduced slender nettle (*Urtica dioica*), has many stiff stinging bristles on its leaves and stem, and has heart-shaped leaves. It usually has male and female flowers on separate plants. Stinging nettle is uncommon in our region. • As with docks (*Rumex* spp.), the juice of the plant can be used to treat the sting its sharp spines impose on the skin! Burns, hives and poisonous stings and bites can be treated with an extract made from the leaves and seeds. Historically, the whole herb, leaves, seeds, flowers, roots or the greenish juice were used in medicine. Roots, seeds and extracts of oils from nettles are used to stimulate hair growth and to treat general hair troubles. • The plants, when picked early, can be cooked as greens that are high in vitamins A, B and D. The heat of cooking causes the spines to collapse. See caution in Introduction. • Stems of older plants provide strong fibres which have been used to make thread, rope and fishing nets.

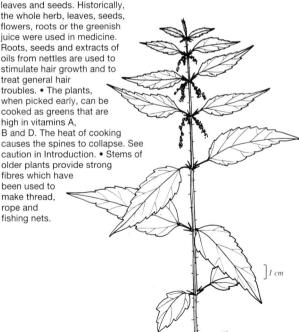

] *1 cm*

E.T.

NORTHERN COMANDRA • *Geocaulon lividum*
COMMANDRE LIVIDE

OMNR

General: Perennial herb; flowering stems erect, 10–25 cm tall, usually unbranched; grow as individuals or in clumps from slender, reddish, creeping underground stems (rhizomes).

Leaves: Alternate, short-stalked, simple, oval, blunt or rounded at tip, narrowed at base, 1–2.5 cm long, 0.5–1 cm wide, thin, pale lead-coloured to purplish.

Flowers: Small, greenish to purplish; 5 petal-like sepals, about 2 mm long; in clusters of 2–4 from leaf axils; May–June.

Fruits: Rounded, fleshy or pulpy, berry-like drupes, fluorescent-orange to scarlet, 6–10 mm in diameter; solitary or rarely 2; ripen August–September.

Habitat: Rock outcrops and moist shorelines.

Notes: This plant is also known as *Comandra livida*. • Some people have considered the berries edible, but they are apparently not very tasty. See caution in Introduction. • Northern comandra is a parasite on the roots of other plants.

] *1 cm*

General: Perennial herb, low-lying, up to 10 cm tall; from long, creeping, branched underground stems (rhizomes), produces 2 leaves and 1 flower annually.

Leaves: 1 pair, long-stalked, heart-shaped, tending to be wider than long, 8–12 cm wide (larger at maturity), hairy.

Flowers: Purplish-brown, 2–4 cm long, no petals; 3 petal-like sepals form tube at base with 3 slender-pointed, spreading lobes; stalk stout, hairy, 2–5 cm long; solitary, from base of leaves; April–May.

Emma Thurley

Fruits: Fleshy, solitary capsules, 1–1.5 cm wide, bursting irregularly, with wrinkled, egg-shaped seeds.

Habitat: Moist to fresh, fine loamy to sandy, intolerant hardwood mixedwood and black ash stands.

Notes: Wild ginger root has a strong, ginger-like odour and taste and it has been used as a spice by aboriginal and other peoples. Wild ginger was often combined with food to protect the food from possible sources of contamination (of the superstitious kind). • Historically, the root was used to treat whooping cough and stomach ills, to ease childbirth, and to dress wounds. The roots were combined with those of spikenard (*Aralia racemosa,* p. 235) to produce a mash which was applied to a fractured arm. A dressing and thin cedar splint were then placed on top. The whole plant was fermented for beverages. See caution in Introduction.

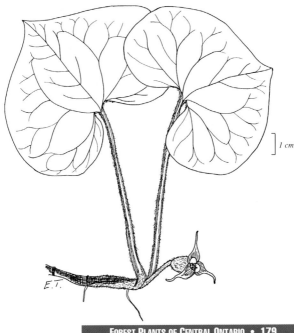

] 1 cm

E.T.

MOUSE-EAR CHICKWEED • *Cerastium fontanum*
CÉRAISTE VULGAIRE   ssp. *triviale*

PINK FAMILY (CARYOPHYLLACEAE)

Karen Legasy

General: Perennial herb; flowering stems 10–65 cm tall, horizontally spreading, ascending or erect, hairy.

Leaves: Opposite, stalkless, simple, 0.5–4 cm long, 1.5–15 mm wide, hairy; basal and lower leaves spoon-shaped to oblong, blunt-tipped; upper leaves oblong, blunt to pointed at tip.

Flowers: White, about 6 mm wide; petals 5, deeply notched; in terminal clusters of 3–60; May–September.

Fruits: Capsules, cylindrical, curved, 7–11 mm long; seeds brown, 0.5–0.7 mm in diameter.

Habitat: Grassy or sandy open areas, fields, roadsides.

Notes: Mouse-ear chickweed was formerly known as *Cerastium vulgatum*. • Historically, the leaves were boiled and eaten as greens. See caution in Introduction. • The common name 'mouse-ear' refers to the fuzzy little leaves. • This plant is often a troublesome weed in gardens and lawns. See notes on northern chickweed (p. 181).

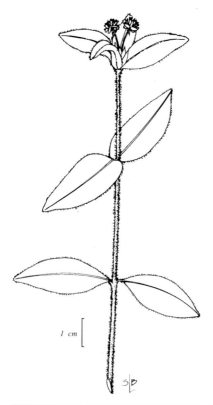

1 cm

PINK FAMILY (CARYOPHYLLACEAE)

General: Perennial; stems 3–50 cm long, hairless or slightly rough, unbranched or loosely branched, reclining, ascending or erect, often matted; from slender underground stems (rhizomes).

Leaves: Opposite, stalkless, simple, narrowly egg-shaped or elliptic to lance-shaped, pointed or slightly pointed at tip, 0.8–4 cm long, 2–8 mm wide, hairless except for a few minute hairs on margin near base.

Flowers: White to greenish, small; petals absent or shorter than sepals; 5 sepals widely lance-shaped to oblong, pointed or blunt-tipped, 2–5.5 mm long; solitary in leaf axils or in open terminal clusters; May–August.

Linda Kershaw

Fruits: Capsules, cone-shaped to inversely egg-shaped, pale brown, thin-walled.

Habitat: Wet areas, meadows, thickets, moist forest openings, wet ditches, streambanks.

Notes: This species was formerly known as *Stellaria calycantha*.
• Northern chickweed varies greatly in its branching patterns and leaf size. This plant is relatively easy to distinguish from mouse-ear chickweed (p. 180). Mouse-ear chickweed is hairy, and its leaves are hairy all over, but northern chickweed is relatively hairless and its leaves have only a few minute hairs along their margins near the leaf base.

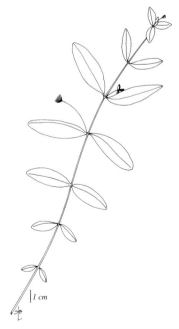

1 cm

General: Perennial herb; flowering stems 7–15 cm tall; from small, slender, creeping, underground stems (rhizomes), distinctively bright yellow or gold.

Leaves: Basal, on long, slender stalks, triangular in outline, 2.5–5 cm wide, compound, with 3 stalkless leaflets; upper surface shiny, dark green; underside paler; margins with rounded teeth.

Flowers: White, star-shaped; 5–7 petals; usually solitary at tip of long, slender, leafless stalks; June–July.

Fruits: Slender capsules (follicles), splitting along 1 side, beaked, 7–12 mm long; in a spreading cluster of 3–7; July–August.

Habitat: Wet organic to moist hardwood and conifer swamps; fresh to moist uplands, in many stand types.

Notes: You can recognize goldthread by its shiny, dark green, trifoliate leafs and by its golden, thread-like roots. • Aboriginal peoples made a yellow dye from the roots. The roots were also used in a mouthwash for cankers, sore throat, sore gums and teething. See caution in Introduction.

Brenda Chambers, top; Linda Kershaw, bottom

Food Use by Wildlife: leaves. • **Mammals:** white-tailed deer. • **Birds:** ruffed grouse.

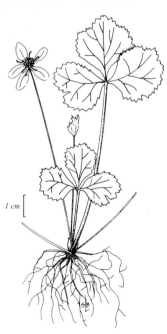

1 cm

CROWFOOT FAMILY (RANUNCULACEAE)

General: Perennial herb; flowering stems slender, 20–70 cm tall; from stout underground stems (rhizomes).

Leaves: Compound, twice divided in 3s, gradually reduced upward on stem with fewer leaflets; leaflets broadly egg-shaped to rounded, cut 1/3 to 1/2 their length, hairless; margins with rounded teeth or lobes.

Flowers: Scarlet-red, 3–4 cm long, showy; 5 petals, funnel-shaped, scarlet-red, yellow inside, prolonged backward into a long, nearly straight, red spur; nodding at branch ends; April–July.

Fruits: Slender capsules (follicles), splitting along 1 side, 1.5–2 cm long, several-seeded; in an erect head of 5.

Habitat: Dry, open rocky areas, forest edges.

Notes: Historically, the roots of this plant were used for food and medicine. Aboriginal peoples used the roots to treat stomach ailments. See caution in Introduction. • The genus name *Aquilegia*, from the Latin *aquila*, 'eagle,' refers to the long spurs, which were said to resemble the talons of an eagle. • The red colour of these flowers attracts hummingbirds, whose long tongues can reach the nectar at the base of the spurs.

Brenda Chambers, both

] *1 cm*

BUTTERCUP FAMILY (RANUNCULACEAE)

General: Perennial herb; flowering stems 20–70 cm tall, hairy, branching from centre of leaf whorls or crowns; from underground stems (rhizomes).

Leaves: Alternate, deeply cut into 3–5, deeply-lobed sections, wider than long; basal leaves long-stalked, 5–15 cm wide; stem leaves stalkless; veins prominent, covered with soft hairs on underside; margins toothed toward tip of lobes.

Flowers: White; 5 unequal, widely oblong, petal-like sepals 1–2.5 cm long; stalks solitary from centre of leaf whorls; June–July.

OMNR

Fruits: Dry, hard, single-seeded (achenes), flat, almost circular, hairy, with a stout style at tip; in rounded heads.

Habitat: Along gravelly shores, in dry, open areas and along roadsides.

Notes: Anemones contain a **toxic compound** which may irritate skin. Aboriginal peoples washed sores and wounds in a remedy made from anemone roots and leaves. Some aboriginal peoples chewed the roots before singing to help clear their throats. See caution in Introduction.

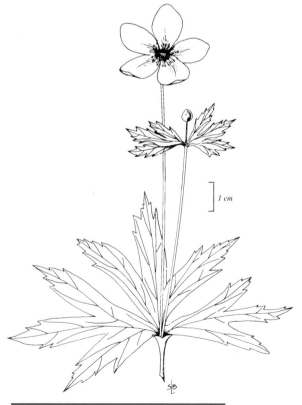

1 cm

BUTTERCUP FAMILY (RANUNCULACEAE)

General: Perennial herb, delicate; stems slightly hairy, 5–30 cm tall; from whitish, crisp, toothed, horizontal, 1–4 mm thick underground stems (rhizomes).

Leaves: Alternate, deeply cut into 3–5 sections, narrowly triangular to egg-shaped or oblong, narrower toward base, 1–5 cm long; basal leaf single, long-stalked; stem leaves in whorls of 3, similar but smaller; margins irregularly and coarsely toothed.

Flowers: Whitish to purplish or pinkish-tinged; 4–9 (usually 5) petal-like sepals 0.6–2.5 cm long; solitary; May–June.

Fruits: Dry, hard, single-seeded (achenes), hairy; in rounded clusters; develop in late June–July.

Habitat: Moist to fresh, sandy to silty, upland, intolerant hardwood stands and jack pine-black spruce stands; hardwood swamps; in northern part of this region.

OMNR

Notes: Wood anemone contains a **toxic compound** which may irritate skin. • Anemones have been called 'wind flowers' because their slender stalks tremble easily in a breeze.

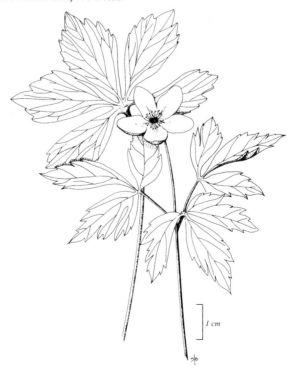

1 cm

BUTTERCUP FAMILY (RANUNCULACEAE)

General: Low perennial herb; flowering stems hairy, 5–15 cm tall, bearing a single flower.

Leaves: Basal, simple, thick, 3-lobed; lobes broadly rounded, cut to near middle of leaf; terminal lobe often wider than long; appear after flowers die and persist through winter.

Flowers: Blue-lavender to pink or white, 12–25 mm wide, with 6 petal-like sepals and 3 sepal-like bracts; bracts blunt, toothless; solitary, appear among previous year's leaves in April–May.

Fruits: Lance- or spindle-shaped, hairy, dry, hard, single-seeded (achenes), 1–1.4 mm thick, tipped with a slender style.

Habitat: Dry to moist, sandy to loamy, tolerant hardwood stands.

Brenda Chambers, both

Notes: This species is also known as *Heptatica triloba*. • Blunt-lobed hepatica (*H. acutiloba*) (inset photo), occasional in the southern part of our region, is distinguished by its pointed leaf lobes. • Historically, round-lobed hepatica was used in medicine and as an ornamental herb in gardens. See caution in Introduction. • The leaf shape resembles that of a mammalian liver, hence the genus name *Hepatica*.

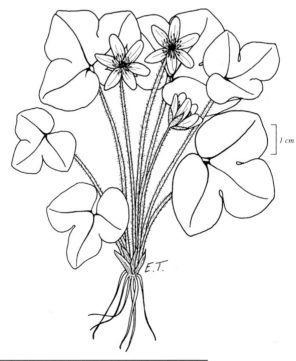

1 cm

E.T.

BUTTERCUP FAMILY (RANUNCULACEAE)

General: Perennial herb; flowering stems 1-several from slender roots, erect, 60–90 cm tall, hairy, branched above, hollow.

Leaves: Basal leaves long-stalked, simple, 2.5–10 cm wide, palmately lobed with 3–7 (usually 5) stalkless, deeply cut sections (lobes) with pointed tips; stem leaves alternate, short-stalked, smaller and scattered.

Flowers: Yellow, glossy, about 2.5 cm wide; 5 petals; 5 sepals small, greenish; stalks hairy; in loose clusters; May–September.

Karen Legasy

Fruits: Dry, hard, single-seeded (achenes), hairless, flattish, margined, 2.5–3.5 mm long with a curved beak; in rounded heads.

Habitat: Dry to moist soils, roadsides, occasional in intolerant hardwood mixedwood stands.

Notes: Tall buttercup leaves were historically used in a headache remedy. The roots contain an antibiotic principle, protoanemonine, which apparently acts against a broad range of bacteria. See caution in Introduction.
• The species name *acris* refers to the acrid stem and leaf juice. This **juice can burn** the mouth and mucous membranes, and can even **blister** sensitive skin.

Food Use of Buttercups (*Ranunculus* spp.) by Wildlife: leaf, seed. • **Mammals:** eastern cottontail, eastern chipmunk, striped skunk.
• **Birds:** ring-necked pheasant, ruffed grouse, wild turkey.

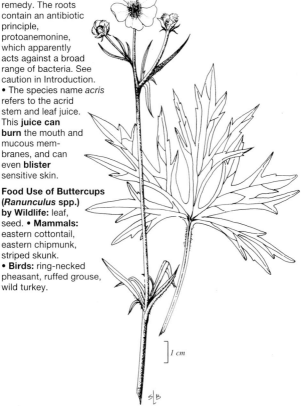

] *1 cm*

BUTTERCUP FAMILY (RANUNCULACEAE)

General: Perennial herb; flowering stems hairless or sparingly hairy, green, erect, 15–60 cm tall, branched; roots slender, fibrous.

Leaves: Alternate; basal leaves long-stalked, simple, kidney-shaped to round, blunt at tip, heart-shaped at base, 1.3–3.8 cm wide, bright green, thick, with rounded teeth, sometimes lobed; stem leaves stalkless, 3- to 5-lobed, mostly toothed.

Flowers: Yellow; petals oblong, 2–3 mm long, shorter than sepals; sepals 5, petal-like; few- to many-flowered; May–August.

Fruits: Dry, hard, single-seeded (achenes), shiny, rounded, 1.2–1.5 mm long, tipped with minute, curved beak; in a rounded to egg-shaped head.

Habitat: Wet organic hardwood swamps; moist to dry upland intolerant hardwood mixedwoods and tolerant hardwood stands.

Brenda Chambers

Notes: With their small petals, the flowers of kidneyleaf buttercup do not look much like buttercups. • Kidneyleaf buttercup was historically used to treat syphilis and to help increase perspiration. See caution in Introduction.

Food Use by Wildlife: See notes under tall buttercup (p. 187).

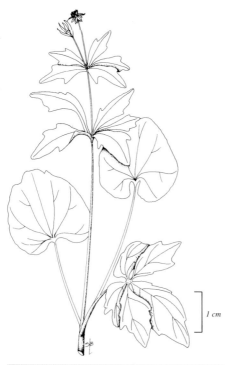

1 cm

General: Perennial herb; flowering stems 15–80 cm tall, erect or reclining, hairless, grooved, branching, thick, hollow, succulent.

Leaves: Alternate, stalked, simple, rounded to heart- or kidney-shaped, 5–17.5 cm wide, glossy, dark green; upper leaves become stalkless; margins with pointed to rounded teeth or nearly toothless.

Brenda Chambers

Flowers: Bright yellow, showy, 2.5–4 cm wide; 5–9 petal-like sepals, widely oval to narrowly egg-shaped; on stalks from leaf axils at top of stem; April–June.

Fruits: Prominently beaked, shiny pods (follicles), split open along 1 side, 1–1.5 cm long, contain numerous small, black seeds; arranged in a small head of 6–12.

Habitat: Wet organic hardwood and conifer swamps to moist, fine-loamy to clayey upland sites; wet areas in meadows, open woods, ditches, marshy creeks.

Notes: The flowers resemble large buttercups. • Marsh marigold leaves contain a toxin that makes them **poisonous when raw.** It is said that the toxin can be destroyed with cooking, but that has not been verified. See caution in Introduction.

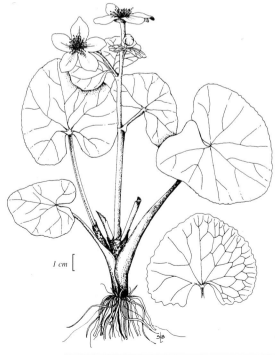

1 cm

General: Perennial herb, bushy, hairy to hairless; flowering stems 30–80 cm tall, erect.

Leaves: Alternate, stalked, compound, divided 2–3 times into groups of 3; leaflets egg-shaped, about 6 cm long, commonly hairy on veins beneath, sharply toothed with irregular, lobed margins.

Flowers: White, about 7–10 mm wide; 4–10 narrow petals; in dense, terminal, egg-shaped or cylindrical clusters on long stalks; May–June.

Fruits: Rounded, red (sometimes white) berries, shiny, on thin stalks; in clusters; ripen July–August.

Habitat: Moist to dry, sandy to clayey upland tolerant hardwood and intolerant hardwood mixedwood stands.

Karen Legasy, top;
Bill Crins, bottom

Notes: White baneberry (*Actaea pachypoda*) (inset photo) is a similar species, but its berries ripen to white instead of red and its flower stalks are thick while those of red baneberry are thin. • After giving birth, aboriginal women drank a tea made from the roots of red baneberry, to 'clear up their systems.' **The berries of red and white baneberry are considered poisonous.** All parts of the red baneberry plant are said to contain a **toxic oil**. See caution in Introduction.

Food Use by Wildlife: fruit. • **Birds:** ruffed grouse.

1 cm

Buttercup Family (Ranunculaceae)

General: Perennial herb; flowering stems erect, 0.5–2.5 m tall, stout, branched and leafy.

Leaves: Alternate, compound, 3–4 times divided in 3s; leaflets usually rounded to oblong, 3-lobed at tip, firm, paler and sometimes hairy on underside; stem leaves stalkless.

Flowers: White, rarely purplish, male or female (rarely with both sexes); no petals; 4–5 sepals 2–3.5 mm long, oblong, blunt, soon falling; stamens in showy clusters with pollen-bearing portions (anthers) 0.7–1.3 mm long, on white, stiff, club-shaped, erect stalks (filaments) 3.5–5 mm long; numerous in loose terminal clusters; June–August.

Fruits: Dry, hard, single-seeded (achenes), 3–5 mm long, about 1 mm wide, narrowed at base and tip, hairless.

Habitat: Wet organic hardwood and conifer swamps; streambanks.

Emma Thurley

Notes: Meadow-rue (*Thalictrum dasycarpum*), a similar species with larger anthers (1.6–3.5 mm long) and larger achenes (4–6 mm long), occurs more commonly in the northern part of this region.

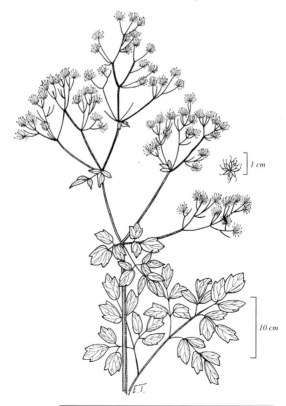

1 cm

10 cm

E.T.

General: Perennial herb, hairless, dark greenish-purple, with a waxy coating (bloom) when young; flowering stems simple, erect, 30–80 cm tall; from thick, long, matted, knotty rootstocks.

Leaves: Compound; main leaf about halfway up stem, divided into 3-stalked sections, each with 3–5 leaflets; leaflets egg-shaped to oblong, 2–5-lobed near tip, 5–8 cm long; smaller leaf below flower cluster.

Brenda Chambers

Flowers: Maroon to yellowish-green, nearly 1 cm wide, stalked; 6 petals much smaller than and opposite to 6 sepals; in loose terminal clusters 3–6 cm long; appear before the leaf is fully open; April–May.

Fruits: 2 round, dark-blue, berry-like seeds, 5–8 mm long, on thick stalks nearly as long as seeds.

Habitat: Moist to fresh, often calcareous upland tolerant hardwood stands.

Notes: The **berries are considered poisonous**. Aboriginal peoples and early white settlers drank a tea made from the roots of this plant to promote menstruation and rapid childbirth. Historically, the root was used in medicine, and the seeds and fruit were used for food. See caution in Introduction. • This plant was also called 'squawroot.'

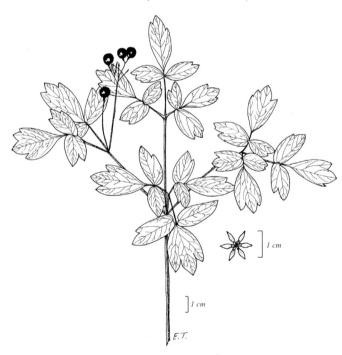

1 cm

1 cm

E.T.

General: Perennial herb, delicate; flowering stems solitary, erect, 5–20 cm tall, usually leafless; from underground stems (rhizomes).

Leaves: Basal, long-stalked, simple, 1–3 cm long, rounded to kidney-shaped with heart-shaped base; upper surface with distinctive, bristly, stiff, whisker-like hairs; margins slightly lobed or deeply toothed.

Flowers: Yellowish to greenish, saucer-shaped; 5 hair-like petals with long fringes; on separate stalks along flowering stems in a long, loose cluster; May–June.

Fruits: Capsules, short, flat; seeds black, smooth, shiny.

Habitat: Wet organic conifer and hardwood swamps; moist sandy to clayey upland sites; upland eastern white cedar mixedwoods.

Frank Boas, both

Notes: You can frequently see just the leaves with their scattered, whisker-like hairs. • The common name 'mitrewort' presumably comes from the seed capsules, which were thought to resemble a bishop's mitre or headdress.

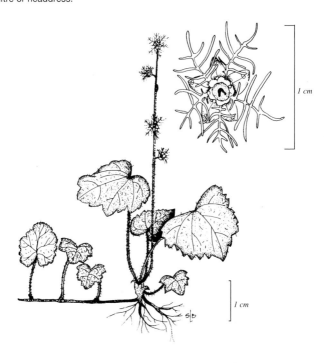

1 cm

1 cm

SAXIFRAGE FAMILY (SAXIFRAGACEAE)

General: Low perennial herb; flowering stems erect, 10–40 cm tall, sparsely hairy below pair of opposite stem leaves, glandular-hairy above; from stout underground stems (rhizomes).

Leaves: Basal leaves long-stalked, heart-shaped, with 3–5 shallow lobes, 3–5 cm long and wide, hairy, with elongated lobe at tip; stem leaves usually 2, opposite, smaller, 3-lobed, with elongated lobe at tip; margins with rounded teeth.

Flowers: White, 5–6 mm wide; 5 petals, slender, deeply fringed, about 2 mm long; in narrow, elongated terminal clusters 5–15 cm long; May–June.

Fruits: Capsules splitting at top and spreading open, cup-like; few seeds, black, 1–1.5 mm long, shiny, smooth.

Habitat: Fresh to moist tolerant hardwood stands in southern part of our region.

Emma Thurley

Notes: The common and genus names refer to the small seed capsules, which are shaped like a bishop's mitre (a tall cap with peaks in front and back). The species name *diphylla* means '2-leaved.'

]*1 cm*

SAXIFRAGE FAMILY (SAXIFRAGACEAE)

General: Perennial herb; flowering stems slender, 15–30 cm tall, glandular-hairy, usually leafless; forms colonies from underground stems (rhizomes) and long runners.

Leaves: Basal, heart-shaped to maple-leaf-shaped with 3–5 shallow lobes, 5–10 cm long; upper surface sparsely hairy; underside downy; margins with rounded teeth.

Flowers: White; 5 petals, lance-shaped, 3–5 mm long, 5 mm wide; in feathery terminal clusters 10 cm long; May–June.

Fruits: Capsules 5–10 mm long, splitting open at tip into 2 unequal parts; seeds several, smooth, round, black.

Habitat: Fresh to moist tolerant hardwood stands.

Notes: Historically, this plant has been used in the treatment of indigestion and diseases of the bladder. See caution in Introduction.
• The species name *cordifolia* means 'heart-leaved.'

Food Use by Wildlife: leaves, flowers.
• **Birds:** ruffed grouse.

Emma Thurley

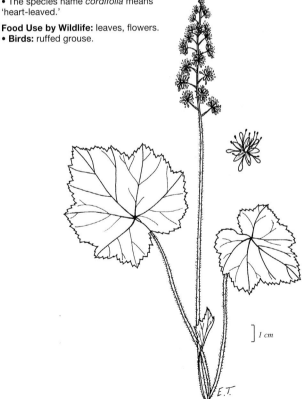

] *1 cm*

E.T.

MARSH GRASS-OF-PARNASSUS • *Parnassia palustris*
PARNASSIE PALUSTRE

SAXIFRAGE FAMILY (SAXIFRAGACEAE)

OMNR

General: Perennial herb, hairless; flowering stems 8–35 cm tall usually with 1 egg-shaped, clasping leaf below middle; from short underground stems (rhizomes).

Leaves: Basal (mainly), slender-stalked, simple, egg-shaped, blunt at tip, usually heart-shaped at base, 3–5 cm long, 1.5–4.5 cm wide.

Flowers: White; petals greenish- or yellowish-veined, oval, 8–15 mm long; solitary; July–August.

Fruits: Capsules, egg-shaped, many-seeded.

Habitat: Wet, rocky shores and clearings; ditches.

Notes: Although called grass-of-Parnassus, this plant is not grass-like.

1 cm

General: Perennial herb; flowering stems 7.5–15 cm tall; often forms large patches from several trailing stems or runners from thick, short rootstocks.

Leaves: Basal, hairy-stalked, 5–15 cm long, compound, with 3 leaflets; leaflets widely oval or egg-shaped, narrower near base; veins straight, prominent; margins coarsely toothed, with tooth at tip **shorter** and narrower than teeth on either side of it.

Flowers: White, 1.5–2.5 cm wide; 5 petals; 2–15, in loose clusters that do not overtop leaves at maturity; May–June.

Brenda Chambers

Fruits: Strawberries, red, pulpy, juicy, rounded to egg-shaped, 0.5–2 cm in diameter with small, dry, pit-like seeds (achenes) embedded in surface; ripen July.

Habitat: All moisture regimes and soil textures; intolerant hardwood mixedwoods, pine stands and conifer swamps, clearings, disturbed areas, roadsides.

Notes: The edible berries often are tastier than domestic strawberries. Aboriginal peoples used the roots in a cure for stomach ache. See caution in Introduction. • See woodland strawberry (p. 198) for notes on distinguishing features.

Food Use of Strawberries (*Fragaria* spp.) by Wildlife: leaf, flower, fruit.
• **Mammals:** opossum, eastern cottontail, snowshoe hare, eastern cottontail, white-footed mouse, red fox, striped skunk, white-tailed deer. • **Birds:** ring-necked pheasant, ruffed grouse, American crow, veery, grey catbird, brown thrasher, rufous-sided towhee, swamp and white-throated sparrows, pine grosbeak.

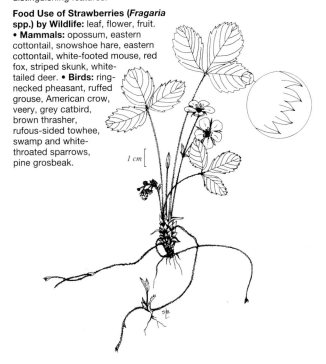

1 cm

OMNR

General: Perennial herb; flowering stems 7.5–15 cm tall, greenish or lightly reddish- or purplish-tinged, hairy, stout, tufted; from long, trailing stems or runners (stolons) growing from a thick, short rootstock.

Leaves: Basal, hairy-stalked, compound, with 3 leaflets; leaflets egg-shaped, somewhat pointed at tip, wedge-shaped at base; upper surface dark green with prominent, straight veins; underside silky-hairy; margins coarsely toothed, with tooth at tip **longer** than 2 on either side.

Flowers: White, 1.0–1.5 cm wide; 5 petals; in loose clusters of 3–15; in clusters that usually extend above leaves at maturity; May–June.

Fruits: Strawberries, red, pulpy, juicy, egg- to cone-shaped, about 1 cm across, above leaves, with small, dry, pit-like seeds (achenes) on smooth surface; ripen July.

Habitat: Moist to dry, clayey to sandy upland tolerant hardwood, pine and intolerant hardwood mixedwood stands.

Notes: You can distinguish woodland strawberry from common strawberry (p. 197) in the following ways: woodland strawberry flowers and fruits usually grow above the leaves, while those of common strawberry are below the leaves; the tooth at the tip of a woodland strawberry leaflet is longer than the teeth to either side, whereas the tooth at the tip of a common strawberry leaflet is shorter than the adjacent teeth; woodland strawberry seeds are not in pits and are on the berry surface, whereas common strawberry seeds are embedded in the berry surface in pits. See common strawberry, (p. 197) for additional notes.

Food Use by Wildlife: See notes under common strawberry (p. 197).

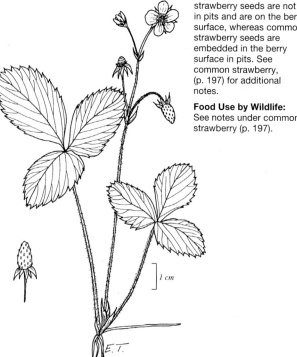

1 cm

E.T.

ROSE FAMILY (ROSACEAE)

General: Low perennial herb; flowering stems leafless, 7–20 cm tall, as long as leaves; spreading by underground stems (rhizomes).

Leaves: Compound, with 3 leaflets; leaflets wedge-shaped to oval, 3–5 cm long, irregularly lobed; lateral leaflets asymmetrical; margins broadly toothed.

Flowers: Yellow; 5 petals, oblong-oval to elliptic, 5–10 mm long, 3–6 mm wide; several in a loose cluster; April–May.

Fruits: Small, dry, hard, single-seeded achenes, few, erect.

Habitat: Dry to moist, sandy to clayey upland pine and intolerant hardwood mixedwood stands; occasional in moist sites along rivers in northern part of our region.

Brenda Chambers

Notes: The species name *fragarioides* means 'like *Fragaria*,' the strawberry.

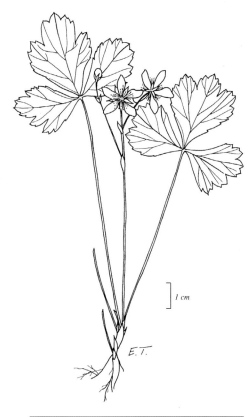

] 1 cm

E.T.

Karen Legasy

General: Perennial herb; flowering stems stout, 10–60 cm tall, hairy above; grow upward from hairless, reclining stems or woody, rooting base.

Leaves: Alternate, stalked, pinnately compound, with 5–7 leaflets; leaflets oval to oblong, blunt to slightly pointed at tips, narrowed at base, 2–10 cm long, 0.7–3.8 cm wide; upper leaves smaller, almost stalkless, in groups of 3–5; margins sharply toothed.

Flowers: Reddish-purple, about 2 cm wide; 5 petals; in loose clusters near top of stem; June–August.

Fruits: Small, dry, hard, single-seeded (achenes), hairless, oval to egg-shaped, numerous, borne on an enlarged, hairy, spongy receptacle.

Habitat: Wet areas, swamps, marshes and peat bogs; shorelines.

Notes: The woody stems root in water. • The common name 'cinquefoil' means 'five leaves,' as there are often 5 leaflets in the compound leaves of these plants.

1 cm

ROSE FAMILY (ROSACEAE)

General: Perennial herb, low, tufted; flowering stems 5–10 cm tall; from slender, creeping stems.

Leaves: Basal, simple, heart-shaped, 3–5 cm long and nearly as wide, sparsely hairy; margins round-toothed; stalks hairy, 3–10 cm long.

Flowers: Of 2 types; first type with 5 showy, white, 4–8 mm long petals, solitary, on 5–10 cm long, downy stalks, usually sterile; second type with no petals on shorter, curved stalks, self-fertilizing; June–September.

Brenda Chambers

Fruits: Dry, seed-like drupes, 3–4 mm long, single-seeded, usually in clusters of 5–10.

Habitat: Wet organic conifer swamps; moist upland tolerant hardwood and eastern white cedar mixedwoods.

Notes: The species name *repens* means 'creeping.'

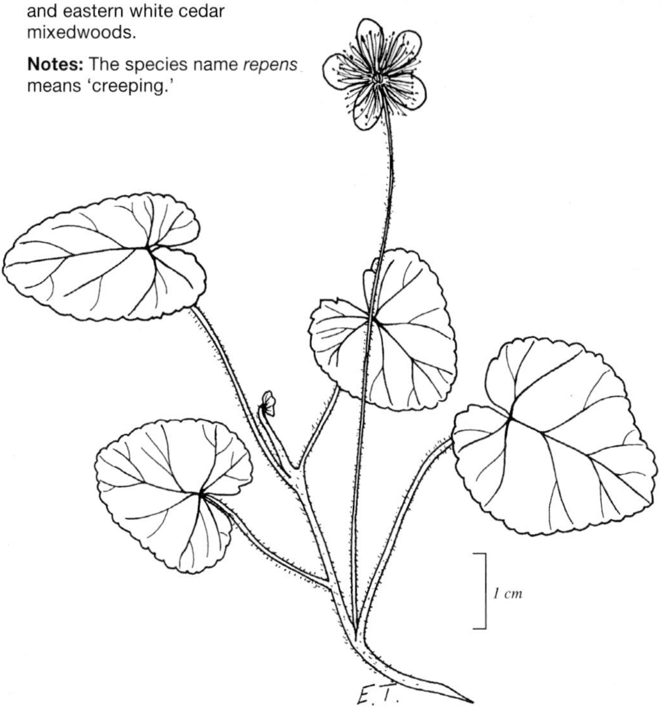

1 cm

E.T.

General: Perennial herb, 0.3–1 m tall; stems erect, bristly-hairy, branched or unbranched.

Leaves: Basal leaves stalked, pinnately compound, with 3–7 leaflets; terminal leaflet largest, rounded to kidney-shaped, toothed, 5–12 cm wide; lateral leaflets 3–6, oval or inversely egg-shaped, interspersed with smaller leaflets; stem leaves short-stalked or stalkless, with 2–4 wedge-shaped lobes or leaflets.

Flowers: Bright yellow, saucer-shaped, short-stalked; petals 5, inversely egg-shaped, 3.5–5 mm long, 3–5 mm wide; single or in few-flowered terminal clusters; June–August.

Fruits: Dry, hard, single-seeded (achenes), 1.2–1.8 mm in diameter, with a slender, hairy style; in rounded clusters.

Habitat: Moist, rich thickets, ditches and other disturbed sites.

Notes: The bright-yellow flowers and rounded terminal leaflets of large-leaved avens distinguish this species from water avens (*Geum rivale*), which has purplish flowers and a more wedge-shaped terminal leaflet. Yellow avens (*Geum aleppicum*) (inset photo) is a similar, yellow-flowering species, but its terminal leaflets are widely egg- to wedge-shaped rather than rounded to kidney-shaped. • Historically, the roots of yellow avens were boiled and used for sore chests or coughs. See caution in Introduction.

Karen Legasy, both

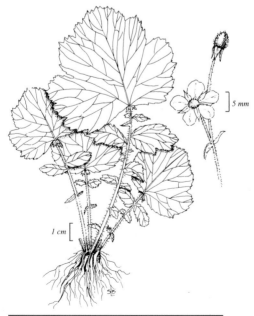

ROSE FAMILY (ROSACEAE)

General: Perennial herb; flowering stems stout, 0.3–1.5 m tall, glandular long-hairy, with a pair of leaf-like, toothed bractlets at base; underground stems (rhizomes) long, fibrous.

Leaves: Alternate, pinnately compound, with 5–9 large leaflets interspersed with smaller leaflets; leaflets oblong to lance-shaped, 1.5–5 cm wide, almost hairless above, glandular-dotted and hairy beneath; pair of somewhat heart-shaped, deeply toothed, leaf-like bracts (stipules) at leaf base, 1–2 cm wide.

Flowers: Yellow, 6–12 mm across, on short, hairy stalks; 5 petals; sepals fused in a 1 cm tube fringed with hooked bristles at mouth; in a slender terminal spike; July–August.

Fruits: Pair of dry, hard, single-seeded (achenes), 6–8 mm long (including beak), enclosed in bristly sepal tube.

Habitat: Wet organic hardwood swamps; moist to fresh upland sites; damp ground at forest edges.

Brenda Chambers

Notes: This plant was used by aboriginal peoples to treat urinary disorders and fevers. See caution in Introduction. • The species name *gryposepala* means 'having hooked bristles,' referring to the bristles on the cup-shaped structure (hypanthium) which bears the petals and sepals.

]*1 cm*

E.T.

COW VETCH • *Vicia cracca*
VESCE JARGEAU

Karen Legasy

General: Perennial; stems climbing or trailing, 60–120 cm long, slender, weak, tufted, finely hairy or sometimes hairless.

Leaves: Alternate, nearly stalkless, thin, pinnately compound, with 5–10 pairs of leaflets and a branched, thread-like, twining tendril at tip; leaflets linear to oblong, blunt to somewhat pointed and bristled at tip, about 2.5 cm long; margins and leaf-like bracts (stipules) toothless.

Flowers: Bluish-purple, pea-like, about 1.3 cm long; in elongated, 1-sided clusters of often 30 or more flowers that bend downward from leaf axils, equal to or just exceeding the leaves; May–August.

Fruits: Pods, brownish, short-stalked, lance-shaped, flattened, 2–3 cm long, 5–7 mm wide.

Habitat: Roadsides, forest trails, fields.

Notes: You can recognize cow vetch by its dense, 1-sided flower clusters on long stalks and by its somewhat flat, brownish seed pods.

Food Use of Vetches (*Vicia* spp.) by Wildlife: leaf, flower, seed.
• **Birds:** ring-necked pheasant, ruffed grouse, wild turkey, song sparrow.

1 cm

General: Perennial; stems up to 1 m long, trailing or climbing by leaf tendrils, often tangled, hairless or with some appressed hairs.

Leaves: Alternate, nearly stalkless, pinnately compound, 4–7 pairs of leaflets and a branched, thread-like, twining tendril at tip; leaflets egg-shaped to oblong, blunt or sometimes shallowly notched at tip with a sharp, abrupt point (mucronate), rounded at base, 1.5–3.5 cm long, 0.6–1.4 cm wide, hairless to hairy; lateral veins prominent and rib-like on underside when dry; leaf-like bracts (stipules) sharply toothed.

Flowers: Bluish-purple, pea-like, 1.5–2 cm long; in loose elongated clusters of 3–9, shorter than leaves; June–July.

Fruits: Pods, short-stalked, 2.5–3.5 cm long, hairless; 4–7 seeds slightly rounded, 4 mm long.

Habitat: Thickets along shores, forest openings.

Linda Kershaw

Notes: Several vetches (*Vicia* spp.) are reported to be **toxic to humans and livestock**.

Food Use by Wildlife: See notes under cow vetch (p. 204).

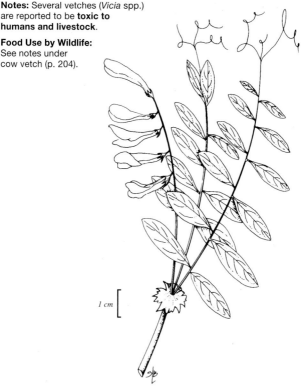

1 cm

OMNR

General: Perennial herb, hairless, slightly waxy; flowering stems slender, angled, 30–100 cm tall, climbing or trailing; rootstocks slender.

Leaves: Alternate, stalked, pinnately compound, with 3–5 pairs of leaflets and a usually branched, thread-like tendril at tip; leaflets egg-shaped to widely oval, blunt at tip, rounded at base, 2.5–5 cm long; lexaf-like bracts (stipules) paired at leaf base, 1.5–3 cm long, slightly heart- to egg-shaped; margins toothless.

Flowers: White to yellowish-white, pea-like, 1–1.5 cm long; in elongated clusters of 5–10 from leaf axils; June.

Fruits: Pods, hairless, 2.5–5 cm long; open August–September.

Habitat: Fresh to moist upland sites, roadsides and shorelines, disturbed areas, clearings and thickets.

Notes: The **seeds are poisonous** and cause headache, difficulty in breathing and partial paralysis, a condition known as 'lathyrism.' • Aboriginal peoples used pale vetchling root as a type of potato. See caution in Introduction.

1 cm

PEA FAMILY (FABACEAE)

General: Annual herb, hairless to slightly hairy; flowering stems erect to ascending, branched, up to 45 cm tall.

Leaves: Alternate, short-stalked, compound, with 3 leaflets; leaflets stalkless or equally short-stalked, oblong to egg-shaped, widest above middle, blunt to notched at tip, narrowed at base, 10–15 mm long; margins finely toothed; stalks only slightly longer than lance-shaped bracts (stipules).

Flowers: Yellow, become brown with age, pea-like, small; in dense, oblong to cylindrical clusters 1–2 cm long, 1–1.5 cm thick, above leaves; June–September.

Fruits: Pods, egg-shaped, usually 1-seeded.

Habitat: Roadsides, fields and waste places.

Karen Legasy

Notes: You can distinguish hop-clover by its oblong to cylindrical clusters of yellow flowers and by its essentially stalkless leaflets. • Hop-clover is rich in protein and has historically been considered edible as cooked greens or in salads, but the leaves and flowers are apparently difficult to digest when raw. The seeds and flowers were reportedly ground into flour. See caution in Introduction.

Food Use of Clovers (*Trifolium* spp.) by Wildlife: leaf, flower, seed. • **Mammals:** opossum, snowshoe hare, eastern chipmunk, raccoon, striped skunk, white-tailed deer. • **Birds:** ring-necked pheasant, ruffed grouse, wild turkey.

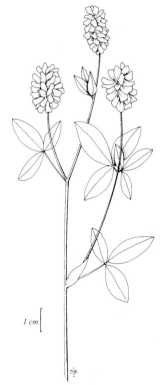

1 cm

ALSIKE CLOVER • *Trifolium hybridum* ssp. *elegans*
TRÈFLE HYBRIDE

OMNR

General: Perennial herb, mostly hairless; flowering stems arched or bent to erect, 30–80 cm tall, hollow, soft, branched.

Leaves: Alternate, long-stalked, compound, with 3 leaflets; leaflets short-stalked, oval to egg-shaped, narrowed toward base, 2.5–4 cm long, 2–3 cm wide; pairs of small leaf-like bracts (stipules) at base of leafstalks are egg- to lance-shaped with pointed tips; margins minutely toothed.

Flowers: Pink to pinkish and whitish, become dull brown with age, pea-like, 8–11 mm long, slender-stalked; clustered in rounded heads 2–3.5 cm in diameter on 2–9 cm long stalks from upper leaf axils; May–October.

Fruits: Pods with 2–4 green to black, lens-shaped seeds.

Habitat: Along roadsides, in fields and waste areas.

Notes: You can distinguish alsike clover by its long-stalked, pinkish flower clusters and its ascending stem. • Historically considered edible in salads or as a cooked green, alsike clover is apparently hard to digest when raw. Clover is rich in protein. The seeds and flower heads have been made into a flour that apparently is nutritious. See caution in Introduction.

Food Use by Wildlife: See notes under hop-clover (p. 207).

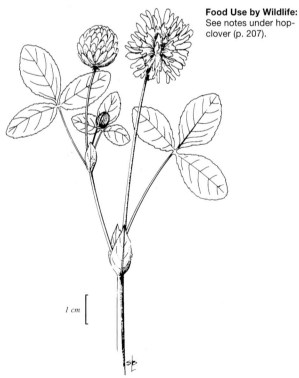

1 cm

PEA FAMILY (FABACEAE)

General: Biennial or short-lived perennial; flowering stems ascending, branched, 5–40 cm tall, usually hairy.

Leaves: Alternate, compound, with 3 (rarely 4) leaflets; leaflets short-stalked, oval to egg-shaped, narrowest at base, blunt or sometimes notched at tip, 1–3 cm long, 0.5–1.5 cm wide, with a lighter, slightly V-shaped marking near middle; small leaf-like bracts (stipules) at base of stalks oval to egg-shaped, prominently veined; lower leaves long-stalked; upper leaves short-stalked to stalkless.

Karen Legasy

Flowers: Reddish or rose to deep pink, rarely white, pea-like, about 1.3 cm long; in dense, rounded to egg-shaped clusters 1.2–3 cm long, stalkless or rarely on short stalks; May–September.

Fruits: Pods containing 1–2 seeds.

Habitat: Roadsides, fields, clearings and waste areas, lawns.

Notes: Introduced from Europe, red clover is planted for hay and pasture in crop rotation. Bacteria in the nodules of clover roots 'fix' nitrogen (convert nitrogen from the air into a form that can be used by the plant). This helps the plant to grow on nitrogen-poor soils and increases soil fertility. See notes under white clover (p. 210).

Food Use by Wildlife: See notes under hop-clover (p. 207).

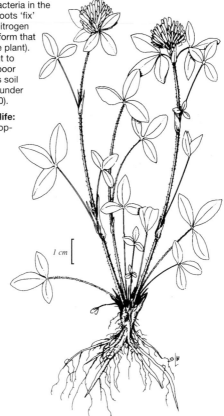

1 cm

WHITE CLOVER • *Trifolium repens*
TRÈFLE RAMPANT

PEA FAMILY (FABACEAE)

General: Perennial; stems 10–25 cm long, creeping, branching at base, often rooting where leaves join stem (at nodes), hairless or with a few hairs.

Leaves: Alternate, long-stalked, compound, with 3 (rarely 4) leaflets; leaflets egg-shaped, narrowest toward base, rounded and slightly notched at tip, 0.5–2 cm long, with a lighter, slightly V-shaped marking near middle; small leaf-like bracts (stipules) at base of stalk egg- to lance-shaped with pointed tips; margins toothed.

Flowers: White to pinkish, turn brown with age, pea-like, 6–13 mm long; in rounded clusters 1.5–3 cm in diameter on long, leafless stalks from creeping stems; May–October.

Karen Legasy

Fruits: Pods with 3–4 rounded to kidney-shaped seeds.

Habitat: In fields, waste places, open areas and lawns.

Notes: You can recognize white clover by its creeping stem, white to pinkish flowers and leaflets with a V-shaped marking near the middle. The creeping stem of white clover distinguishes it from alsike clover (p. 208) and red clover (p. 209), which have ascending stems; also, white clover flowers are on long, leafless stalks. White clover is the typical clover of lawns—the one people examine hoping to find a 'lucky' four-leaf clover.

Food Use by Wildlife: See notes under hop-clover (p. 207).

1 cm

WOOD-SORREL FAMILY (OXALIDACEAE)

General: Perennial herb, creeping and low-growing; flowering stems leafless (scapes), 7.5–15 cm tall, equal to or taller than leaves; from pale, scaly underground stems (rhizomes).

Leaves: Basal, stalked, clover-like, compound, with 3 leaflets; leaflets stalkless, widely heart-shaped at tip, narrowed at base, 1.2–3 cm wide, shiny with a few hairs; margins hairy.

Flowers: Whitish to pinkish with deep-pink or red veins, about 2 cm wide; 5 petals oblong, notched; solitary; June–July.

Fruits: Capsules, cylindrical to awl-shaped; seeds with distinctive, white-crested, horizontal ridges.

Habitat: Wet organic conifer swamps; moist to fresh upland eastern white cedar mixedwoods and tolerant hardwood stands with yellow birch and eastern hemlock; all soil textures.

Notes: A yellow-flowering species, upright wood-sorrel (*Oxalis stricta*) grows in roadside habitats. • Some common wood-sorrel flowers are produced later in the season on curved stems at the base of the plant. These flowers do not open, but produce self-fertilized seeds and capsules. The leaves close at night. •The sour-tasting leaves have historically been used in salads, but it has been found that the consumption of large amounts can cause calcium depletion in the body. The leaves were also used to brew a beverage. See caution in Introduction.

Food Use by Wildlife: leaf, flower, seed. • **Mammals:** eastern chipmunk, white-tailed deer. • **Birds:** dark-eyed junco.

Steve Williams, top;
Brenda Chambers, bottom

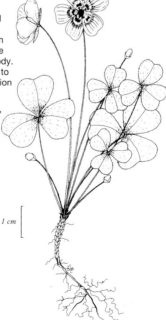

1 cm

OMNR

General: Perennial herb, hairless to finely hairy, usually erect at first, later develops slender, numerous, spreading stems; flowering stems leafy, up to 10 cm tall; from woody, jagged underground stems (rhizomes).

Leaves: Alternate; basal and lower leaves long-stalked, egg-shaped, slightly heart-shaped at base, gradually narrowing to a blunt tip; upper stem leaves narrower, shorter-stalked, less rounded at tips; small, leaf-like bracts (stipules) at stem base deeply toothed; margins finely scalloped.

Flowers: Purple to violet-blue, 5–15 mm long; 5 unequal petals; lateral petals bearded; lower (lip) petal with fine, dark lines, spurred; spur long, often hooked or curved upward; usually above leaves on 3.5–10 cm long stalks; May–July.

Fruits: Capsules, small, rounded; seeds dark brown.

Habitat: In sandy or rocky forest openings.

Notes: Hooked-spur violet can be distinguished by its numerous spreading stems which have both basal and stem leaves, and by the distinctive toothed, leaf-like bracts (stipules) at the base of its leaves.
• The leaves and flowers have historically been used in salads. See caution in Introduction.

Food Use of Violets (*Viola* spp.) by Wildlife: leaf, flower, seed. • **Mammals:** eastern cottontail, white-footed mouse. • **Birds:** ruffed grouse, woodcock, wild turkey, dark-eyed junco.

1 cm

General: Perennial, more or less hairy except for earliest leaves; flowering stems leafless, 7.5–12.5 cm tall; from stout, branching rootstocks.

Leaves: Basal, stalked, simple, egg- to kidney-shaped, somewhat pointed but blunt at tip, heart-shaped at base, 3–7 cm wide, pale green, often purplish beneath; fringed with hairs on margins, veins and stalks; margins with 10–26 teeth; stalks slender, wiry, often purplish at base.

Flowers: Deep violet to pale lilac, 0.8–1.3 cm wide; 5 unequal petals; 3 lower petals bearded at base; lowermost (lip) petal with a spur; solitary on leafless flowering stems; May–June.

Fruits: Capsules, rounded, purple or sometimes green, 5–8 mm long; seeds dark brown.

Habitat: Damp, open forests, streambanks, shore thickets.

Notes: You can recognize northern blue violet by the fringe of hairs along its leaf margins.

Food Use by Wildlife: See notes under hooked-spur violet (p. 212).

Brenda Chambers

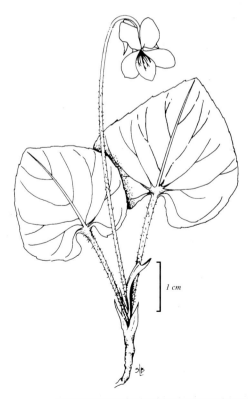

1 cm

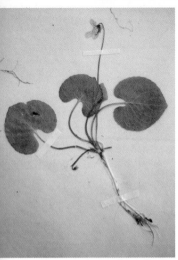

Brenda Chambers

General: Perennial herb; flowering stems leafless (scapes), hairless, usually tinged red, 7.5–12.5 cm tall, taller than leaves; from underground stems and slender runners (stolons).

Leaves: Basal, stalked, simple, rounded, sharp-pointed at tip, heart-shaped at base, up to 6.3 cm wide, dark green, hairless except for scattered, minute hairs on upper surface, shiny; basal sinus narrow with lobes almost overlapping; margins toothed; stalks hairless, usually tinged red.

Flowers: White, about 1.3 cm wide, fragrant, hairless; 5 unequal petals; upper 2 petals narrow, bent strongly backward; lower 3 petals with brown-purple veins; lowermost (lip) petal with a spur; solitary; April–June.

Fruits: Capsules, purplish, egg-shaped, 4–6 mm long; seeds dark brown, minutely wrinkled.

Habitat: All moisture regimes and soil textures (except rocky sites); tolerant hardwood stands.

Notes: Sweet white violet can be distinguished from the very similar northern white violet (p. 215) by its reddish leafstalks and flower stems and its purplish seed capsules. This species is also known as *V. incognita*.

Food Use by Wildlife: See notes under hooked-spur violet (p. 212).

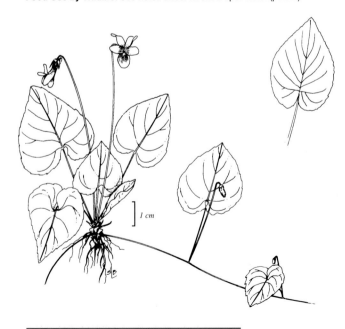

1 cm

VIOLET FAMILY (VIOLACEAE)

General: Perennial herb; flowering stems leafless, often red-dotted in summer and slightly hairy, 2.5–12.5 cm tall, taller than leaves; from slender, thread-like, rooting runners (stolons).

Leaves: Basal, stalked, simple, widely egg-shaped to rounded or heart-shaped, blunt or rounded at tip, 1–5 cm long and wide, hairless; margins have low, rounded teeth; stalks often red-dotted in summer and slightly hairy.

Flowers: White, fragrant, 7–10 mm long; 5 unequal petals; upper 2 petals widely egg-shaped; lower 3 petals with purple veins; 2 lateral petals have small tuft of hairs; lower (lip) petal with a spur; May–June.

Brenda Chambers

Fruits: Capsules, green, rounded to cylindrical; seeds mature black.

Habitat: Low, wet areas, thickets, open areas; along streams or by springs.

Notes: You can recognize northern white violet by the red dots on its leaf stalks and flower stems during the summer. The flowers and leaves grow on separate basal stalks and the flowers are taller than the leaves. See notes on sweet white violet (p. 214).

Food Use by Wildlife: See notes under hooked-spur violet (p. 212).

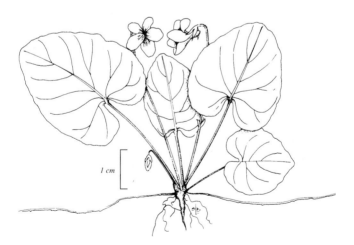

1 cm

Karen Legasy

General: Perennial herb; flowering stems leafless, 2.5–10 cm tall, slender; from underground stems (rhizomes) that become scaly and stout with age.

Leaves: Basal, long-stalked, simple, 1.5–8 cm wide, kidney-shaped to circular, rounded or blunt at tip, heart-shaped at base; upper surface soft, white hairy, occasionally hairless; underside of young leaves and stalk soft, white hairy; margins have broad, flat or rounded teeth.

Flowers: White with brownish-purple stripes or veins on 3 lower petals; 5 unequal petals; lower petal with short, rounded spur; stalks erect, usually shorter than leaves; May–June.

Fruits: Purplish capsules containing brown, rounded seeds; July–August.

Habitat: Moist to dry, clayey to sandy upland eastern white cedar mixedwoods and tolerant hardwood stands; hardwood swamps.

Notes: Kidney-leaved violet can be recognized by its kidney-shaped leaves. • The leaves and flowers are edible. The flowers have been used to scent teas, vinegars and perfumes. The roots were apparently used to induce vomiting. See caution in Introduction.

Food Use by Wildlife: See notes under hooked-spur violet (p. 212).

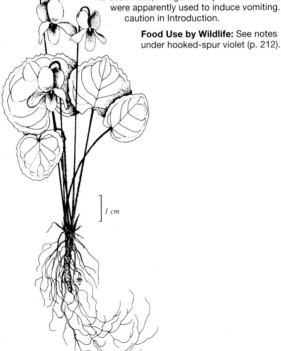

1 cm

VIOLET FAMILY (VIOLACEAE)

General: Creeping perennial, woolly; flowering stems leafless; from stout, oblique, creeping underground stems (rhizomes).

Leaves: Basal, simple, egg-shaped to rounded, heart-shaped at base, up to 7 cm wide at maturity, hairy (especially on underside); margins round-toothed; stalks hairy.

Mike Oldham

Flowers: Violet, lavender or white; 5 unequal petals; 2 lateral petals bearded; lower (lip) petal spurred; stalks more or less hairy, as tall as leaves, with 2 small, leaf-like bracts near middle; solitary; April–June.

Fruits: Egg-shaped, usually purple capsules; seeds dark-brown, 1.8–2.5 mm long, 1.2–1.5 mm thick.

Habitat: Moist to dry, clayey to sandy upland tolerant hardwood stands, hardwood and conifer swamps.

Notes: The showy, spring flowers of many violets do not always produce seed. When this happens, the plants produce inconspicuous, short-stemmed, egg-shaped summer flowers that lack petals and lie on or under the ground. These 'cleistogamous' flowers never open, but rather fertilize themselves. The stalks then elongate in fruit, and the seeds are shot out from the capsules under pressure. Many Violets (*Viola* spp.) hybridize freely when growing near one another. • The species name *sororia* means 'sisterly, resembling other species.'

Food Use by Wildlife: See notes under hooked-spur violet (p. 212).

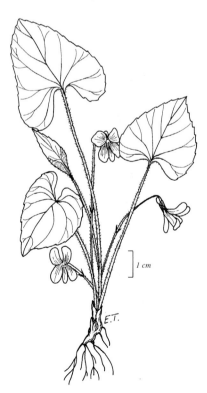

1 cm

E.T.

VIOLET FAMILY (VIOLACEAE)

General: Perennial herb; flowering stems leafy, slender, 10–25 cm tall, leafy; from thick, short, woody underground stems (rhizomes) without runners (stolons).

Leaves: Simple, 5–10 cm long; basal leaves several, long-stalked, heart-shaped; stem leaves numerous, crowded towards top, heart-shaped to broadly wedge-shaped at base; margins toothed; leaf-like bracts (stipules) thin, lance-shaped, slender-pointed.

Flowers: White with yellowish eyespot and dark veins near base, often violet-tinged on underside; 5 unequal petals; 2 lateral petals bearded; lower (lip) petal spurred; solitary; April–July.

Fruits: Rounded to elliptic capsules, 5–7 mm long; seeds brown.

Brenda Chambers

Habitat: Fresh to moist tolerant hardwood stands; southern part of our region.

Notes: Canada violet roots, seeds, leaves and flowers have been used in medicine. Aboriginal peoples used a medicine made from the roots to ease pain in the area of the bladder. See caution in Introduction.

Food Use by Wildlife: See notes under hooked-spur violet (p. 212).

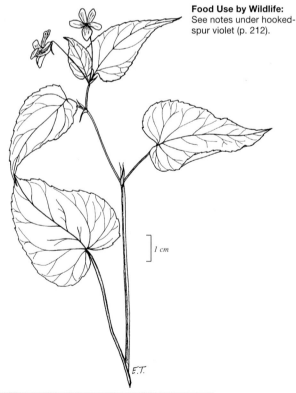

1 cm

E.T.

VIOLET FAMILY (VIOLACEAE)

General: Perennial herb, soft-hairy; flowering stems leafy, usually solitary (sometimes 2), 10–45 cm tall; from stout, brown, woody underground stems (rhizomes) with coarse, fibrous roots.

Leaves: Basal leaves single (or absent), rounded to heart-shaped; stem leaves near top of stem, broadly egg-shaped to round, pointed at tip, truncate at base, 4–10 cm long, about 1 cm wider than long, densely soft-hairy beneath; leaf-like bracts (stipules) at base of stalk broadly egg- to lance-shaped and toothed.

Flowers: Yellow, with brown-purple veins near base; 5 unequal petals; 2 lateral petals bearded; lower (lip) petal spurred; stalks short, hairy, above stem leaves; solitary; May-June.

Brenda Chambers

Fruits: Egg-shaped, hairless to white-woolly capsules, 10–12 mm long; seeds pale, 2.5–2.9 mm long, 1.5–1.8 mm thick.

Habitat: Moist to dry, clayey to sandy upland tolerant hardwood stands; hardwood swamps.

Notes: Also known as *Viola pensylvanica*. • Smooth yellow violet (*Viola pubescens* var. *scabriuscula*) is not hairy and has stem leaves with heart-shaped (rather than truncate) bases, typically has 1–3 (rather than 0–1) basal leaves and has stems in groups of 2 or more (rather than 1–2). • Historically, the flowers, leaves, roots and seeds were used in medicine. Aboriginal peoples used a medicine made from the roots to ease sore throats. See caution in Introduction. • The species name *pubescens* means 'pubescent,' or hairy.

Food Use by Wildlife: See notes under hooked-spur violet (p. 212).

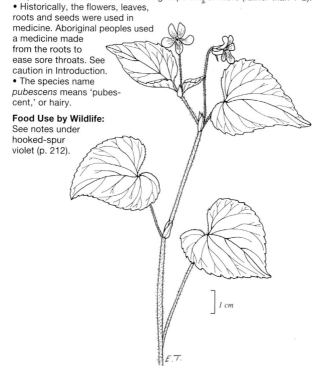
1 cm
E.T.

Brenda Chambers

General: Low perennial herb; flowering stems erect, 8–15 cm tall, with several scale-like leaves below main leaf cluster; from slender underground stems (rhizomes).

Leaves: Alternate, 3–6 crowded at stem top, appear whorled, simple, elliptic to oval, blunt-tipped, 1.5–4 cm long; margins toothless.

Flowers: Showy, rose-purple to white, 1–1.5 cm long; 3 petals fused into tube; lower (lip) petal with finely fringed crest; 2 petal-like sepals wing-like; 1–4 on short stalks near top of stem, in leaf axils; May–June.

Fruits: Rounded capsules, notched at tip, about 6 mm across; ripen in June and July.

Habitat: Dry to moist, sandy to clayey pine stands; conifer swamps.

Notes: Non-flowering fringed polygala plants are similar to those of wintergreen (p. 75), although their taste and smell are less pungent. • The whole plant, but mainly the root, has a mild wintergreen taste and scent. Historically, the whole plant was used in medicine. See caution in Introduction. • The species name *paucifolia* means 'few-leaved.' • These plants also produce small, self-fertilizing flowers scattered on short, underground branches.

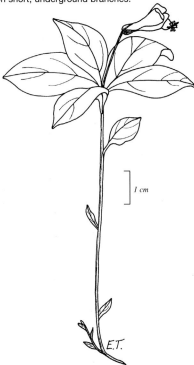

1 cm

E.T.

TOUCH-ME-NOT FAMILY (BALSAMINACEAE)

General: Annual, hairless; flowering stems 60–150 cm tall, semi-transparent, branched, soft and watery.

Leaves: Alternate, stalked, simple, egg-shaped, 3–9 cm long, thin; underside pale, with fine, whitish, waxy powder (glaucous); margins with coarse, often sharp teeth; stalks slender, 1–10 cm long.

Flowers: Orange with crimson or reddish-brown spots, about 2.5 cm long, with 3 petals and 3 sepals; 2 lower (lip) petals lobed; lower sepal forming an orange, cone-shaped sac,

Karen Legasy

longer than wide with a slender, hooked tip (spur); horizontal, hanging on slender stalks; July–October.

Fruits: Capsules, fragile, swollen, burst at maturity to eject seeds.

Habitat: Wet organic hardwood swamps; along wet, shaded streams and in wet ditches along the forest edge.

Notes: Spotted touch-me-not was formerly known as *Impatiens biflora* and is also commonly known as 'spotted jewelweed.'

Food Use By Wildlife: leaf, flower, seed. • **Mammals:** snowshoe hare, white-footed mouse. • **Birds:** ring-necked pheasant, ruffed grouse.

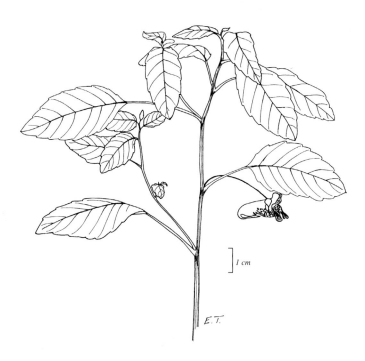

1 cm

E.T.

POPPY FAMILY (PAPAVERACEAE)

General: Low perennial herb; flowering stems 5–15 cm tall; from elongate, simple or forking, underground stems (rhizomes), thick with abundant red-orange, acrid juice.

Leaves: Single, basal, stalked, simple, round, 3- to 9-lobed, 10–28 cm across, thin, soft, pale green on underside; margins wavy to coarsely toothed; fully expand after flower has bloomed.

Flowers: White, occasionally pink, 2–5 cm wide, 4-sided (4 petals usually longer than others); petals usually 8 but as many as 16; solitary; April–May.

Fruits: Elliptic to spindle-shaped capsules, 3–5 cm long; seeds smooth with a large crest.

Habitat: Fresh to moist tolerant hardwood stands; uncommon.

Notes: Stomach cramps were treated with a combination of this plant and the roots of blue cohosh (p. 193). **The roots are considered poisonous. Death has resulted from overdose.** The juice has also been used as an insect repellent. See caution in Introduction. • Aboriginal peoples used the red juice of bloodroot as a dye for facial decoration, clothing, tools and weapons. • This plant gets its name from the colour of its juice; *sanguinaria* means 'bleeding.'

Daisy Wannamaker, top;
Linda Kershaw, bottom

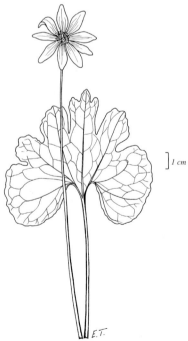

] *1 cm*

E.T.

FUMEWORT FAMILY (FUMARIACEAE)

General: Lacy, delicate perennial herb, hairless; flowering stems 10–30 cm tall; from a cluster of grain-like tubers, crowded together like a scaly bulb.

Leaves: Alternate, long-stalked, broadly triangular in outline, finely divided into many linear segments, bluish-green.

Flowers: White, tipped with pale yellow, nodding, 1.5–2 cm wide; 4 unequal petals; 2 outer petals joined at base, widely spurred; 2 inner petals small, minutely crested; 3–12 in hanging, one-sided, terminal clusters; May–June.

Fruits: Brown pods, spindle-shaped, with 10–20 crested seeds.

Habitat: Fresh to moist tolerant hardwood stands.

Brenda Chambers

Notes: The stems and leaves wither and disappear soon after the flowers. • A related species, squirrel corn (*Dicentra canadensis*), has short, rounded spurs. The 2 species often grow together. • Historically, the roots were used in medicine. See caution in Introduction. • The common name, 'Dutchman's breeches,' refers to the flowers, which resemble a pair of breeches hung up to dry.

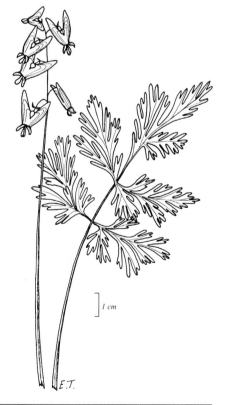

1 cm

E.T.

FUMEWORT FAMILY (FUMARIACEAE)

Brenda Chambers

General: Annual or biennial, pale green, covered with fine, whitish, waxy powder (glaucous), hairless, soft and watery; flowering stems erect or ascending, 30–60 cm tall, freely branching; from taproot.

Leaves: Alternate, short-stalked near base, nearly stalkless above, compound, with leaflets in opposite pairs; leaflets again divided into 3-lobed, inversely egg- or wedge-shaped segments, blunt or rounded and often abruptly pointed (mucronate) at tips, about 1.3 cm long.

Flowers: Pink with yellow tips, 1.3 cm long, tubular, sac-like with a rounded spur; in loose clusters at branch ends; May-September.

Fruits: Capsules, pod-like, erect, narrowly linear, 3–5 cm long, 1.5–2 mm wide, knobby when mature; seeds black, shiny, with raised 'bumps' (tubercles).

Habitat: Dry rocky outcrops.

Notes: Pale corydalis is also called *Capnoides sempervirens*.
• The bitter roots were historically used in a remedy for parasitic intestinal worms and to promote menstrual discharge. See caution in Introduction.

1 cm

SB

MUSTARD FAMILY (BRASSICACEAE)

General: Perennial herb; flowering stems 10–25 cm tall; from horizontal underground stems (rhizomes).

Leaves: Basal and 2 nearly opposite on stem, up to 10 cm long and wide, compound with 3 egg-shaped leaflets about 1/2 as long as wide; margins with coarse, blunt teeth; stalks long at base, short on stem.

Flowers: White to pink (when fading); 4 petals, egg-shaped to oblong, 11–17 mm long; in few-flowered, open terminal clusters; May–June.

Fruits: Erect pods (siliques), lance-shaped, flattened, 2–4 cm long with a slender, 6–8 mm long beak; seeds in 1 row, wingless, flattened; rarely mature.

Habitat: Fresh to dry, sandy to loamy tolerant hardwood stands in the southern part of our region; moist, clayey intolerant mixedwood stands in the northern part of our region; uncommon.

Brenda Chambers

Notes: This species is also known as *Dentaria diphylla*. • The rhizome has a pleasant, pungent taste. Aboriginal peoples ate the roots raw or cooked. Historically, the roots were used for food and medicine, and in spring the whole plant was used for medicine. See caution in Introduction. • The species name *diphylla* means 'two-leaved.'

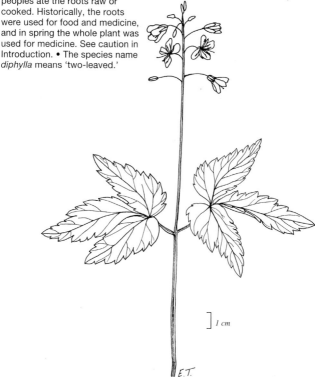

] *1 cm*

E.T.

COMMON ST. JOHN'S-WORT • *Hypericum perforatum*
MILLEPERTUIS COMMUN

ST. JOHN'S-WORT FAMILY (HYPERICACEAE)

Karen Legasy

General: Perennial herb; flowering stems 30–90 cm tall, 2-edged, tough, many-branched, usually clustered from flattened base.

Leaves: Opposite, stalkless, simple, oblong to linear, blunt-tipped, 2.5–5 cm long on main stem, half as long on branches, more or less black-dotted.

Flowers: Yellow, 2–2.5 cm wide; 5 petals, with black dots on margins; June–September.

Fruits: Capsules, brown, egg-shaped, pointed, 3-celled, glandular, with a conspicuous network of raised veins and pits.

Habitat: Dry roadsides, fields, waste places.

Notes: The glandular dots on the leaves contain a **phototoxin** that can make some people susceptible to sunburn and dermatitis. • St. John's-wort is said to bloom on June 24th, the Feast of St. John. • This plant was introduced from Europe and is often a troublesome weed.

1 cm

General: Annual or biennial; flowering stems 15–60 cm tall, hairy, spreading, usually much-branched but may be unbranched.

Leaves: Opposite, slender-stalked, 2–7 cm wide, generally rounded in outline, deeply cut into 5 wedge-shaped segments and further cut into oblong lobes, hairy along veins.

Flowers: Pinkish to purplish, 2 per stalk; 5 petals, notched, 5–7 mm long; July–September.

Fruits: Beaked, 1.5–2.3 cm long, glandular-hairy, splitting open from below with 5 segments curling upward; seeds thick, cylindrical, clearly veined (reticulate).

Habitat: Disturbed areas, clearings, open woods and roadsides; often locally abundant following fire.

Notes: Herb robert (*Geranium robertianum*), has a similar pair of flowers, but its leaves are more intricately dissected and their terminal segment is well-stalked. • The roots were used in a gargle for canker sores in the mouth and throat and in remedies for diarrhea and hemorrhoids. See caution in Introduction. • Bicknell's cranesbill grows abundantly for a few years after a wildfire or a clearcut followed by slash burning.

Food Use by Wildlife: leaf, flower, seed. • **Mammals:** least chipmunk, white-tailed deer.

Karen Legasy, top; Linda Kershaw, bottom

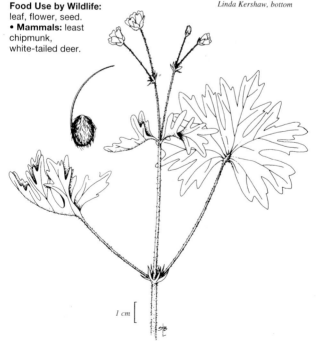

1 cm

EVENING PRIMROSE • *Oenothera biennis*
ONAGRE MURIQUÉE

EVENING-PRIMROSE FAMILY (ONAGRACEAE)

General: Biennial, 0.5–1.5 m tall; stems erect, usually stout, unbranched and 'wand-like' or branched, more or less hairy, rarely hairless, green to purple-tinged; from stout taproot.

Leaves: Alternate, forming a leafy rosette in the first year, stalkless or lowest short-stalked, simple, lance-shaped to oblong, pointed at tip, narrowed at base, 10–20 cm long; margins slightly toothed or toothless.

Flowers: Yellow, large; 4 sepals form a tube at first but eventually separate and bend back; 4 petals 1.2–2.5 cm long; in a leafy terminal spike; June–September.

Fruits: Capsules, oblong, narrowed toward tip, hairy, erect, about 2.5 cm long.

Habitat: Clearings, roadsides, fields.

Notes: Evening primrose is so-named because its flowers open in the evening. • The roots have historically been considered edible. See caution in Introduction.

Karen Legasy

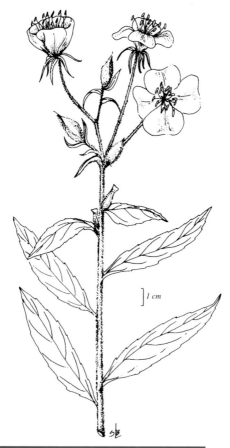

] *1 cm*

General: Perennial herb; flowering stems up to 2 m tall, erect, usually unbranched, upper part often purplish.

Leaves: Alternate, very short-stalked, simple, narrowly lance-shaped, pointed at tips, narrowed at base, 3–20 cm long, numerous; upper surface green; underside paler with a network of veins; margins toothless to slightly toothed.

Flowers: Deep pink, purplish or magenta, rarely white, about 2.5 cm wide; 4 petals; numerous in elongated terminal clusters; July–August.

Fruits: Linear capsules, green to red or purplish, 3–7 cm long, split open to release seeds; seeds numerous, 1–1.3 mm long, spindle-shaped, with fluffy, white tufts of soft hairs 9–14 mm long.

Brenda Chambers

Habitat: All moisture regimes and soil textures; in upland pine, pine-black spruce and intolerant hardwood mixedwood stands; roadsides and clearings; locally abundant following fire or other disturbance.

Notes: The stem pith was used to make an ale. Aboriginal peoples used the roots and leaves to make a variety of poultices for bruises, swellings and sores. See caution in Introduction. • The silky hairs from fireweed seeds were mixed with cotton or fur to make stockings.

Food Use by Wildlife: seeds.
• **Mammals:** moose.

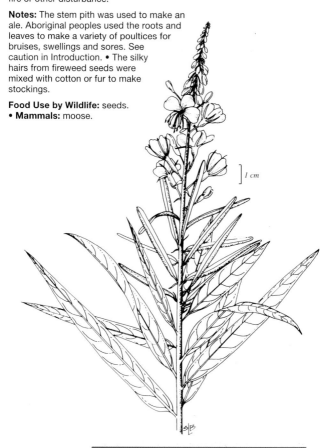

] 1 cm

NORTHERN WILLOW-HERB • *Epilobium ciliatum*
ÉPILOBE GLANDULEUX

Evening-primrose Family (Onagraceae)

General: Perennial herb; flowering stems erect, 30–90 cm tall, usually much-branched.

Leaves: Alternate, stalkless to short-stalked, simple, lance- to egg-shaped, bluntish or sometimes pointed at tip, up to 6 cm long; margins sparingly, minutely toothed.

Flowers: Pink or white, 5–15 mm wide, 4–6 mm long; sepals with glandular hairs; usually nodding at first; July–September.

Fruits: Linear capsules with glandular hairs, 4–8 cm long; stalks 6–15 mm long; seeds about 1 mm long, inversely egg-shaped with abrupt, short beak and tuft of tawny, silky hairs 3.5–6.5 mm long.

Habitat: Moist to wet areas.

Notes: Northern willow-herb was formerly known as *Epilobium adenocaulon* and *E. leptocarpum*.

Frank Boas

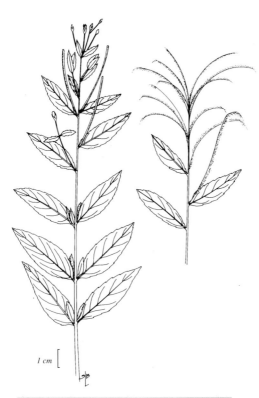

1 cm

EVENING-PRIMROSE FAMILY (ONAGRACEAE)

General: Perennial herb; flowering stems weak, soft, 10–25 cm tall; from underground stems (rhizomes).

Leaves: Opposite, stalked, simple, egg- to heart-shaped, pointed at tip, 1–6.5 cm long, thin; upper surface hairless to short-haired; underside more prominently hairy; margins coarsely toothed; stalks narrowly winged.

Julie Hrapko

Flowers: White, about 2 mm long; 2 petals, deeply lobed; stalks 5–6 mm long; in sparse, elongated terminal clusters; July–September.

Fruits: Small, club- or pear-shaped capsules about 2 mm long, covered with soft, hooked hairs.

Habitat: Wet organic hardwood swamps; moist to fresh, sandy to fine loamy tolerant hardwood stands with eastern hemlock and yellow birch.

Notes: The name *Circaea* comes from the Greek goddess Circe, who was a mythological enchantress. Some sources say she made a powder from the plant and used it to enchant people to love; other sources claim she was a sorceress who used a poisonous member of the *Circaea* genus in her sorcery, hence the name 'enchanter's nightshade.'

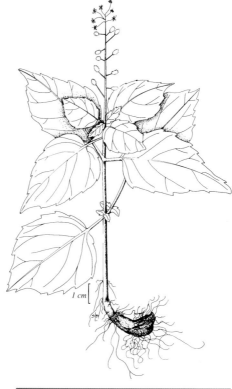

1 cm

EVENING-PRIMROSE FAMILY (ONAGRACEAE)

General: Perennial herb; flowering stems 20–60 cm tall; from slender, horizontal underground stems (rhizomes) with thread-like runners (stolons).

Leaves: Opposite, simple, oblong to egg-shaped, tapered to point at tip, rounded or nearly heart-shaped at base, 6–12 cm long, usually half as wide, hairless; margins shallow-toothed; stalks slender, rounded or angled on lower side.

Flowers: White, small; 2 petals, deeply notched, 2.5–3.5 mm long; sepals fused in a hairy, 2-lobed tube; in many-flowered terminal clusters up to 20 cm long; June–August.

Brenda Chambers

Fruits: Bur-like with stiff, hooked bristles, egg-shaped with 3–5 rounded ridges and furrows on each half, 3.5–5 mm long (including bristles), 2-seeded; remain closed.

Habitat: Fresh to moist tolerant hardwood stands in the southern part of our region; uncommon.

Notes: This species is also known as *Circaea quadrisiculata*.

1 cm

E.T.

General: Perennial herb, up to 50 cm tall (leaf); stems short (barely reach soil surface); from extensive, creeping underground stems (rhizomes).

Leaves: Basal, solitary, long-stalked, taller than flower stalk, twice compound, with 3 sections, each with 3–5 leaflets; leaflets egg-shaped or oval, pointed at tips, 5–12.5 cm long; margins finely toothed.

Flowers: Greenish-white, small; 5 petals, about 2.5 mm long; numerous, usually in 3 rounded clusters at top of leafless stalk.

Karen Legasy

Fruits: Rounded, purplish-black berries 6–8 mm long; in clusters; ripen July–August.

Habitat: All moisture regimes, soil and stand types.

Notes: During wars and on hunting excursions, aboriginal peoples subsisted on the nutritious rhizomes of wild sarsaparilla, which were also used in remedies for nosebleeds, wounds and sores. The berries were used to make wine, and to add flavour to beer made from the rhizomes. See caution in Introduction.

Food Use by Wildlife: seed.
• **Mammals:** eastern chipmunk, black bear. • **Birds:** Swainson's thrush, wood thrush, white-throated sparrow.

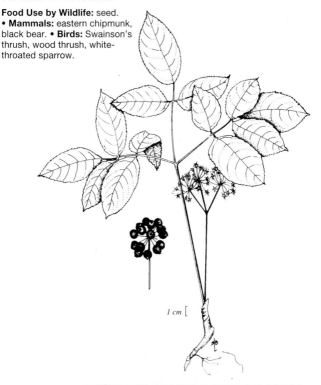

1 cm

BRISTLY SARSAPARILLA • *Aralia hispida*
ARALIE HISPIDE

General: Perennial herb; flowering stems erect, 20–90 cm tall, leafy, with sharp, slender spines near base; from stout underground stems (rhizomes).

Leaves: Alternate, several, short-stalked, twice palmately compound, with 3 sections, each with 3–5 leaflets; leaflets oblong to egg-shaped, pointed at tips, 2.5–5 cm long, hairless or with hairs on underside along veins; margins sharply toothed; stalks shorter than blades.

Flowers: Greenish-white, small; numerous, in 2–10 rounded terminal clusters on slender stalks.

Fruits: Dark-purple, rounded berries, 5-lobed when dry.

Brenda Chambers

Habitat: Sandy or rocky areas, clearings or open forests.

Notes: Bristly sarsaparilla is also commonly known as 'dwarf-elder' or 'wild elder.' • The roots and bark were historically used in a remedy for kidney and urinary problems. See caution in Introduction.

Food Use by Wildlife: See notes under sasparilla (p. 233).

]*1 cm*

General: Stout perennial herb; flowering stems smooth, blackish, up to 1 m tall, with many branches; from large, aromatic root.

Leaves: Alternate, few, wide-spreading, up to 80 cm long, twice palmately compound, usually with 3 sections, each with 3–5 leaflets; leaflets, egg- to heart-shaped, pointed, up to 15 cm long, pale beneath, slightly hairy; margins double-toothed.

Flowers: Greenish-white, about 2 mm across; in numerous, rounded, loose clusters along main stem; June–August.

Fruits: Hard, dark-purple berries about 5 mm across, numerous.

Habitat: Fresh to moist tolerant hardwood stands; wet organic hardwood swamps; uncommon.

Brenda Chambers

Notes: The roots are large and have a spicy aroma. Aboriginal peoples combined spikenard roots with those of wild ginger (p. 179) and pounded them into a mash for the treatment of broken bones (see notes under wild ginger, p. 179). Historically, the ground roots have been used to treat asthma, coughs and rheumatism. The roots, root bark, leaves, berries and seed oil were also used in medicine. See caution in Introduction.
• The young tips and roots provided food, and the berries were used for flavouring. The roots are still used to make tea and root beer. See caution in Introduction.

Food Use by Wildlife: See notes under sasparilla (p. 233).

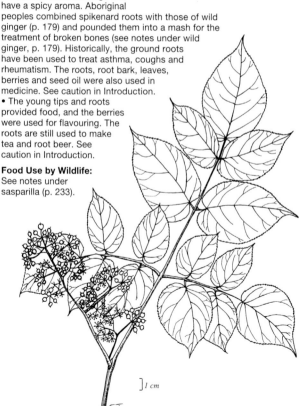

] *1 cm*

E.T.

DWARF GINSENG • *Panax trifolius*
GINSENG À TROIX FOLIOLES

GINSENG FAMILY (ARALIACEAE)

Brenda Chambers

General: Perennial herb; stems erect, 10–20 cm tall; from deep, rounded root, about 1.5 cm in diameter.

Leaves: In a single whorl of 3, stalked, palmately compound, with 3–5 leaflets; leaflets narrowly oblong to egg-shaped, blunt-tipped, 4–8 cm long, stalkless; margins finely toothed.

Flowers: White or pink-tinged, small, bisexual or with male and female flowers on separate plants; 5 petals; in a single, rounded terminal cluster; April–June.

Fruits: Yellowish, drupe-like berries, fleshy, about 5 mm in diameter.

Habitat: Fresh to moist tolerant hardwood stands, mainly in the southern part of our region; uncommon.

Notes: Ginseng (*Panax quinquefolia*), a close relative, is occasional in the southern part of our region and is highly valued in Chinese medicine. • Dwarf ginseng tubers are edible and were used as a source of starch. See caution in Introduction. • The species name *trifolius* means 'three-leaved.'

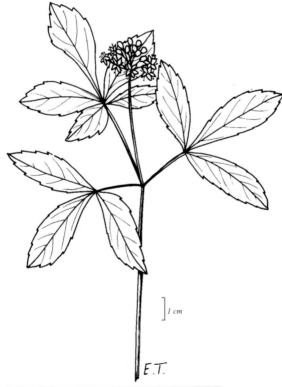

1 cm

E.T.

CARROT FAMILY (APIACEAE)

General: Erect perennial herb, soft, hairy; flowering stems 30–50 cm tall; from thickened, fibrous, aromatic roots.

Leaves: Opposite, fern-like, compound, 2–3 times divided in 3s; segments egg- to lance-shaped, hairy on both sides; lower leaves stalked, often 20 cm long and wide; upper leaves stalkless, reduced; margins deeply toothed.

Flowers: White or greenish-white; styles, even in fruit, not more than 1.0 mm; in short-stalked, hairy-stemmed, terminal and lateral clusters with few branches, usually surpass leaves; May–June.

Emma Thurley

Fruits: Long, thin, brown pods, 1–1.3 cm long (excluding tapered base); slender, bristly, tail-like appendages at base 5–7 mm long; June–August.

Habitat: Fresh to moist tolerant hardwood stands in the southern part of our region; uncommon.

Notes: The leaves of this species are similar to those of smooth sweet cicely (*Osmorhiza longistylis*), and mature fruits are needed to differentiate the 2 species. The style and its base are 2–4 mm long in smooth sweet cicely, and only 1 mm long in sweet cicely. • The crushed leaves and carrot-like roots of sweet cicely have a pleasant licorice- or anise-like odour. However the root is said to be rank-tasting. Aboriginal peoples chewed the root to treat sore throats, and applied a mixture of pounded roots and water to sores. See caution in Introduction.

Food Use by Wildlife: plant. • **Mammals:** deer.

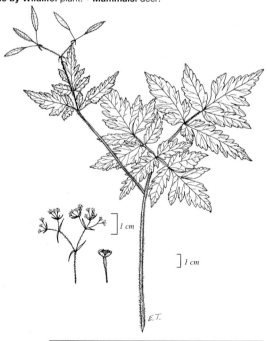

] *1 cm*

] *1 cm*

E.T.

Brenda Chambers

General: Perennial herb; flowering stems 30–80 cm tall; from thick cluster of fibrous roots.

Leaves: Palmately compound; basal leaves long-stalked, 10–30 cm wide, thick, divided into 5 (often appear as 7) lance- to egg-shaped, pointed leaflets; stem leaves smaller upwards, stalkless or short-stalked; margins singly to doubly sharp-toothed.

Flowers: Greenish-white, bisexual or male only, about 1 mm long; male flowers stalked; bisexual flowers stalkless; in several flat-topped, branched, 12–25-flowered clusters; June–August.

Fruits: Rounded to elliptic, 4–6 mm long, thickly covered with stout, hooked bristles.

Habitat: Fresh to moist tolerant hardwood stands in the southern part of our region; uncommon.

Notes: Aboriginal peoples used the crushed leaves of black snakeroot to treat bruises and swellings. The roots were pounded and used to treat snake bites, and a tea was made to cure fevers. See caution in Introduction. • The genus name comes from the Latin *sanare*, 'to heal,' referring to the medicinal properties of these plants.

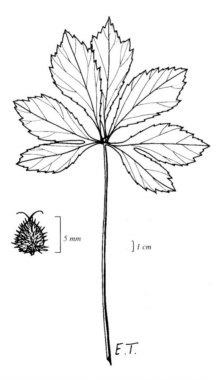

5 mm

] 1 cm

E.T.

CARROT FAMILY (APIACEAE)

General: Perennial herb, large; flowering stems 1.2–2.7 m tall, stout, hairy or woolly, rank-smelling; stems single, ridged, often 5 cm thick at base, hollow; from a thick taproot or cluster of fibrous roots.

Leaves: Alternate, widely oval, compound, with 3 maple-leaf-like leaflets 5–15 cm wide, thin, very hairy on underside; margins lobed, coarsely toothed; stalks inflated, winged, with a dilated sheath at base.

Flowers: White, often purple-tinged, about 1.3 mm wide; petals notched; numerous in flat-topped, umbrella-like, compound terminal clusters 10–20 cm wide, with 1–4 smaller clusters from lateral shoots on main stem; June–August.

Fruits: Widely oval to heart-shaped, flat, about 10 mm long and nearly as wide, finely hairy, 1-seeded.

Karen Legasy

Habitat: Wet organic to moist hardwood swamps; roadsides and streambanks.

Notes: Caution: Similar-looking plants such as water hemlock are extremely poisonous. • The stems and leafstalks have historically been considered an edible green. Aboriginal peoples ate cow parsnip. The taste is apparently similar to celery but the texture is more rhubarb-like. See caution in Introduction.

Food Use by Wildlife: plant.
Mammals: American black bear.

1 cm

1 cm

General: Perennial herb; flowering stems ascending to erect, up to 30 cm tall; from thick, spongy, scaly, submerged rootstocks scarred by bases of former leafstalks.

Leaves: Alternate, stalked, crowded near base of flowering stem, compound, with 3 leaflets; leaflets oblong or inversely egg-shaped, blunt-tipped, narrowed at base, stalkless, 3–8 cm long; stalks 10–30 cm long, sheathing at base.

Flowers: White, usually purplish- or pinkish-tinged, 1–1.3 cm long, funnel-shaped, 5-lobed; lobes spreading, covered with long, white hairs on inner surface; stalks 5–15 mm long; few to many in dense terminal clusters; May–July.

Fruits: Capsules, egg-shaped, blunt-tipped, about 1 cm long, irregularly rupturing at maturity; seeds flattened, egg-shaped, smooth, shiny and brown.

Habitat: Bogs and conifer swamps; wet, mucky shores.

Notes: Buckbean flowers have a foul odour. • The roots were historically used in a tonic which, when taken in small doses, was said to add vigour to the stomach and strengthen digestion, but large doses apparently caused vomiting. The entire plant is said to be bitter. See caution in Introduction.

*Linda Kershaw, top;
Bill Crins, bottom.*

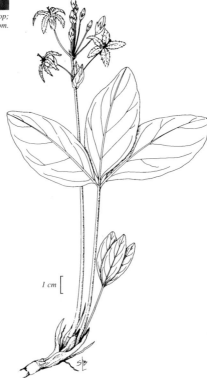

1 cm

DOGWOOD FAMILY (CORNACEAE)

General: Perennial herb; flowering stems 10–20 cm tall, erect; often growing in large patches from creeping underground stems (rhizomes).

Leaves: Opposite, but crowded at top of stem and appearing whorled, 4–6, essentially stalkless, simple, egg-shaped to oblong, taper to point at tip and base, 2–9 cm long, with 7–9 prominent major veins parallel to leaf margin; normally 1–2 pairs of small leaves on stem below main leaves.

Karen Legasy, top; Brenda Chambers, bottom.

Flowers: Tiny, greenish or purplish; in a tight cluster surrounded by 4 white, 1–2 cm long, spreading, petal-like bracts (the inflorescence resembles a flower); June–July.

Fruits: Bright scarlet, round, berry-like drupes, single-seeded, pulpy; in clusters; ripen July–August.

Habitat: All moisture regimes, soil textures and stand types.

Notes: The fruits are apparently not very palatable, although they have been considered edible. The plant was used in a cold remedy, and its roots were used in a colic remedy. See caution in Introduction.

Food Use by Wildlife: leaf, flower, fruit. • **Mammals:** white-tailed deer. • **Birds:** spruce grouse.

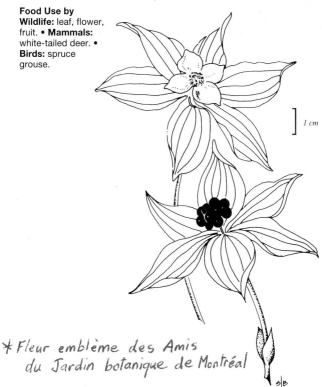

] *1 cm*

❋ Fleur emblème des Amis du Jardin botanique de Montréal

s/b

PRIMROSE FAMILY (PRIMULACEAE)

General: Perennial herb; flowering stems single, erect, up to 25 cm tall, unbranched; from slender underground stems (rhizomes).

Leaves: 5–10 in single terminal whorl, stalkless, simple, lance-shaped, tapering to slender point at tip and base, 4.5–10 cm long, hairless, shiny, prominently veined; tiny, scale-like leaf or leaves on stem below whorl.

Flowers: White, star-shaped, 10–12 mm across; 7 petals; stalks long, slender; 1–3 from centre of leaf whorl; May–June.

OMNR

Fruits: Rounded capsule with tiny, black seeds; July.

Habitat: All moisture regimes, soil and stand types.

Notes: Aboriginal peoples included starflower root in a smoke-making mixture used for attracting deer to hunters.

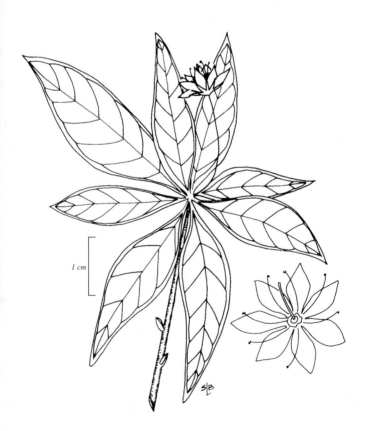

1 cm

WINTERGREEN FAMILY (PYROLACEAE)

General: Perennial herb; flowering stems 3–13 cm tall, leafless above except for 1–2 small, scale-like bracts, leafy near base; from slender underground stems (rhizomes).

Leaves: In 1–3 pairs or whorls near stem base, short-stalked, simple, rounded to inversely egg-shaped, blunt to rounded at tip, narrowed to rounded or slightly heart-shaped at base, 1–3 cm long, thin, veiny; margins toothed or nearly toothless.

Flowers: Waxy, white, rarely pinkish, 1–2 cm wide; 5 petals; solitary, terminal, nodding; July–August.

Fruits: Capsules, round, 6–8 mm across, brown, erect; August.

Habitat: All moisture regimes and soil types; conifer and hardwood mixedwood stands; uncommon.

Notes: One-flowered wintergreen was historically used in a tea to remedy such ailments as diarrhea, cancer, smallpox and sore throat. This tea was also believed to bring power and good luck. See caution in Introduction.

Linda Kershaw

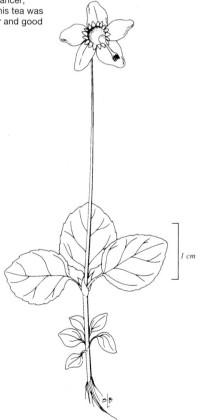

1 cm

ONE-SIDED WINTERGREEN • *Orthilia secunda*
PYROLE UNILATÉRALE

WINTERGREEN FAMILY (PYROLACEAE)

Brenda Chambers

General: Perennial herb; flowering stems 5–25 cm tall, with scattered, scale-like bracts above, often woody and leafy near base; from slender, creeping underground stems (rhizomes).

Leaves: Alternate, on lower part of stem but not in a basal rosette, stalked, simple, mainly thin, egg-shaped to oval or nearly round, usually pointed at tip and rounded to narrowed at base, 1.5–3 cm long; margins wavy to finely toothed.

Flowers: Whitish to greenish, bell-shaped, about 6 mm wide, nodding, short-stalked; 5 petals; style straight, 3–4 mm long, protruding; 6–20 in 1-sided, elongated, erect to nodding terminal clusters; July–August.

Fruits: Capsules, rounded, 3–5 mm wide; style persisting.

Habitat: Moist to dry, sandy to coarse loamy upland pine-black spruce and intolerant hardwood mixedwood stands; conifer swamps.

Notes: One-sided pyrola was formerly known as *Pyrola secunda*. • Its 1-sided flower cluster helps to distinguish this species from the *Pyrola*s.

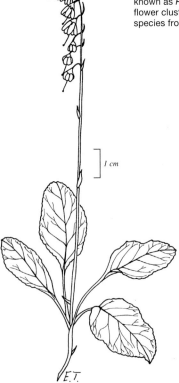

1 cm

E.T.

General: Perennial herb; flowering stems single, 12–30 cm tall, leafless except for 1–2 small, scale-like bracts; from creeping underground stems (rhizomes).

Leaves: In a basal rosette, stalked, simple, thin, dark green, dull; blade widely oval to oblong or egg-shaped with narrow end toward base and extending slightly down stalk, rounded or bluntly pointed at tip, 2–8 cm long, normally longer than stalk; margins wavy or with small, rounded teeth.

Flowers: Greenish-white or creamy, about 16 mm wide, nodding, short-stalked; 5 petals, spreading, egg-shaped with narrow end toward base; style long, protruding, bent; 3–21 in loose, cylindrical, elongated terminal clusters of; July–August.

Fruits: Capsules, rounded, 5-chambered; style persisting; August.

Habitat: Dry to fresh, sandy to clayey upland intolerant hardwood mixedwood and tolerant hardwood stands; wet organic hardwood swamps.

Notes: You can recognize shinleaf by its creamy-coloured flowers and its dull, thin leaf blades which are usually longer than

Brenda Chambers

and extend slightly down their stalks. • Aboriginal peoples used the roots in a remedy for weakness and back ailments. See caution in Introduction.

Food Use of Wintergreens (*Pyrola* spp.) by Wildlife: leaves, seeds. • **Birds:** ruffed grouse.

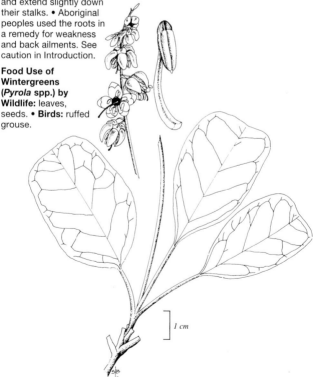

1 cm

Ray Demey

General: Perennial herb; flowering stems single, 5–25 cm tall, leafless or with 1–2 scale-like bracts; from slender underground stems (rhizomes).

Leaves: In a basal rosette or slightly scattered, stalked, simple, thin, dull, dark green; blades widely oval to rounded, blunt to slightly pointed at tip, slightly heart-shaped or narrowed to rounded at base, 1–4.5 cm long, shorter than or equal to stalk; margins have small, rounded teeth.

Flowers: White to pinkish, nodding; 5 petals 3–5 mm long; styles short, straight, not protruding from flower; 6–17 in loose, cylindrical or elongated terminal clusters 1.3–7 cm long; June–August.

Fruits: Capsules, rounded; style short, persisting.

Habitat: Moist coniferous, mossy or mixedwood forests.

Notes: Lesser pyrola is our only pyrola that does not have a style protruding from its flowers. • The leaves have historically been used in a salve or poultice for sores and wounds. See caution in Introduction.

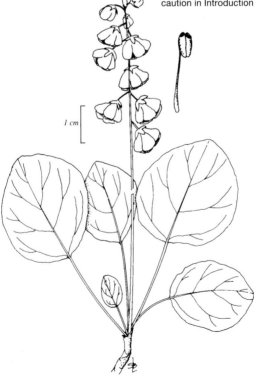

1 cm

General: Perennial herb; flowering stems 10–30 cm tall, leafless except for 1–3 small, scale-like bracts; from extensively creeping underground stems (rhizomes).

Leaves: In a basal rosette, simple, shiny, leathery; blades kidney- to heart-shaped or rounded with blunt tip, equal to or greater in width than length, 2–6.5 cm long; margins wavy or with rounded teeth; stalks narrowly margined, longer than blades.

Flowers: Pink to pale purple, nodding, short-stalked; 5 petals; style protruding; 4–22 in loose, cylindrical, elongated terminal clusters 3–20 cm long; July–August.

Linda Kershaw

Fruits: Capsules, rounded, nodding, with persisting style; August.

Habitat: Wet organic hardwood and conifer swamps; fresh upland pine and intolerant hardwood mixedwood stands.

Notes: You can distinguish pink pyrola by its purplish to pinkish flowers (the flowers of other pyrolas are whitish), and by its shiny, leathery leaves, with blades that are usually shorter than their stalks.
• Aboriginal peoples used the leaves in remedies for a variety of ailments. See caution in Introduction.

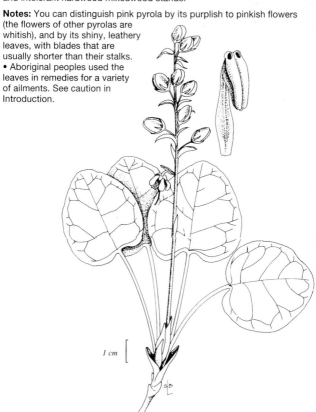

1 cm

OMNR

General: Perennial herb; flowering stems single, 10–30 cm tall, leafless except for 1–7 brownish, scale-like bracts; from creeping underground stems (rhizomes).

Leaves: In a basal rosette, stalked, simple, thick, leathery, shiny; blades rounded, 1.8–5 cm long, extend slightly down stalk; margins wavy or with small, rounded teeth; stalks winged.

Flowers: Creamy to white, nodding, fragrant; 5 petals, 5–7 mm long, 4–6 mm wide, thick, firm, scarcely veiny; styles protruding, downward-arching; 3–13 in loose, cylindrical, elongated terminal clusters; July–August.

Fruits: Capsules, nodding; style persisting.

Habitat: All moisture regimes, sandy to loamy upland intolerant hardwood mixedwood and tolerant hardwood stands; wet organic conifer swamps.

Notes: This species was previously called *Pyrola rotundifolia* • You can recognize round-leaved pyrola by its rounded leaves, winged leafstalks and leaf blades that extend slightly down their stalks. • Round-leaved pyrola was historically used to wash tumours and cancerous sores. The leaves were also used on wounds and bruises to help reduce pain. See caution in Introduction.

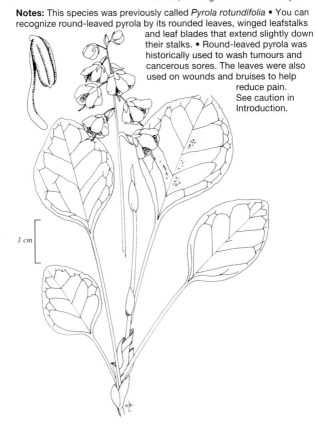

1 cm

WINTERGREEN FAMILY (PYROLACEAE)

General: Perennial herb; flowering stems single, 7–30 cm tall, leafless or rarely has 1–2 small, scale-like bracts; from creeping underground stems (rhizomes).

Leaves: In a basal rosette, simple, leathery, dull, numerous; blades rounded or widely oval, usually blunt or rounded at both ends, occasionally narrowed at base, 1–3 cm long, 1.5–3.5 cm wide; margins have small, obscure and rounded teeth; stalks longer than blades.

Flowers: Greenish-white (white with green veins), short-stalked, nodding; 5 petals, 5–7 mm long; style bent downward, upturned at tip, protruding from flower; 2–13 in loose, cylindrical, elongated terminal cluster; July–August.

Fruits: Capsules, nodding, 5.5–9 mm long; style persisting, 4–7 mm long.

Habitat: Wet organic hardwood swamps; dry, sandy upland conifer and intolerant hardwood mixedwood stands.

Brenda Chambers

Notes: You can distinguish greenish-flowered pyrola by its greenish-white flowers with protruding styles and its long-stalked, dull, leathery leaves. This plant is also called *Pyrola virens*. • The leaves contain acids and have historically been used to treat skin sores. See caution in Introduction.
• The genus name *Pyrola* is said to be derived from the Latin word *pyrus*, 'pear.'

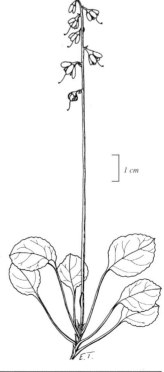

1 cm

E.T.

General: Perennial herb, hairy or downy, brownish-orange to yellowish, sometimes red, turns black when dried, fleshy; flowering stems 10–40 cm tall, unbranched; from fleshy roots.

Leaves: Alternate, simple, scale-like, up to 15 mm long, thick, more crowded toward stem base, same colour as stem.

Flowers: Brownish-orange to yellowish, same colour as stem, nodding when young, erect at maturity, about 1.3 cm long; 4 petals on lateral flowers, 5 petals on terminal flower; many in dense terminal cluster; June–November.

OMNR

Fruits: Capsules, egg-shaped to rounded, turn brown and split open when dry, erect.

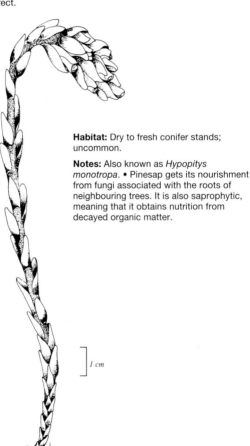

Habitat: Dry to fresh conifer stands; uncommon.

Notes: Also known as *Hypopitys monotropa*. • Pinesap gets its nourishment from fungi associated with the roots of neighbouring trees. It is also saprophytic, meaning that it obtains nutrition from decayed organic matter.

1 cm

INDIAN-PIPE FAMILY (MONOTROPACEAE)

General: Perennial herb, hairless, waxy, white or rarely pinkish, turns black when dried; flowering stems 5–30 cm tall, unbranched; roots brittle, in matted mass.

Leaves: Alternate, lance-shaped to oval, scale-like, often overlap, up to 10 mm long.

Flowers: White, urn- or bell-shaped, solitary, terminal, nodding at first, become erect as fruiting capsule matures; 4–5 petals; late July–September.

Fruits: Capsules, erect, brown, split open when mature, often persist through winter.

Habitat: All moisture regimes and soil textures; in many stand types.

Brenda Chambers

Notes: Indian pipe gets its nourishment through fungal connections between its roots and those of nearby trees. • The root was historically used as a remedy for epileptic seizures. Aboriginal peoples used the plant in a remedy to soothe and heal sore eyes. See caution in Introduction. • Indian pipe is also referred to as 'corpse plant' because of its colour, and 'ice plant' because if it is rubbed, it appears to melt like ice.

1 cm

NORTHERN BULGEWEED • *Lycopus uniflorus*
LYCOPE UNIFLORE

General: Perennial herb; flowering stems 10–40 cm tall, square; from runners (stolons) and short, fleshy tuber just below soil.

Leaves: Opposite, stalkless or nearly so, simple, lance-shaped to slightly oblong, tapered at both ends, 2–11 cm long, 0.5–3.5 cm wide, light green; margins coarsely toothed.

Flowers: White, tiny, about 3 mm long; calyx lobes broad-triangular; in dense clusters in leaf axils.

Fruits: Nutlets, smooth with thicker margins, squared at top, narrowed at base.

Habitat: Wet organic to moist hardwood swamps; along streams and shorelines.

Brenda Chambers

Notes: Northern bulgeweed belongs to the mint family and resembles field mint (p. 253), but it does not have a mint fragrance. Cut-leaved water horehound (*Lycopus americanus*) is a similiar species, but it has deeply and sharply lobed leaf margins.

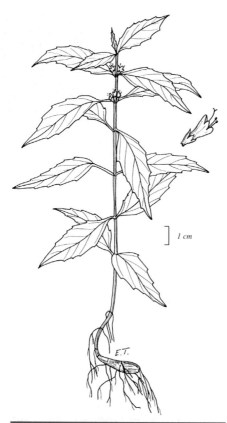

1 cm

E.T.

General: Perennial herb; flowering stems erect, 15–80 cm tall, unbranched or branched, slender, square, with downward-pointing hairs on angles, hairy or hairless on sides; aromatic, with a distinctive mint scent; from spreading underground stems (rhizomes).

Leaves: Opposite, stalked, simple, 3–7 cm long, smaller towards top of stem, egg- to lance-shaped or oblong, pointed to blunt at tip, rounded at base; margins toothed.

Flowers: Pale lilac to purplish, pinkish or sometimes white, bell-shaped, about 6 mm long and 3 mm wide; in dense, rounded clusters at leaf axils; July–September.

Fruits: Nutlets, egg-shaped, smooth.

Habitat: Moist to wet areas, shorelines, wet forest openings.

Notes: You can recognize field mint by its strong minty aroma. Field mint resembles northern bugleweed (p. 252), but that plant does not have a minty smell. • Field mint leaves have historically been used to flavour beverages, jellies and sauces. Aboriginal peoples used the leaves or the top of the plant in a fever remedy. See caution in Introduction.

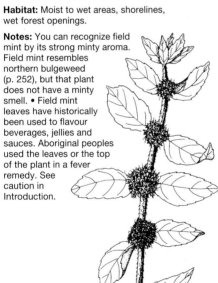

Linda Kershaw, top;
Karen Legasy, bottom.

1 cm

Linda Kershaw

General: Perennial herb, finely hairy to hairless; flowering stems erect, 30–50 cm tall, square, usually branched.

Leaves: Opposite, short-stalked, simple, oblong to lance-shaped, pointed at tip, slightly heart-shaped to rounded at base, 3–5 cm long, thin; margins toothed; stalks 1–4 mm long; upper leaves reduced, much smaller, may be stalkless, and sometimes toothless.

Flowers: Blue, tubular, slender, slightly wider at tip, gradually tapering to slender base, 1.5–2.5 cm long, 2-lipped; 2 sepals form a short tube, with a conspicuous bump on upper sepal; in pairs from leaf axils; July–August.

Fruits: Nutlets, rounded to flattish, covered with minute bumps.

Habitat: Moist to wet areas; swamps, marshes, clearings, ditches and wet shorelines.

Notes: Marsh skullcap is a member of the mint family, but it does not have a minty fragrance. • Mad-dog skullcap (*Scutellaria lateriflora*), has longer leafstalks (5–20 mm) and its flowers are in slender, 1-sided clusters from leaf axils. • Skullcaps were historically used in remedies for nervous disorders, hysteria, convulsions and cases of severe hiccupping. See caution in Introduction.

1 cm

General: Perennial herb, hairy; flowering stems slender, erect or ascending, 20–45 cm tall, square, usually branched; from short runners (stolons).

Leaves: Opposite, simple, egg-shaped to widely lance-shaped, 2–4 cm long; stalks up to 1 cm long, shorter on upper leaves; margins toothless or with a few rounded teeth.

Flowers: Pale purple or pink to white; petals 9–10 mm long, fused in a 2-lipped tube, protruding 2–5 mm beyond hairy tubular calyx; in rounded, dense clusters at stem tips or in upper leaf axils, with hairy, leaf-like bracts throughout; June–September.

Fruits: Nutlets, 4, small, smooth, 1-seeded.

Habitat: Woods, thickets, rocky shores.

Notes: Also known as *Satureja vulgaris*.
• European settlers used the leaves of this plant in cooking. A tea made from the leaves was used to relieve sore throats and treat kidney ailments. See caution in Introduction.

Emma Thurley

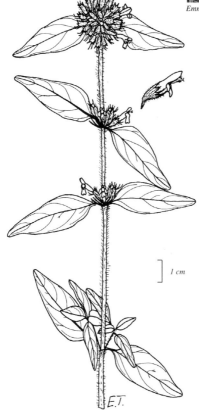

1 cm

HEAL-ALL • *Prunella vulgaris*
PRUNELLE VULGAIRE

General: Perennial herb; flowering stems 15–30 cm tall, square, hairy or almost hairless, slender, trailing to erect, usually unbranched; from running underground stems (rhizomes).

Leaves: Opposite, stalked, simple, egg-shaped to oblong, blunt to slightly pointed at tip, usually narrowed at base, 2–5 cm long; margins slightly toothed or toothless.

Flowers: Purple, 10–16 mm long, 2-lipped; upper lip arched; lower lip 3-lobed, pointing outward; greenish, hairy, small, leaf-like bract under each flower; in dense, terminal, cylindrical spikes 2–5 cm long; June–September.

Fruits: Nutlets, egg-shaped, smooth.

Habitat: Roadsides and shorelines; fields, clearings, waste areas and lawns.

Notes: Heal-all is named for its historical use in healing. It was used to stop internal and surface bleeding, to heal wounds and to cure hemorrhoids. It was also used as a remedy for sore throat and mouth ulcers. See caution in Introduction.

Emma Thurley

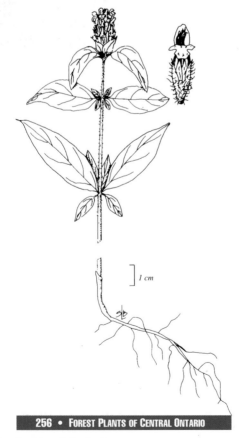

] *1 cm*

DOGBANE FAMILY (APOCYNACEAE)

General: Perennial herb; flowering stems 10–70 cm tall; branches numerous, wide-spreading, mostly hairless; stems and leaves release a milky juice when broken; from horizontal underground stems (rhizomes).

Leaves: Opposite, stalked, usually spreading or hanging, simple, egg-shaped to oval, with short, abrupt point at tip, rounded to narrowed at base, 2–7 cm long, 2.5–6.3 cm wide; upper surface hairless; underside paler, usually covered with downy hairs; stalks short, slender.

Flowers: Pinkish, fragrant, about 8 mm wide, bell-shaped with 5 spreading lobes, stalked, upwardly spreading to nodding; in showy terminal clusters, sometimes from leaf axils; June–July.

Fruits: Pods (follicles), slender, cylindrical, 8–12 cm long, in pairs, open along 1 side; seeds numerous, elongated, with tufts of long, cottony hair.

Habitat: Dry to moist, sandy to clayey, pine and intolerant hardwood mixedwood stands; edges of dry forest areas or roadsides.

Notes: Spreading dogbane can be recognized in the field by its milky juice, which is released when the stems or leaves are broken. The juice may irritate sensitive skin. • Aboriginal peoples used the roots in remedies for headache, nervousness, heart palpitations and kidney problems, and as an oral contraceptive. See caution in Introduction.

Brenda Chambers, both

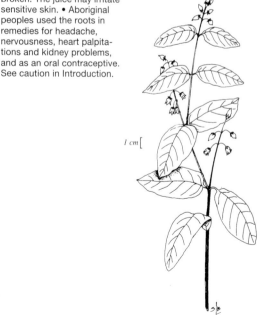

1 cm

Karen Legasy

General: Perennial herb, usually hairy, rough to touch; flowering stems erect, 20–80 cm tall, unbranched or loosely branched above; branches slender.

Leaves: Alternate, simple, thin, prominently veined, coarsely hairy, rough to touch; basal leaves long-stalked, egg-shaped, rounded to heart-shaped at base, 5–20 cm long; stem leaves on winged stalks near base, almost stalkless near top, egg- to lance-shaped, pointed at tip, tapered at base.

Flowers: Blue (pinkish in bud), 1–1.5 cm long, bell- or funnel-shaped with 5 lobes, nodding; in loosely branched, open clusters; June–July.

Fruits: Nutlets, small, rounded; July–September.

Habitat: Moist to fresh intolerant hardwood mixedwood stands; wet organic hardwood swamps; uncommon.

Notes: Northern bluebells is sometimes called 'lungwort' because it resembles a European lungwort (*Pulmonaria officinalis*). European lungwort was considered useful for treating lung diseases. This common name is also used for the leaf lichen, *Lobaria pulmonaria* (p. 418), which can lead to some confusion.

1 cm

BLUEBELL FAMILY (CAMPANULACEAE)

General: Slender perennial herb; flowering stems simple to freely branched, 10–50 cm tall; from slender underground stems (rhizomes).

Leaves: Alternate, hairless; basal leaves heart- to broadly egg-shaped, long-stalked, soon withering mostly toothed on margins; stem leaves linear or narrowly lance-shaped, 2–10 cm long, smaller above, toothless on margins.

Flowers: Purplish-blue, rarely white, 12–20 mm long, bell-shaped, with 6 short lobes; in loose terminal clusters of 1–15 stalked flowers; June–August.

Fruits: Capsules, short-egg-shaped, nodding, open by pores at base; seeds numerous.

Habitat: Open rocky areas, meadows and shores.

Notes: The species name *rotundifolia* means 'round-leaved' and refers to the basal leaves.

Brenda Chambers

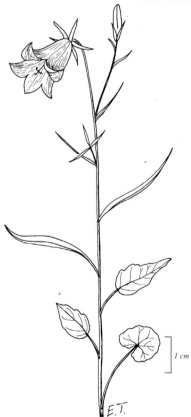

1 cm

E.T.

Brenda Chambers

General: Perennial herb; flowering stems slender, 30–60 cm tall, reclining or spreading, 3-angled with stiff, downward-pointing hairs on angles, leafy.

Leaves: Alternate, stalkless, simple, narrowly linear to linear-lance-shaped, gradually tapering, pointed at tip, 2–8 cm long, up to 8 mm wide; short, stiff hairs bent downward on margins and midribs; margins toothless or minutely toothed.

Flowers: Bluish, bell-shaped, 4–12 mm long, lobed for 1/3 of length; stalks thread-like; in a loose, widely branching, few-flowered cluster; June–August.

Fruits: Capsules, slightly rounded, 5–10 mm long, slender-stalked, open near base.

Habitat: Wet fields, marshes and alder thickets.

Notes: This plant was formerly known as *Campanula uliginosa*.
• *Campanula* means 'bell-shaped,' and refers to the flowers of these plants.

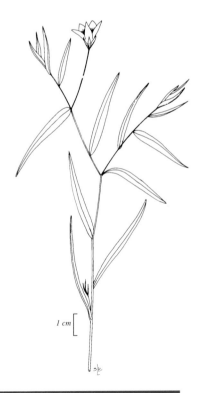

1 cm

FIGWORT FAMILY (SCROPHULARIACEAE)

General: Perennial herb, hairy, trailing, mat-forming, 20–30 cm long; flowering stems erect or curved upwards.

Leaves: Opposite, elliptic to narrowly egg-shaped, tapered to a short stalk, usually 2–4 cm long and 1–2 cm wide, hairy; margins uniformly toothed, except near base.

Flowers: Lilac-blue to pale violet with darker lines, 5–7 mm wide, 4-lobed; in several many-flowered, narrow, terminal clusters 3–6 cm long; May–July.

Fruits: Heart-shaped capsules, 4.5–5 mm wide and nearly as long, glandular-hairy; seeds few to many.

Habitat: Dry to fresh tolerant hardwood stands; roadsides.

Emma Thurley

Notes: The young leaves and stems of common speedwell are edible raw or cooked, and are high in vitamin C. See caution in Introduction.

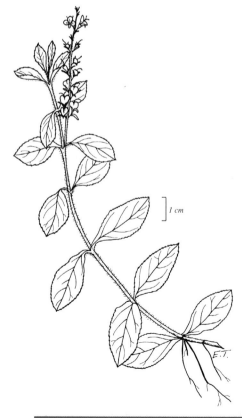

1 cm

CANADA WOOD BETONY • *Pedicularis canadensis*
PÉDICULAIRE DU CANADA

FIGWORT FAMILY (SCROPHULARIACEAE)

Brenda Chambers

General: Perennial herb; flowering stems several, erect or ascending, hairy, 15–40 cm tall, tufted.

Leaves: Alternate, sparsely hairy; basal leaves long-stalked, fern-like, lance-shaped to narrowly oblong, cut more than halfway to midvein into many oblong or egg-shaped segments; stem leaves similar, reduced upwards, short-stalked to stalkless; margins toothed.

Flowers: Yellow to purple, 15–20 mm long, tubular, 2-lipped; upper lip rounded, hood-like, with 2 teeth near tip; lower lip shorter than upper; in dense, solitary, terminal clusters 3–5 cm long; May–June.

Fruits: Capsules, oblong to lance-shaped, flattened, brown, about 15 mm long, 2–3 times as long as calyx, in elongated clusters 12–20 cm long; seeds many.

Habitat: Dry to fresh, sandy to loamy upland pine and intolerant hardwood mixedwood stands; in the southern part of our region.

Notes: Another common name for this genus is 'lousewort.' • Aboriginal peoples cooked the leaves of this plant in the same way that we cook spinach. Wood betony is reported to have aphrodisiac qualities. Historically, the whole plant was used in medicine and the root was included in a food for ponies. Wood betony is considered **poisonous to humans or animals if eaten in quantity**. See caution in Introduction.

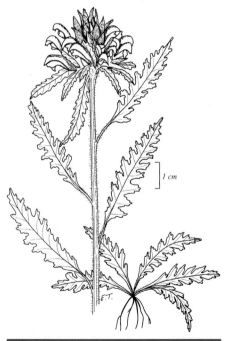

1 cm

General: Annual herb; flowering stems slender, 5–50 cm tall, slightly squared, unbranched to branched.

Leaves: Opposite, short-stalked, simple, lance- to linear-lance-shaped, 2–6 cm long, pointed at tip, tapered at base; margins toothless, or with 1 or more bristle-like teeth near base of upper leaves.

Flowers: Whitish with yellow tip and throat, tubular, 2-lipped, about 6–12 mm long, short-stalked; solitary or in pairs from leaf axils, often in leafy clusters; June–July.

Fruits: Capsules, 3–5 mm wide; seeds 1–4, white, ripen to blackish with white to brown tips; July–August.

Karen Legasy

Habitat: Dry to fresh pine, pine-oak and pine-black spruce stands; occasional in wet organic conifer swamps.

Notes: The genus name *Melampyrum* combines Greek words for 'wheat' and 'black' and apparently was chosen because cow wheat seeds are often black and resemble wheat. Some people find that the flowers resemble a snake's head.

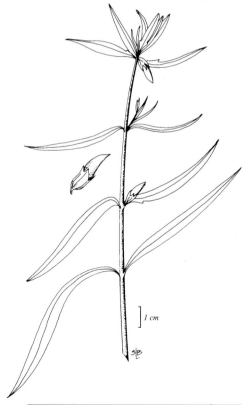

1 cm

FRAGRANT BEDSTRAW • *Galium triflorum*
GAILLET À TROIS FEUILLES

MADDER FAMILY (RUBIACEAE)

OMNR

General: Perennial herb, 20–80 cm long; stems square, hairless, weak, often trailing or leaning on other plants, sometimes rough to touch.

Leaves: Usually in whorls of 6, stalkless, simple, narrowly oblong, with short bristle-point at tip, 2–8.5 cm long, single-veined; underside often with coarse hairs; margins rough with stiff hairs.

Flowers: Greenish-white, small; 4 petals; stalks long, spreading; in groups of 3 in loose clusters from leaf axils or terminal; June–August.

Fruits: Pairs of 2–3 mm, joined balls, densely covered with hooked bristles; August.

Habitat: Moist to fresh, clayey to sandy upland tolerant hardwood and intolerant hardwood mixedwood stands; wet organic hardwood and conifer swamps; absent in very dry conifer stands.

Notes: Fragrant bedstraw has a strong vanilla fragrance and contains a substance which apparently prevents blood from clotting. • The dried and roasted fruit has historically been used as a coffee substitute. See caution in Introduction.

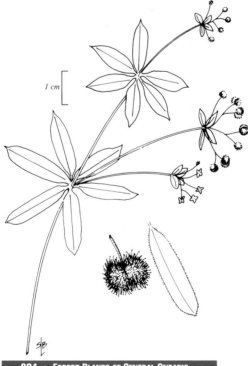

1 cm

General: Perennial herb, 50–180 cm long; stems much-branched, leaning on bushes or sometimes erect and climbing, square, rough with downward-pointing, stiff hairs.

Leaves: In whorls of 6 (4–5 on branchlets), stalkless, simple, narrowly oval to lance-shaped, with sharp, firm, pointed tip (cuspidate), narrowed at base, 1–2 cm long, 2–4 mm wide; midrib and margins rough with stiff hairs.

Flowers: White, 2–3 mm wide; 4–6 (usually 4) petals; stalks slender, spreading; numerous in clusters, terminal or from leaf axils; June–August.

Fruits: Smooth nutlets in pairs, shiny, about 2 mm long.

Habitat: Moist clayey to sandy upland intolerant hardwood mixedwood and tolerant hardwood stands; wet organic hardwood and conifer swamps; open wet forest edges.

Brenda Chambers

Notes: Rough bedstraw is distinctly rough to touch and can stick to clothing. • This plant was historically used in a remedy for kidney ailments. See caution in Introduction.

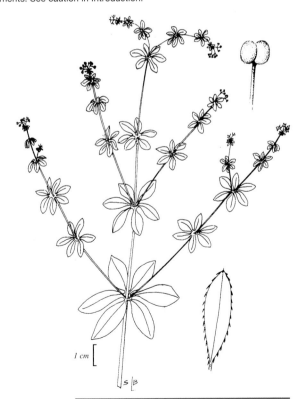

1 cm

MADDER FAMILY (RUBIACEAE)

General: Perennial herb, hairless; flowering stems erect, 20–60 cm tall, straight, unbranched or branched, leafy, square.

Leaves: In whorls of 4, lance-shaped to linear, blunt to pointed at tip, 2.5–6 cm long, 2–6 mm wide, 3-nerved; margins sometimes fringed with hairs.

Flowers: White, 3 mm wide, 4-lobed; in dense, showy terminal clusters; May–August.

Fruits: Nutlets in pairs, about 2 mm wide, usually densely hairy, sometimes almost hairless when mature.

Habitat: Gravelly or rocky areas, shorelines, clearings, ditches.

Notes: Northern bedstraw is usually hairless; this distinguishes it from rough bedstraw (p. 265) which feels rough because of its stiff, backward-pointing hairs. Small bedstraw (*Galium trifidum*), a smaller, reclining species, also has leaves in whorls of 4, but its flowers are 3-lobed. • Aboriginal peoples used northern bedstraw roots to make a red dye.

Robert Norton

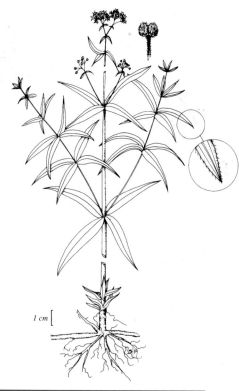

1 cm

PLANTAIN FAMILY (PLANTAGINACEAE)

General: Perennial herb, hairless to slightly hairy; flowering stems 15–45 cm tall; from short, thick, erect underground stems (rhizomes).

Leaves: Basal, long-stalked, simple, egg-shaped, 5–30 cm long, firm, minutely hairy, usually slightly rough when dry, ascending or lying on ground; lengthwise ribs prominent, almost parallel.

Flowers: Numerous, greenish, about 2 mm long; in narrow, elongated terminal spike 5–25 cm long; June–October.

Fruits: Capsules, 2–4 mm long, brown or purplish; top comes off like a lid; 6–15 seeds, 1 mm long.

Habitat: Roadsides, waste areas, lawns, open shores, forest clearings.

Notes: This introduced species is a weed in lawns. • The mature leaves were considered too stringy to eat, but the young leaves have historically been eaten in salads and as cooked greens. See caution in Introduction.

Karen Legasy, top;
Linda Kershaw, bottom

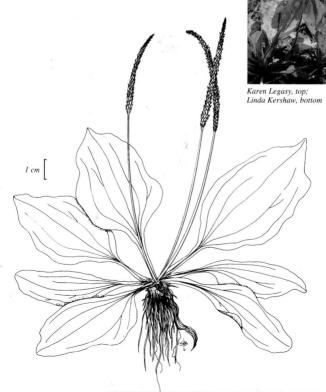

1 cm

Karen Legasy

General: Perennial herb, with milky juice; flowering stems hollow, slender, 5–45 cm tall, leafless; from thick, deep taproot, often about 25 cm long.

Leaves: Basal, stalked, simple, oblong to spoon-shaped, pointed to blunt at tip, narrowing to slender, slightly winged stalk at base, 5–40 cm long, often slightly hairy; margins with irregular, downward-pointing teeth or lobes.

Flowerheads: Yellow, 2–5 cm long; florets strap-like (ray); small sepal-like bracts at base (involucre) narrow, pointed; outer bracts bend backward; solitary on leafless stalk; April–September.

Fruits: 3–4 mm long, not splitting open, dry, hard, single-seeded (achenes), tipped with long beak and tuft of white bristles (pappus); mature heads silky or downy, round.

Habitat: Fields, waste places, disturbed sites, roadsides.

Notes: Dandelions are related to lettuce and their young leaves have historically been used in salads. The roots were used as a coffee substitute. The flowers are used to make dandelion wine and the entire plant has been used to brew beer. See caution in Introduction. • The irregular teeth of the leaves are said to resemble the teeth of a lion, thus the name *dent de lion* or dandelion. • This is an introduced species from Europe.

Food Use by Wildlife: leaf, flower, seed.
• **Mammals:** eastern cottontail, snowshoe hare, least chipmunk, white-tailed deer. • **Birds:** ring-necked pheasant, ruffed grouse, American goldfinch.

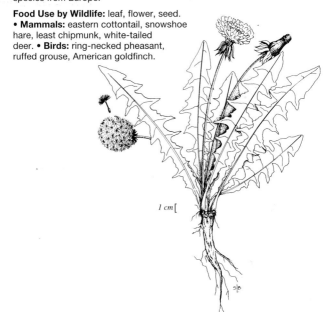

1 cm [

ASTER FAMILY (ASTERACEAE)

General: Perennial herb with milky juice; flowering stems upright, up to 2 m tall, leafy, hairless or often hairy towards the base; from thick, spindle-shaped, tuberous root.

Leaves: Alternate, simple, long-stalked, triangular to arrowhead-shaped, unlobed to deeply lobed, gradually reduced upwards and becoming stalkless, variable in size and shape, 4–15 cm long, 2.5–16 cm wide; upper surface hairless, underside often hairy; margins toothless to coarsely toothed.

Flowerheads: Greenish-white to yellowish-white; florets strap-like (ray), few (8–15 florets in *P. alba* and 5–6 in *P. altissima*); sepal-like bracts (involucre) few (7–10 in *P. alba* and typically 5 in *P. altissima*), 11–14 mm long; in loose terminal clusters; August–September.

Fruits: Small, elongate, dry, hard, single-seeded (achenes), with tuft of white to brownish-white, hair-like bristles (pappus) at tip.

Habitat: Dry to moist tolerant hardwood stands.

Notes: The flower characteristics noted above distinguish white lettuce (*P. alba*) and rattlesnake root (*P. altissima*), the 2 *Prenanthes* species typical of our region. • After giving birth, aboriginal women drank a broth made from wild lettuce roots to induce the production of milk. Historically, the root has been used as an antidote to the bites of poisonous snakes. See caution in Introduction.

Emma Thurley, top;
Brenda Chambers, bottom

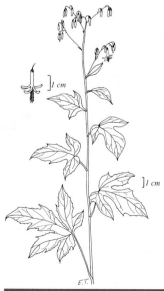

ORANGE HAWKWEED • *Hieracium aurantiacum*
ÉPERVIÈRE ORANGÉE

ASTER FAMILY (ASTERACEAE)

General: Perennial herb, hairy, glandular; flowering stems slender, 20–70 cm tall, leafless (rarely has 1–2 stalkless leaves); from long underground stems (rhizomes) and runners (stolons).

Leaves: In a basal rosette, simple, oblong to spoon-shaped, blunt-tipped, narrowed at base, 4–20 cm long, 1–3.5 cm wide, coarsely hairy; margins toothless or sometimes slightly toothed.

Flowerheads: Orange, about 2 cm wide; florets all strap-like (ray); small, green, sepal-like bracts at base (involucre) covered with blackish, glandular hairs; in rounded terminal clusters; June–August.

Karen Legasy

Fruits: Dry, hard, single-seeded (achenes), oblong, with dirty-white, hair-like bristles (pappus) at tip.

Habitat: Sandy roadsides, fields and clearings, rocky shores.

Notes: Orange hawkweed is also called 'devil's paintbrush.'
• This weedy species was introduced from Europe. • See notes on yellow hawkweed (p. 271) for ways to distinguish these 2 species.

Food Use of Hawkweeds (*Hieracium* spp.) by Wildlife: leaf, flower, seed. • **Mammals:** white-tailed deer. • **Birds:** ruffed grouse, wild turkey.

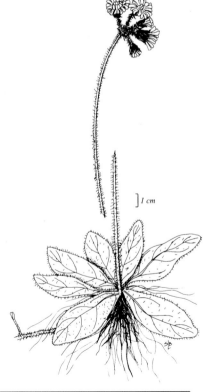

] *1 cm*

General: Perennial herb, covered with long hairs; flowering stems leafless or with 1–2 leaves below middle, 30–60 cm tall, blackish hairy.

Leaves: In a basal rosette, on winged stalks or stalkless, simple, narrowly oblong, blunt-tipped, tapered at base, 6–25 cm long, 10–20 mm wide, hairy on both sides; margins toothless or with a few distantly spaced, minute, glandular-tipped teeth.

Flowerheads: Yellow, about 1.3 cm wide; florets all strap-like (ray); small, green, sepal-like bracts at base (involucre) covered with gland-tipped, black hairs; in loose, irregular and slightly flat-topped clusters; May–August.

Karen Legasy

Fruits: Dry, hard, single-seeded (achenes), oblong, with hair-like bristles at tip (pappus).

Habitat: Fields, clearings, roadsides.

Notes: Yellow hawkweed is very similar to orange hawkweed (p. 270). You can differentiate them by their flower colours when they are in bloom, and by the shapes of their leaves when they are not. Yellow hawkweed leaves are narrowly oblong, whereas those of orange hawkweed are more spoon-shaped.

Food Use by Wildlife: See notes under orange hawkweed (p. 270).

1 cm

WILD LETTUCE • *Lactuca* spp.
LAITUE

ASTER FAMILY (ASTERACEAE)

General: Annual, biennial or perennial herb, with milky juice; flowering stems leafy, up to 2.5 m tall.

Leaves: Alternate, dandelion-like, usually deeply lobed, sometimes without lobes, variable.

Flowerheads: Yellow, white, pink or blue; florets all strap-like (ray), 5-toothed at tip; sepal-like bracts (involucre) arranged in a cone or cylinder; in spreading clusters, rarely solitary; summer and fall.

Ray Demey

Fruits: Small, dry, hard, single-seeded (achenes), oval, oblong or linear, flat, 3–5-ribbed on each side, tipped with a thread-like beak (or beakless) and a cluster of soft hairs (pappus) which falls separately.

Habitat: Moist thickets and woods, shorelines.

Notes: Due to the large amount of variation in this genus, it is often difficult to identify wild lettuce to the species level in the field.
• The genus name *Lactuca* comes from the Latin word *lac*, which means 'milk' and refers to the milky juice the plants release when their foliage is crushed.

Food Use by Wildlife: leaf, flower, seed. • **Mammals:** white-tailed deer.
• **Birds:** ring-necked pheasant, rufous-sided towhee, American goldfinch.

1 cm

General: Perennial herb; flowering stems 60–180 cm tall, purple or purple-spotted, usually with longitudinal lines; from horizontal underground stems (rhizomes).

Leaves: In whorls of 3–5, stalked to almost stalkless, simple, egg- to lance-shaped, tapered at tip and base, 6–20 cm long, 2–9 cm wide, thick; upper leaves smaller; upper surface slightly rough; underside minutely hairy to hairless and rough; margins coarsely toothed.

Karen Legasy

Flowerheads: Pinkish to purplish, cylindrical, thick, fluffy, about 8 mm wide; florets all tubular (disc); cluster of sepal-like bracts (involucre) 6–9 mm high, hairy, often purplish; in flat-topped clusters 10–14 cm wide; July–September.

Fruits: Dry, hard, single-seeded (achenes), 3–4 mm long, glandular-dotted, tipped with a cluster of fluffy, brownish hairs (pappus).

Habitat: Moist soils, thickets, wet clearings, shorelines.

Notes: Purple Joe-Pye weed (*Eupatorium purpureum*) is similar, but has dark purple coloration at the nodes, leaves in whorls of 3 or 4 and dome-shaped, fewer-flowered clusters. • The roots of spotted Joe-Pye weed were historically used in a remedy for urinary and kidney problems. See caution in Introduction. • The species name *maculatum* means 'spotted.'

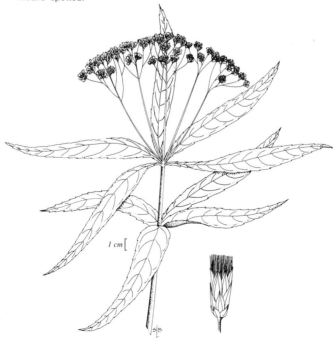

1 cm

ASTER FAMILY (ASTERACEAE)

Karen Legasy

General: Biennial, 0.5–2 m tall; stems woolly or hairy when young, becomes hairless, slender, leafy, hollow; from small, forking, slightly turnip-shaped root.

Leaves: Alternate, simple, densely white-woolly beneath when young, sometimes hairless on both sides when mature; basal leaves long-stalked, egg-shaped to elliptic, deeply lobed, 10–20 cm long; stem leaves stalkless, thin, smaller, unlobed, toothed or deeply cut (cleft) almost to midrib in lance-shaped to oblong lobes; spines usually tipped with slender prickles.

Flowerheads: Purple to rose-purple, about 3.5 cm wide, lacking prickles, long-stalked; florets all tubular (disc); small sepal-like bracts at base (involucre) cottony, sticky; several to many, in loose or crowded inflorescence; July–September.

Fruits: Small, dry, hard, single-seeded (achenes), oblong, flattish, not ribbed.

Habitat: Moist to wet sites; hardwood swamps; clearings and ditches, thickets; streambanks.

Notes: Aboriginal peoples used thistle roots in a remedy for back pain. See caution in Introduction.

Food Use of Swamp Thistle (*Cirsium* spp.) by Wildlife: seeds.
• **Birds:** dark-eyed junco, American goldfinch.

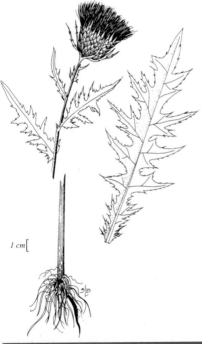

1 cm

General: Perennial herb; flowering stems 30–100 cm tall, hairless to finely hairy, erect, slender, lacking branches or forked near top.

Leaves: Alternate, simple; basal leaves on long, slender stalks, spoon- or egg-shaped with narrower end at base, 4–15 cm long; stem leaves usually stalkless and partly clasping stem, oblong, smaller; margins coarsely and regularly toothed or lobed.

Flowerheads: White with yellow disc in centre, 2.5–5 cm wide, solitary; 20–30 outer strap-like (ray) florets; tubular florets yellow, forming central disc; sepal-like bracts (involucre) overlapping (imbricate), with dark-brown margins; June–August.

Karen Legasy

Fruits: Small, dry, hard, single-seeded (achenes), black with about 10 white ribs.

Habitat: Roadsides, fields, meadows, clearings.

Notes: Ox-eye daisy is also known as *Leucanthemum vulgare*, and is a weed species introduced from Europe. • It gets its common name 'daisy' from the yellow central disc of the flower, which resembles the sun or 'day's eye.'

1 cm

YARROW • *Achillea millefolium*
ACHILLÉE MILLEFEUILLE

ASTER FAMILY (ASTERACEAE)

Karen Legasy

General: Perennial herb, covered with woolly or cobwebby hairs to nearly hairless; flowering stems erect, 30–70 cm tall, branched near top; from horizontal, spreading rootstocks.

Leaves: Alternate, fern-like, lance-shaped to narrowly oblong, twice or thrice deeply cut into narrow segments (bi- or tri-pinnatifid); stem leaves stalkless, 2–10 cm long, up to 1 cm wide; basal leaves longer, usually stalked.

Flowerheads: White, 4–6 mm wide, with 4–6 strap-like (ray) florets surrounding 10–30 tiny tubular (disc) florets; sepal-like bracts (involucre) 4–5 mm high, overlapping, with dry, membranous margins; in slightly rounded to flat-topped clusters; June–September.

Fruits: Dry, hard, hairless, single-seeded (achenes).

Habitat: Dry to fresh roadsides, fields.

Notes: Some people have an allergic skin reaction when they come into contact with yarrow. • Yarrow was historically used for ailments such as colds, fevers, stomach cramps, bleeding from the lungs and diabetes. The leaves were used to stop bleeding and reduce swelling, but they were also said to cause nosebleed when put into the nostrils. Some people believed that washing your head with yarrow could prevent baldness. See caution in Introduction.

Food Use by Wildlife: leaf, flower, seed. • **Mammals:** eastern cottontail, white-tailed deer. • **Birds:** ruffed grouse.

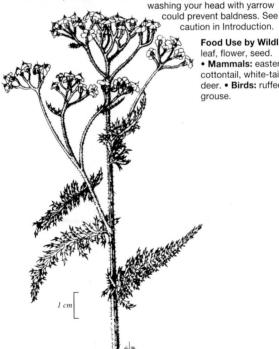

1 cm

ASTER FAMILY (ASTERACEAE)

General: Perennial herb; flowering stems erect, 10–90 cm tall, white-woolly, leafy; branched near top to form flower cluster; from spreading underground stems (rhizomes).

Leaves: Alternate, stalkless, simple, narrowly lance-shaped, 7.5–13 cm long, up to 1.5 cm wide, whitish-green; upper surface slightly hairy; underside densely woolly; midvein visible.

Flowerheads: Pearly-white with yellowish centres, rounded, about 6 mm wide; florets tubular (disc), all of 1 sex or nearly so; sepal-like bracts (involucre) overlapping, 5–7 mm high, scarious, pearly-white; in dense, broad terminal clusters; July–September.

Karen Legasy

Fruits: Dry, hard, single-seeded (achenes) with tuft of white, hair-like bristles at tip (pappus).

Habitat: Dry to fresh cutovers, roadsides, fields.

Notes: Pearly everlasting was historically used as a substitute for tobacco. The flowers were used in a remedy for paralysis and as a charm to keep evil spirits away. They were also used to help heal burns. See caution in Introduction. • These flowers are often dried and used in flower arrangements.

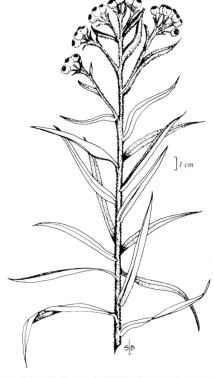

] 1 cm

Frank Boas

General: Perennial herb; flowering stems 15–40 cm tall; loosely spreading by runners or rooting branches (stolons) which develop terminal rosettes of leaves.

Leaves: Alternate, simple; basal leaves in a rosette, wedge- to spoon-shaped with widest part toward tip, 1.5–6.5 cm long, 0.5–1.3 cm wide, rounded or with an abrupt, sharp point at tip, tapered at base; upper surface green, hairless to thinly woolly; underside white-woolly, usually with a single prominent vein; stem leaves stalkless, linear, 3–8, smaller towards stem top.

Flowerheads: Whitish, rounded, male and female on separate plants, with tubular (disc) florets only; sepal-like bracts (involucre) 7–10 mm high, overlapping, sharply pointed at tips, dry, membranous; female heads about 1 cm wide; male heads smaller, 2–8 in flat-topped clusters; May–June.

Fruits: Dry, hard, single-seeded (achenes), rough, many fine, with tuft of white hairs (pappus) at tip.

Habitat: Dry rocky areas in forest openings.

Notes: Many people pick and dry field pussytoes for ornamental bouquets.

Food Use by Wildlife: leaf, flower, seed. • **Mammals:** eastern cottontail, snowshoe hare, white-tailed deer. • **Birds:** wild turkey.

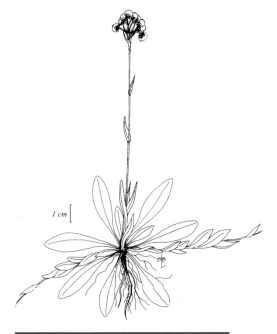

1 cm

ASTER FAMILY (ASTERACEAE)

General: Perennial herb; flowering stems 10–30 cm tall, with alternate, scale-like bracts with parallel veins, appear before leaves emerge; from extensively creeping, thick, cord-like underground stems (rhizomes).

Leaves: Basal, on long, slender stalks, simple, circular to kidney-shaped, 5–25 cm wide, with 5–7 lobes cut over halfway to base; upper surface green, hairless; underside white-woolly; margins coarsely toothed.

Flowerheads: Creamy white, fragrant, mainly male or female, with outer strap-like (ray) and inner tubular (disc) florets; in a terminal cluster; May–June.

Karen Legasy

Fruits: Dry, hard, single-seeded (achenes), 1.5–2.5 mm long, with a tuft of long white hairs (pappus) at tip, forming soft, fluffy seed heads; late May–June.

Habitat: Wet organic hardwood and conifer swamps; moist intolerant hardwood mixedwood stands.

Notes: Sweet coltsfoot was formerly known as *Petasites palmatus*.
• *Palmatus* means 'palmate' or 'shaped like a palm' and refers to the leaves.

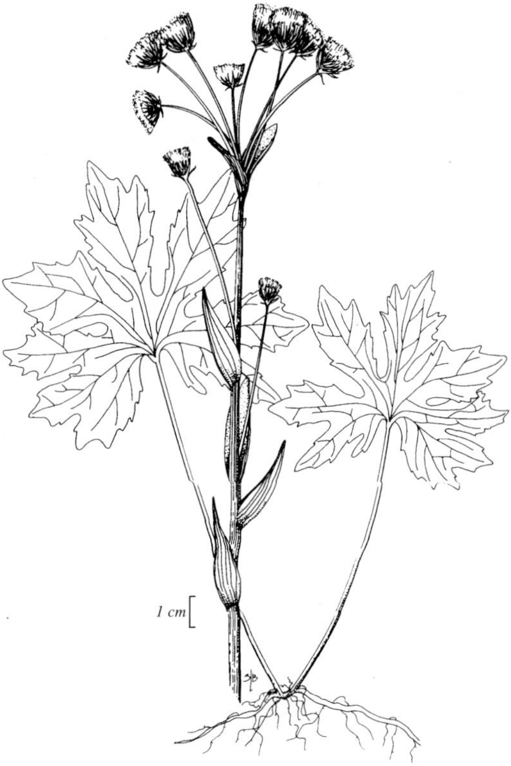

1 cm

CALICO ASTER • *Aster lateriflorus*
ASTER LATÉRIFLORE

General: Perennial herb; flowering stems ascending to erect, 20–120 cm tall, sparsely to densely hairy, slender, purplish or green, with wide-spreading to upwardly arched branches from near or below the middle; rootstocks woody and branched.

Leaves: Alternate, stalkless, simple; upper surface has short hairs pointing in 1 direction; underside usually hairless; margins toothed; stem leaves broadly lance-shaped to oblong lance-shaped, gradually tapered and pointed at tip, tapered at base, 5–15 cm long, 5–30 mm wide; basal leaves small, spoon- or lance-shaped, widest toward tip.

Flowerheads: White to pale-purple with pale-yellow to purplish or brownish disc or centre, less than 1.3 cm wide; 8–15 outer strap-like (ray) florets, 4.5–7.5 mm long, 0.7–2 mm wide; 8–16 central tubular (disc) florets, 3–5 mm long; sepal-like bracts (involucre) overlapping, 5–7.5 mm high, hairless to sparsely hairy; usually crowded on 1 side of straggly lateral branches; August–October.

Fruits: Small, dry, hard, single-seeded (achenes), compressed, cone-shaped with wider part toward tip, 0–1 rib per side, minutely hairy, with tuft of sparse, long, flat hairs (pappus) at tip.

Habitat: Dry to moist upland sites.

Notes: The floral discs range from pale yellow to purplish or brownish on the same plant, and even a single flower may have these colours at the same time.

Food Use of Asters (*Aster* spp.) by Wildlife: leaf, flower, seed.
• **Mammals:** eastern cottontail, snowshoe hare, white-footed mouse, white-tailed deer, moose.
• **Birds:** wild turkey, black-capped chickadee, swamp sparrow.

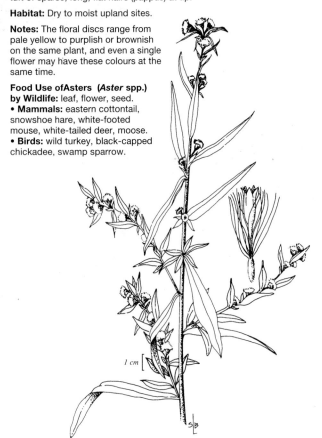

1 cm

ASTER FAMILY (ASTERACEAE)

General: Perennial herb; flowering stems sparsely hairy, single, erect, 20–70 cm tall, slender, unbranched or branched near top at flower cluster; from elongated, creeping underground stems (rhizomes).

Leaves: Alternate, stalkless, simple, oblong to lance-shaped, both ends pointed, 1–6 cm long, 2–12 mm wide, hairy, stiff, numerous and crowded on upper stem; margins toothless or toothed, often curled downward.

Flowerheads: Purplish to rose-pink, with yellow disc at centre (eventually becoming purple), daisy-like, 2.5–4 cm wide; 15–25 outer strap-like (ray) florets, 11–18 mm long and 1.2–2.6 mm wide; 20–35 central tubular (disc) florets, 5–7 mm long; sepal-like bracts (involucre) overlapping, 5–7.5 mm high, purplish to greenish, hairless to sticky-hairy; solitary or few in clusters; August–September.

Karen Legasy

Fruits: Small, dry, hard, single-seeded (achenes), glandular-hairy, with tuft of cottony, bristle-like hairs (pappus) at tip.

Habitat: Sphagnum bogs; edges of open bogs and boggy ditches.

Notes: You can distinguish bog aster from other asters by its unbranched or few-branched stem, its small, stiff, crowded leaves with downward-curled margins and its solitary or few flowers. • Aboriginal peoples used bog aster in an earache remedy. See caution in Introduction.

Food Use by Wildlife: See notes under calico aster (p. 280).

1 cm

PURPLE-STEMMED ASTER • *Aster puniceus*
ASTER PONCEAU

Karen Legasy

General: Perennial herb; flowering stems erect, 40–170 cm tall, hairless to (usually) covered with spreading, bristle-like hairs, purplish to reddish; from thick, woody, branched underground stems (rhizomes).

Leaves: Alternate, stalkless, clasping, simple, narrowly to widely lance-shaped, pointed at tips, up to 20 cm long and 1–4 cm wide; upper surface hairless to rough with sharp, stiff hairs; underside hairless or with hairs along midrib; margins toothless to sharply toothed.

Flowerheads: Blue-violet to purplish with yellow disc at centre (eventually turns purple), daisy-like, 2.5–4 cm wide; 30–50 outer strap-like (ray) florets, 7–15 mm long, 0.9–1.3 mm wide; 30–90 central tubular (disc) florets, 3.5–6.5 mm long; sepal-like bracts (involucre) overlapping, 6–10 mm high, hairless to sparsely hairy; usually numerous in loose to dense and slightly flat-topped cluster; July–September.

Fruits: Small, dry, hard, single-seeded (achenes), with tuft of cottony, bristle-like hairs (pappus) at tip.

Habitat: Wet organic hardwood swamps; open wet fields, thickets, shorelines.

Notes: Purple-stemmed aster is a highly variable species. Its distinguishing features are its large, clasping leaves and purplish, bristly stem.

Food Use by Wildlife: See notes under calico aster (p. 280).

1 cm

General: Stout perennial herb, 0.5–2 m tall; from creeping underground stems (rhizomes).

Leaves: Alternate, stalkless, simple, linear or lance-shaped, usually 10–20 cm long and 2–4 cm wide, hairless to sparsely hairy; lower leaves deciduous at flowering time; upper leaves 3–10 cm long; margins toothless to sparsely toothed.

Flowerheads: About 2 cm across, with white to pale-pink or pale blue-violet rays and a yellow, button-like disc at centre; 16–47 outer strap-like (ray) florets 3–10 mm long; 20–40 central tubular (disc) florets 3–6 mm long, yellow, becoming purple; basal cluster of overlapping sepal-like bracts (involucre) 3–8 mm high; several to many in loose, leafy clusters, on sparsely to densely hairy, forked stalks; August–October.

Brenda Chambers

Fruits: More or less flattened, minutely hairy, hard, dry, single-seeded (achenes); tuft of hair-like bristles (pappus) at tip, 3.6–7 mm long, whitish to pale yellow.

Habitat: Dry to fresh upland sites.

Notes: Also known as *A. simplex*.

Food Use by Wildlife: See notes under calico aster (p. 280).

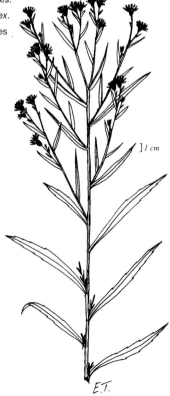

] *1 cm*

E.T.

ASTER FAMILY (ASTERACEAE)

Bill Crins

General: Perennial herb; flowering stems 0.3–2 m tall, hairless to densely hairy, leafy and branching at top; from numerous fibrous roots and creeping underground stems (rhizomes).

Leaves: Alternate, simple, lance-shaped, tapering to tip and base, up to 15 cm long; upper surface dark green; underside lighter with fine cross-veins; margins fringed with fine hairs; upper leaves reduced.

Flowerheads: 1.2–2 cm in diameter, with white outer rays and a yellow, button-like disc at centre; 4–15 outer strap-like (ray) florets, 5–9 mm long, 1.2–2.1 mm wide; 10–40 central tubular (disc) florets, yellow, 5.5–6.5 mm long; basal cluster of overlapping, sepal-like bracts (involucre) sparsely glandular, 3–7 mm high; numerous, in flat-topped clusters, on hairy stalks; July–September.

Fruits: More or less flattened, dry, hard, single-seeded, (achenes), slightly hairy, with 4–5 (sometimes 6–7) main nerves and double crown of hair-like bristles (pappus) at tip; inner bristles numerous, 3–4 mm long; outer bristles few, less than 1 mm long.

Habitat: Wet organic hardwood swamps; fresh to moist upland pine and intolerant hardwood mixedwood stands; cutovers, roadsides and shorelines, meadows.

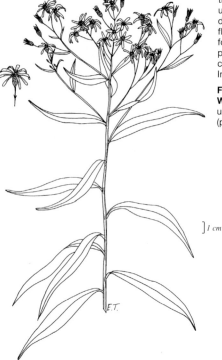

Notes: Historically, the leaves were used in food and drinks, and the flowers were used for medicinal purposes. See caution in Introduction.

Food Use by Wildlife: See notes under calico aster (p. 280).

] *1 cm*

E.T.

ASTER FAMILY (ASTERACEAE)

General: Perennial herb; flowering stems erect, 20–100 cm tall, hairless or with slightly hairy lines, sparingly leaved; may appear as a cluster of basal leaves without flowering stem; from long, creeping underground stems (rhizomes).

Leaves: Alternate, simple; lower leaves on long, winged, slender stalks fringed with fine hairs, narrowly egg-shaped, slightly heart-shaped or rounded at base, taper to pointed tip, firm, hairless to sparsely hairy, sharply toothed; upper stem leaves stalkless, lance-shaped, sparsely hairy, toothless.

Karen Legasy

Flowerheads: Bluish (rarely pink), with yellow disc at centre (turns purple), daisy-like; 15–25 outer strap-like (ray) florets 7–11 mm long, 1.2–2 mm wide; 20–35 central tubular (disc) florets, 4–6 mm long; sepal-like bracts (involucre) overlapping, 4–6 mm high, hairy on margins near tips, glandless; usually few in open, elongated clusters; August–September.

Fruits: Dry, hard, single-seeded (achenes), hairless, with tuft of white, hair-like bristles (pappus); seed heads fluffy white.

Habitat: Dry to fresh upland forests, cutovers.

Notes: Ciliolate aster can be distinguished from large-leaved aster (p. 287) by its leaves. The basal leaves of ciliolate aster are only slightly heart-shaped, while those of large-leaved aster have a more noticeably heart-shaped base. Also, the winged leafstalks of ciliolate aster have a fringe of fine hairs along their margins, but those of large-leaved aster do not.

Food Use by Wildlife: See notes under calico aster (p. 280).

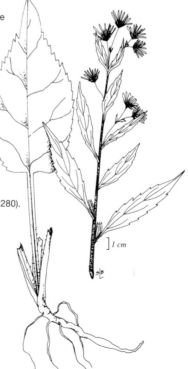

] *1 cm*

Emma Thurley

General: Perennial herb, nearly hairless; flowering stems 20–120 cm tall; from a persistent, branched base and many fibrous roots, may also have long creeping underground stems (rhizomes).

Leaves: Basal leaves heart-shaped with deep sinuses, 3.5–12 cm long, 2.5–7 cm wide, with long, winged stalks; stem leaves smaller, egg- to lance-shaped, stalkless upwards; upper surface rough to touch near margins; underside hairy; margins sharply toothed.

Flowerheads: Blue or purple (rarely white) rays with a yellow, button-like central disc; 10–16 outer strap (ray) florets, 6–8 mm long, 1.4–1.8 mm wide; 14–20 central tubular (disc) florets, 3–5 mm long, yellow becoming purple; basal cluster of overlapping, sepal-like bracts (involucre) 4–5.5 mm high; stalks sparsely to densely hairy; many, in a loose, much-branched inflorescence 15–70 cm high and 10–30 cm wide; August–October.

Fruits: Pale, smooth, dry, hard, single-seeded (achenes), 3–5-nerved, with tuft of white, hair-like bristles (pappus) at tip.

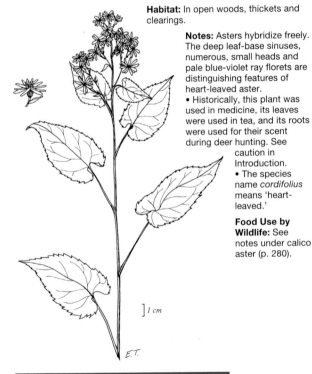

Habitat: In open woods, thickets and clearings.

Notes: Asters hybridize freely. The deep leaf-base sinuses, numerous, small heads and pale blue-violet ray florets are distinguishing features of heart-leaved aster.
• Historically, this plant was used in medicine, its leaves were used in tea, and its roots were used for their scent during deer hunting. See caution in Introduction.
• The species name *cordifolius* means 'heart-leaved.'

Food Use by Wildlife: See notes under calico aster (p. 280).

] 1 cm

E.T.

ASTER FAMILY (ASTERACEAE)

General: Perennial herb; flowering stems zigzag, up to 100 cm tall, reddish; from long, thick rootstocks.

Leaves: Alternate, thick; 1–4 basal leaves stalked, heart-shaped pointed at tip, up to 20 cm long and 15 cm wide, usually hairy and rough on upper surface, broadly toothed; stem leaves smaller, narrower, oblong, stalkless near top.

Karen Legasy

Flowerheads: Pale blue to purplish or lilac with yellowish disc at centre (turning reddish-brown), daisy-like; 9–20 outer strap-like (ray) florets 8–15 mm long, 1.5–2 mm wide; 20–40 central tubular (disc) florets 6–7.5 mm long; sepal-like bracts (involucre) overlapping, 7–10 mm high, often purple-tinged, hairy-glandular; in loose, open, flat-topped clusters with sticky hairs; August–September.

Fruits: Narrow or linear dry, hard, single-seeded (achenes) with fluffy tuft of white hairs (pappus) at tip; September.

Habitat: All moisture regimes, soil textures and stand types.

Notes: Aboriginal peoples ate the young, tender leaves and brewed a tea from the young roots to bathe the head as a remedy for headaches. See caution in Introduction.

Food Use by Wildlife: See notes under calico aster (p. 280).

1 cm

Karen Legasy

General: Perennial herb; flowering stems slender, 30–150 cm tall, hairless at base to hairy toward top, unbranched to clustered and branched; from long, creeping underground stems (rhizomes).

Leaves: Alternate, stalkless or lowest ones stalked, simple, lance-shaped, about 6–13 cm long and 0.5–1.8 cm wide, slightly to roughly hairy, with 3 parallel veins (triple-nerved), numerous and crowded on stem; margins sharply toothed to toothless.

Flowerheads: Yellow; 6–12 outer strap-like (ray) florets, about 2 mm long, 0.2–0.3 mm wide; 2–7 central tubular (disc) florets, about 2.5 mm long; sepal-like bracts (involucre) overlapping, 2–3 mm high, yellowish; numerous on upper side of flowering branches that often are curved, in a somewhat triangular cluster; August–September.

Fruits: Small, dry, hard, single-seeded (achenes), slightly hairy, and with a tuft of hair-like, white bristles (pappus) at tip.

Habitat: Dry to moist areas, roadsides, fields, clearings, thickets, streambanks.

Notes: A yellow dye was made from these flowers, and goldenrod was historically used as a tea substitute. See caution in Introduction.

Food Use of Goldenrods (*Solidago* spp.) by Wildlife: leaf, flower, seed. • **Mammals:** opossum, eastern cottontail, porcupine, white-tailed deer. • **Birds:** spruce and ruffed grouse, American tree sparrow, swamp sparrow, dark-eyed junco, pine siskin, American goldfinch.

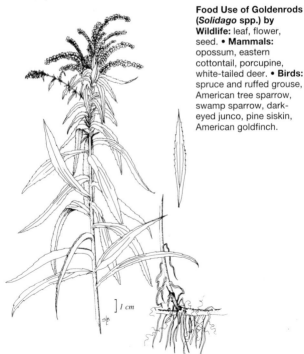

] 1 cm

General: Perennial herb; flowering stems solitary or in small clumps, 30–180 cm tall, leafy, rough, hairy; branches diverging or arching, with flowerheads on upper side; from long, creeping underground stems (rhizomes).

Bill Crins

Leaves: Alternate, stalkless, simple, oval or lance-shaped, pointed (rarely blunt) at tip, tapered to base, 3.8–12.5 cm long, 1.3–4.0 cm wide, hairy, wrinkled, rough; margins sharply toothed.

Flowerheads: Yellow; 6–11 outer strap-like (ray) florets, 2.5–2.8 mm long, 0.4–0.7 mm wide; 4–8 central tubular (disc) florets, 3–3.3 mm long; sepal-like bracts (involucre) overlapping, 2–4 mm high; in 1-sided clusters on spreading or arching branches of usually large, compound flower clusters; July–October.

Fruits: Dry, hard, single-seeded (achenes), with minute, sharp, stiff hairs, and tuft of soft, white, hair-like bristles (pappus) at tip.

Habitat: Thickets and fields, roadside ditches, forest edges and open shores.

Notes: Rough-stemmed goldenrod varies greatly in size, flowerhead shape and hairiness.

Food Use by Wildlife: See notes under Canada goldenrod (p. 288).

1 cm

BOG GOLDENROD • *Solidago uliginosa*
VERGE D'OR DES MARAIS

ASTER FAMILY (ASTERACEAE)

Brenda Chambers

General: Perennial herb; flowering stems solitary or few, 30–120 cm tall, hairless; branches hairy in flower cluster; from thickened, elongated, branched underground stems (rhizomes).

Leaves: Alternate, simple, thick, hairless, not strongly veined, oblong to lance-shaped, pointed at tip, narrowed and sheathing at base; lower leaves stalked, 10–35 cm long, 0.5–6 cm wide, toothed; upper stem leaves stalkless, 2–5 cm long, 0.5–1 cm wide, toothless.

Flowerheads: Yellow, 4–6 mm long; 1–8 outer strap-like (ray) florets, 3.2–3.7 mm long; 6–8 central tubular (disc) florets, 4.5–5 mm long; sepal-like bracts (involucre) 3–5 mm high, overlapping, with outer bracts egg-shaped and inner bracts linear to lance-shaped, hairy on margins at tips; in elongated clusters varying from narrow, 1-sided and few-headed to wide, not 1-sided, and with branches strongly ascending; August–September.

Fruits: Dry, hard, single-seeded (achenes), hairless, with tuft of soft, white hairs (pappus) at tip.

Habitat: Marshy and boggy sites around lakes, sandy shores and openings, damp, open thickets.

Notes: Bog goldenrod's appearance is highly variable, but you can recognize it by its sheathing basal and lower stem leaves.

Food Use by Wildlife: See notes under Canada goldenrod (p. 288).

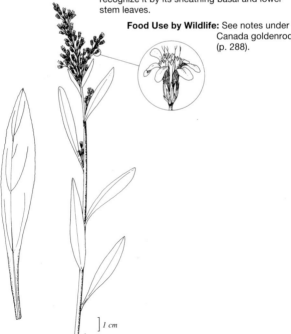

] *1 cm*

ASTER FAMILY (ASTERACEAE)

General: Perennial herb; stems stout, up to 100 cm tall, densely hairy, branchless or sometimes branched.

Leaves: Alternate, simple; lower leaves on winged stalks, oval, pointed to blunt at tips, 5–12.5 cm long, 2.5–5 cm wide, hairy on both sides, usually toothed; upper leaves stalkless, oblong, pointed at tips, smaller, hairy, slightly toothed or toothless.

Flowerheads: Yellow, short stalked; outer strap-like (ray) florets white to pale cream-coloured, 3.5–4 mm long, about 1 mm wide; 9–12 central tubular (disc) florets, 3–3.5 mm long; sepal-like bracts (involucre) overlapping, 3–5 mm high, white-edged, with green, conspicuous midrib; densely crowded in somewhat narrow, elongated clusters toward top of stem; August–September.

Fruits: Small, dry, hard, single-seeded (achenes), hairless or with a few flattened hairs, with a tuft of soft, white hairs (pappus) at tip.

Habitat: Dry, sandy and rocky areas, sandy pine stands.

Karen Legasy

Notes: Hairy goldenrod is almost identical to silverrod (*Solidago bicolor*), but silverrod flowers are white to pale cream while those of hairy goldenrod are yellow; however, the yellow flowers of hairy goldenrod sometimes fade, making it difficult to distinguish between the 2 plants. Silverrod occurs in parts of southern Ontario.

Food Use by Wildlife: See notes under Canada goldenrod (p. 288).

] *1 cm*

GRAMINOIDS

To assist in the identification of graminoids (grasses, sedges and rushes), illustrations and pictorial keys are provided below which highlight the main identifying features of these often difficult-to-identify plants.

PARTS OF A GRASS

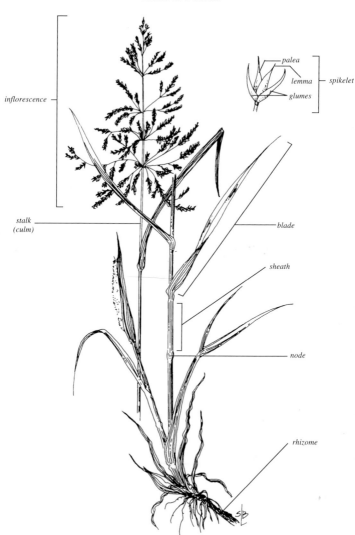

PARTS OF A SEDGE

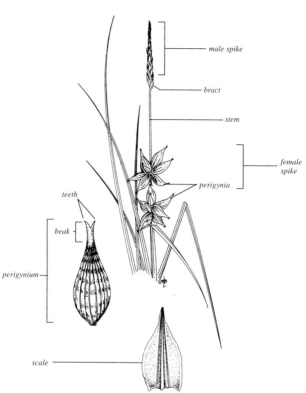

- male spike
- bract
- stem
- female spike
- perigynia
- teeth
- beak
- perigynium
- scale

KEY TO GRASSES, SEDGES AND RUSHES

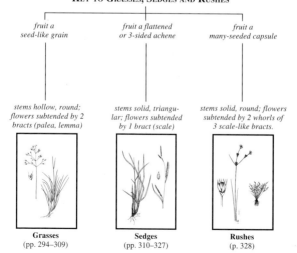

fruit a seed-like grain	fruit a flattened or 3-sided achene	fruit a many-seeded capsule
stems hollow, round; flowers subtended by 2 bracts (palea, lemma)	stems solid, triangular; flowers subtended by 1 bract (scale)	stems solid, round; flowers subtended by 2 whorls of 3 scale-like bracts.
Grasses (pp. 294–309)	**Sedges** (pp. 310–327)	**Rushes** (p. 328)

KEY TO THE GRASSES (POACEAE)

A. Compact inflorescence

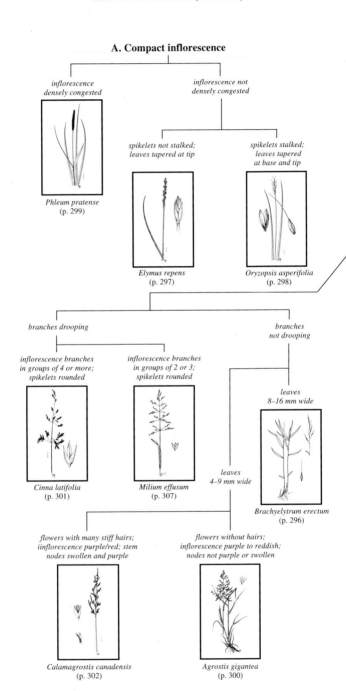

*inflorescence
densely congested*

Phleum pratense
(p. 299)

*inflorescence not
densely congested*

*spikelets not stalked;
leaves tapered at tip*

Elymus repens
(p. 297)

*spikelets stalked;
leaves tapered
at base and tip*

Oryzopsis asperifolia
(p. 298)

branches drooping

*inflorescence branches
in groups of 4 or more;
spikelets rounded*

Cinna latifolia
(p. 301)

*inflorescence branches
in groups of 2 or 3;
spikelets rounded*

Milium effusum
(p. 307)

*branches
not drooping*

*leaves
8–16 mm wide*

Brachyelytrum erectum
(p. 296)

*leaves
4–9 mm wide*

*flowers with many stiff hairs;
iinflorescence purple/red; stem
nodes swollen and purple*

Calamagrostis canadensis
(p. 302)

*flowers without hairs;
inflorescence purple to reddish;
nodes not purple or swollen*

Agrostis gigantea
(p. 300)

Les herbes graminées

B. Open inflorescence

spikelets 1- flowered

spikelets 2 or more flowered

lemmas awned

lemmas not awned

glumes without lines; leaves with boat-shaped tip

Poa saltuensis
(p. 305)

glumes with prominent lines; leaves without boat-shaped tip

Glyceria striata
(p. 306)

leaves 1–5 mm wide; spiklets few

Schizachne purpurascens
(p. 303)

leaves 5–10 mm wide; spiklets many

Bromus ciliatus
(p. 308)

leaves 1–2 mm wide and tufted at the base; spiklets many

inflorescence branched

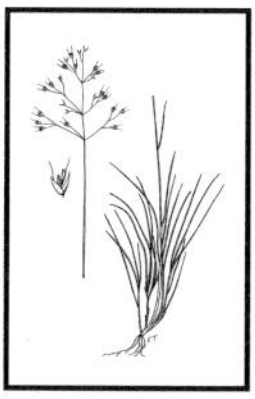

Deschampsia flexuosa
(p. 309)

inflorescence not branched

Danthonia spicata
(p. 304)

BEARDED SHORTHUSK • *Brachyelytrum erectum*
BRACHYÉLYTRUM DRESSÉE

GRASS FAMILY (POACEAE)

General: Perennial grass; flowering stems erect, 50–100 cm tall, hairless or hairy; in clumps from short, knotty rhizomes.

Leaves: Flat, broad, rough to hairy, 8–18 cm long, 8–16 mm wide.

Inflorescence: Slender, 5–15 cm long, loose, open, erect or nodding panicle; spikelets 1-flowered, 1 cm long (excluding awn); flower scales (lemmas) 6–10 mm long, rough hairy on nerves, with 12–25 mm long bristle-tip (awn), soon deciduous; June–August.

Habitat: Moist to dry, clayey to rocky tolerant hardwood and intolerant hardwood mixedwood stands; wet organic hardwood swamps.

Notes: The genus name, *Brachyelytrum*, comes from the Greek, *brachys*, 'short,' and *elytron*, 'husk,' referring to the persistent outer scales (glumes).

Karen Legasy

Food Use of Grasses by Wildlife: leaf, seed. • **Mammals:** eastern cottontail, snowshoe hare, red fox, black bear, raccoon, striped skunk, white-tailed deer, moose. • **Birds:** spruce grouse, black-capped chickadee.

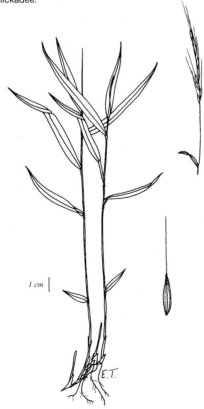

1 cm

E.T.

GRASS FAMILY (POACEAE)

General: Perennial grass, erect, 30–120 cm tall; from tough, extensively creeping, whitish or yellowish underground stems (rhizomes).

Leaves: Flat, green, somewhat soft, rough-hairy on edges, upper surface sparsely hairy; underside hairless, with fine, white lines, 7.5–30.5 cm long, 5–10 mm wide; sheaths usually shorter than internodes, with small, slender teeth at upper edge.

Inflorescence: Narrow, tight spike, 5–25 cm long; spikelets flattened, 0.6–2.2 cm long, 2–9-flowered; flower scales (lemmas) oblong to lance-shaped with a sharp point or bristle (awn) at tip, hairless, with 5–7 prominent, slender nerves; July–September.

Habitat: Sandy roadsides and disturbed habitats.

Notes: Also called *Agropyron repens*.
• Quack grass has been used for grazing and hay production, but it is also a troublesome weed. It persists in ploughed areas and spreads very rapidly.

Karen Legasy

] *1 cm*

Bill Crins

General: Perennial grass; flowering stems erect or abruptly bent at lowest node, smooth or rough, 25–50 cm tall.

Leaves: Basal leaves dark, evergreen, erect, rough on surface and edges, 40 cm long, 4–10 mm wide, often equalling or exceeding flowering stem, tapering gradually to tip and base, flat or curved inward on margins; sheaths purple; stem leaves shorter, usually less than 1 cm long.

Inflorescence: Narrow, short-branched panicle, 5–12 cm long; spikelets alternate (sometimes paired) on stem, 6–8 mm long (excluding 6–14 mm long hair-like awn); 2 large, translucent outer scales (glumes) remain after seeds are shed; May–June.

Habitat: Dry to fresh rocky, sandy to loamy upland tolerant hardwood, pine, pine-oak and intolerant hardwood mixedwood stands.

Notes: White-grained mountain rice is not related to cultivated rice and does not even look like it, but its grains vaguely resemble those of rice and are about the same size.
• The genus name *Oryzopsis* combines words for 'rice' and 'appearance.'

1 cm

GRASS FAMILY (POACEAE)

General: Perennial grass; flowering stems stiffly erect, hairless, 30–100 cm tall, bulb-like at base; in clumps with fibrous roots.

Leaves: Wide, flat, tapering to tip, hairless, rough-margined, 8–22 cm long, 5–10 mm wide.

Inflorescence: Narrow, cylindrical, very dense, short-branched panicle, green, becomes drab, 1–22 cm long, 7–10 mm thick, stiff, harsh, rough-textured; 2 outer scales (glumes) of each spikelet have short, stiff bristle-tips (awns); July–August.

Habitat: Fields, clearings, roadsides.

Notes: Timothy is an important hay grass. Older plants are too rough to be eaten by livestock, but they make good winter food when dried.

Karen Legasy

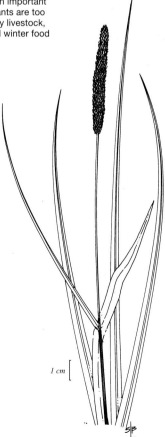

1 cm

REDTOP • *Agrostis gigantea*
AGROSTIS GÉANT

GRASS FAMILY (POACEAE)

Karen Legasy

General: Perennial grass; stems erect or abruptly bent at base, up to 1.5 m tall; basal shoots sterile, reclining, often form creeping stems (stolons); forms large colonies from numerous prolonged, scaly, underground stems (rhizomes).

Leaves: Flat, deep green, 5–9 mm wide.

Inflorescence: Loose, irregular, open, pyramid-shaped panicle 15–30 cm long, purple to green or reddish; branches spread when in fruit; spikelets 1-flowered, 2–3.5 mm long, with outer scales (glumes) rough on midvein only.

Habitat: Dry sandy or wet grassy edges of roads, in meadows and clearings.

Notes: Redtop was introduced from Europe. You can recognize it by its often reddish-tinged inflorescence. • Tickle grass (*A. scabra*) is also a common red-topped grass, often forming large colonies in disturbed habitats. It has delicate panicles and lacks rhizomes.

1 cm

General: Perennial grass; flowering stems erect, tall, hairless, slender, 60–120 cm tall, pale green; from creeping underground stems (rhizomes), solitary or in loose tufts.

Leaves: Flat, thin and wide, 10–25 cm long, 7–15 mm wide, spreading at right angles to stem; margins rough.

Inflorescence: Nodding, loose, open panicle, pale green, shiny, 12.5–25 cm long; branches hair-like, usually spreading and often drooping; lowest branches 3.8–12.5 cm long; spikelets 2–4.5 mm long, 1 flowered; flower scales (lemmas) with short, straight bristle-tip (awn); August–September.

Habitat: Moist to dry, fine loamy to sandy upland tolerant hardwood and intolerant hardwood mixedwood stands; grassy roadsides.

Notes: This grass is also called 'drooping woodreed.' The species name *latifolia* means 'broad-leaved.'

Bill Crins

1 cm

Karen Legasy, top;
Linda Kershaw, bottom

General: Robust perennial grass; flowering stems erect, smooth or somewhat rough, not hairy, often bluish to purplish around joints, 5–15 cm tall; in dense clumps or patches from numerous creeping underground stems (rhizomes).

Leaves: Flat, numerous, gradually tapering to a long, pointed tip, 15–30 cm long, 4–8 mm wide, rough (especially on edges and upper surface); sheaths shorter than internodes.

Inflorescence: Loose, irregular, open panicle, widely lance- to egg-shaped, occasionally nodding, 10–20 cm long, 2–10 cm wide, purplish to greenish or straw-coloured; spikelets 2–6 mm long, 1-flowered; flower scale (lemma) with a slender hair (awn) on back and a dense tuft of long hairs at base; outer scales (glumes) lance-shaped to narrowly egg-shaped; July–September.

Habitat: Wet organic hardwood swamps and moist to dry, sandy to coarse loamy upland sites; wet ditches.

Notes: This grass spreads quickly in disturbed areas.

] *1 cm*

General: Perennial grass; flowering stems erect from reclining base, 0.3–1 m tall, slender, hairless; loosely tufted; from underground stems (rhizomes).

Leaves: Flat, upright, narrowed at base, 1–5 mm wide, shorter than flowering stem; sheaths purplish at base of stem, closed at first, splitting with maturity.

Inflorescence: Loose, lax panicle, 5–15 cm long, slender with few drooping branches; spikelets bronze to purplish, 1.3–2.3 cm long, 1–3 per branch, 3–5-flowered;

Bill Crins

flower scale (lemma) with an 8–15 mm long, hair-like bristle (awn) at tip and a dense tuft of short hairs at base; June-August.

Habitat: Dry to fresh upland tolerant hardwood, pine, cedar mixedwood and intolerant hardwood mixedwood stands.

Notes: The species name *purpurascens* means 'becoming purple,' referring to the colour of the spikelets.

] *1 cm*

GRASS FAMILY (POACEAE)

General: Perennial grass; flowering stems erect, wiry, smooth, hairless, leafless for much of length, 10–60 cm tall, 0.5–1.5 mm thick at base; in clumps, from shallow, fibrous roots.

Leaves: Basal leaves in tufts, rolled inward or flat, slender, often curved or twisted, much shorter than flowering stems; stem leaves few, 3–18 cm long, smaller upwards, remote from long-stalked panicle; sheaths shorter than internodes, usually with long, soft hairs and a tuft of long hairs at upper edge.

Inflorescence: Stiff, erect panicle 2.5–5 cm long, with 2–13 spikelets; branches short, usually with 1 spikelet; flower scales (lemmas) hairy on back, 3.4–5.2 mm long, with a 4–7 mm long bristle (awn); July–September.

Habitat: Dry sandy open sites; roadsides.

Notes: Also known as 'danthonia.'
• Because the plants pull up by their roots so easily, poverty oat grass is not particularly palatable to livestock.

Karen Legasy

1 cm

GRASS FAMILY (POACEAE)

General: Perennial grass, hairless; flowering stems slender, 20–85 cm tall; loosely to densely clumped.

Leaves: 2–4, widely spaced on stem, soft, 2–5 mm wide, keeled at tips.

Inflorescence: Loose, nodding panicle, 5–10 cm long; branches slender, typically in pairs; spikelets few, borne above middle of branches, 3–5-flowered, egg-shaped, 3.5–5.5 mm long; flower scales (lemmas) egg- to lance-shaped, 2.4–3.9 mm long, distinctly nerved, with a tuft of cobwebby hairs at base; May–August.

Habitat: Hardwood and mixedwood stands; forest clearings and roadsides.

Notes: The pairs of branches at each node are a diagnostic character of bushy pasture spear grass. • Other common blue grasses (*Poa* spp.) in the region include swamp bluegrass (*Poa palustris*), Kentucky bluegrass (*Poa pratensis*), and Canada bluegrass (*Poa compressa*). • The species name *saltuensis* means 'of bushy pastures.'

Brenda Chambers

] *1 cm*

E.T.

GRASS FAMILY (POACEAE)

Brenda Chambers

General: Slender perennial grass, pale green, 0.3–1.5 m tall; often in large clumps from long, creeping underground stems (rhizomes).

Leaves: Flat, firm, 2–10 mm wide, with tiny, pointed tips, rough on surface; uppermost leaves 10–20 cm long; sheaths smooth or slightly rough, closed nearly to the top, finely nerved and cross-veined.

Inflorescence: Nodding panicles 10–20 cm long; spikelets crowded near branch tips, greenish to purplish, 2–4 mm long, oblong to egg-shaped, 4–7-flowered; flower scales (lemmas) distinctly nerved, overlapping; June–September.

Habitat: Wet organic conifer swamps; wet to moist forest edges.

Notes: Fowl manna grass's distinguishing features are its small, usually purplish panicles, distinctly nerved scales, nodding branches and slender stems.

1 cm

General: Perennial grass; flowering stems erect, smooth, slender, often with whitish bloom, 0.7–1.2 m tall; in small tufts.

Leaves: Flat, broad, sometimes rough-margined, greyish-green, 10–30 cm long, 0.7–1.5 cm wide.

Inflorescence: Open, drooping, egg-shaped panicle, 10–25 cm long; branches slender, wide-spreading or drooping; spikelets 1-flowered, 3–3.5 mm long, borne above middle of branches; spikelets hard and shiny; flower scales (lemmas) about 3 mm long, firm and shiny when mature; June–August.

Habitat: Dry to fresh tolerant hardwood stands, with sugar maple, red oak, ironwood and white ash; roadsides; uncommon.

Notes: The soft, succulent growth of millet-grass makes it a good forage plant. Its abundance has been reduced by grazing and clearing. • The species name *effusum* means 'spread out,' referring to the open inflorescence.

Bill Crins

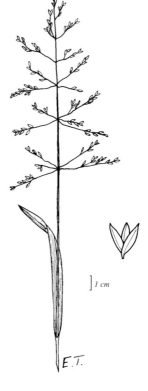

] *1 cm*

E.T.

GRASS FAMILY (POACEAE)

General: Perennial grass; flowering stems erect, hairless or soft-hairy, 0.3–1.2 m tall; often in tufts.

Leaves: Flat, lax, taper to long tip, 10–20 cm long, 5–10 mm wide; upper surface hairless or sparsely hairy and rough; underside smooth, with white midrib; sheaths often soft-hairy, shorter than internodes.

Inflorescence: Open panicle 10–20 cm long, at top of stem, with branches nodding to 1 side; spikelets 5–10-flowered, 1.5–3 cm long, 4–10 mm wide, greenish to bronze or purplish-tinged; flower scales (lemmas) overlapping, rounded on back, lance-shaped with a 3–4 mm bristle-tip (awn), slender-nerved, long-hairy along margins; July–August.

Habitat: Shorelines, roadsides, clearings.

Notes: Fringed brome grass is Ontario's most common native brome grass.

Karen Leagsy, top;
Derek Johnson, bottom

1 cm

General: Perennial grass; flowering stems erect, slender, 30–80 cm tall; in dense clumps.

Leaves: Mostly basal, bristle-shaped, rolled inward, hairless, 1–2 mm wide, 5–20 cm long.

Inflorescence: Open loosely branched panicle, up to 15 cm long; lowest branches in groups of 2–5; spikelets 2-flowered, shining, often purplish above middle, 4–6 mm long; flower scales (lemmas) minutely rough hairy, with twisted bristle (awn) 1–3 mm longer than scale; June–August.

Habitat: On bedrock and in dry to fresh, sandy to coarse loamy open pine and pine-oak sites.

Notes: This species is abundant on dry, rocky, acidic sites throughout our region. It has persisted in the Sudbury area, where smelter fumes have had widespread effects on vegetation.

Brenda Chambers

1 cm

KEY TO THE SEDGES (CYPERACEAE)

A. Fruiting clusters not woolly; seeds surrounded by perigynia in axils of scales

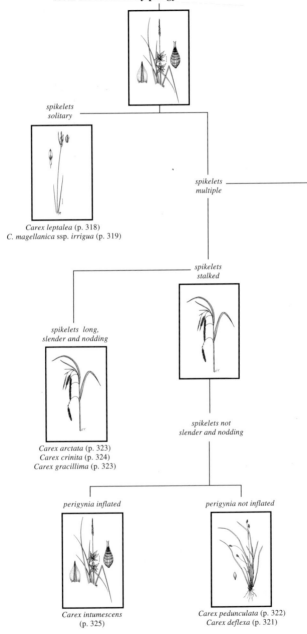

spikelets solitary

Carex leptalea (p. 318)
C. magellanica ssp. *irrigua* (p. 319)

spikelets multiple

spikelets stalked

spikelets long, slender and nodding

Carex arctata (p. 323)
Carex crinita (p. 324)
Carex gracillima (p. 323)

spikelets not slender and nodding

perigynia inflated

Carex intumescens (p. 325)

perigynia not inflated

Carex pedunculata (p. 322)
Carex deflexa (p. 321)

Les carex ou les laîches

**B. Fruiting clusters appearing woolly;
seeds naked in axils of scales**

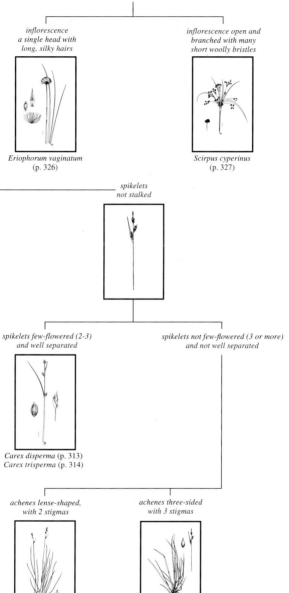

*inflorescence
a single head with
long, silky hairs*

*inflorescence open and
branched with many
short woolly bristles*

Eriophorum vaginatum
(p. 326)

Scirpus cyperinus
(p. 327)

*spikelets
not stalked*

*spikelets few-flowered (2-3)
and well separated*

*spikelets not few-flowered (3 or more)
and not well separated*

Carex disperma (p. 313)
Carex trisperma (p. 314)

*achenes lense-shaped,
with 2 stigmas*

*achenes three-sided
with 3 stigmas*

Carex brunnescens (p. 315)
Carex deweyana (p. 316)

Carex communis (p. 320)
Carex lucorum (p. 317)
Carex deflexa (p. 321)

Karen Legasy

General: Perennial; flowering stems triangular in cross-section, solid (i.e., filled with pith rather than hollow); usually from underground stems (rhizomes) with fibrous roots.

Leaves: In 3 vertical rows (3-ranked) along stem angles ; sheaths closed, usually unclear exactly where leaves attach to stem (i.e., no distinct joints).

Inflorescence: Tiny male or female flowers borne in narrow, unbranched, tight spikes; female flowers produce single, minute, hard fruits (achenes) in tiny sacs called perigynia; perigynia inflated and easy to see on some species, such as bladder sedge (*Carex intumescens*), flat or thin on others; fruits angled (3-sided in species with 3 slender stigmas, 2-sided [lens-shaped] in those with 2 stigmas).

Notes: You can usually distinguish sedges by their triangular stems ('sedges have edges').

Food Use by Wildlife: leaf, seed. • **Mammals:** grey squirrel, black bear, raccoon, white-tailed deer, moose. • **Birds:** ring-necked pheasant, spruce grouse, ruffed grouse, woodcock, wild turkey, American crow, northern cardinal, rufous-sided towhee, American tree sparrow, song sparrow, swamp sparrow and common redpoll.

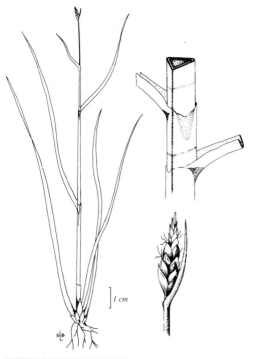

1 cm

SEDGE FAMILY (CYPERACEAE)

General: Perennial sedge, light green; stems almost thread-like, rough, often reclining, 5–60 cm tall; solitary or in loose tufts from long, slender underground stems (rhizomes).

Leaves: Flat, soft, weak, usually shorter than flowering stem, 1–2 mm wide; margins usually rough to base; nodding or weakly ascending.

Bill Crins

Inflorescence: Slender, interrupted, linear cluster of 1–5 small, stalkless spikes; spikes distantly spaced or upper few close together, with 1–5 (typically 1–2) female flowers at base and 1–2 male flowers at tip; seed sacs (perigynia) egg-shaped to elliptic, 2–2.8 mm long, hard, with minute beak and many fine nerves, green, ripen dark brown or black; bract at base of inflorescence either very short or missing; June–August.

Habitat: Wet organic conifer swamps.

Notes: The very short bract at the base of soft-leaved sedge's inflorescence distinguishes it from the similar-looking three-fruited sedge (p. 314), which has a long, bristle-like bract that overtops the flowers.

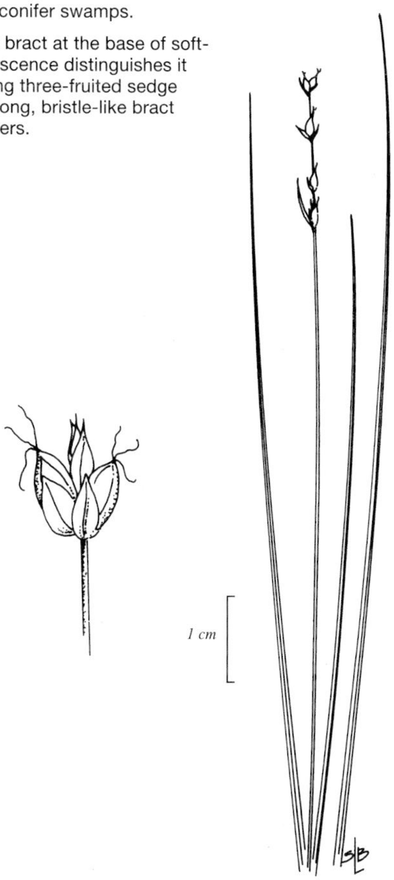

1 cm

SEDGE FAMILY (CYPERACEAE)

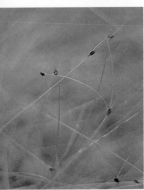

Karen Legasy

General: Perennial sedge, bright green; flowering stems thread-like, weak, slightly roughened, usually reclining or spreading, 20–70 cm long; solitary or in loose tufts from slender, often elongated underground stems (rhizomes).

Leaves: Soft, flat, narrow, 1–2 mm wide, shorter than flowering stem, usually drooping; margins rough to base; numerous dead leaves at base of plant.

Inflorescence: Slender, interrupted linear cluster of 1–3 (usually 3), stalkless spikes spaced about 1–4 cm apart; spikes have 1–5 (usually 2–3) female flowers at tip and several male flowers at base; seed sacs (perigynia) greenish, 2–4 mm long, flattened on 1 side, short-beaked, finely many-nerved; bristle-like bract at base of inflorescence 2–4 cm long, overtops upper spikes; June–August.

Habitat: Wet organic conifer and hardwood swamps.

Notes: Three-fruited sedge may be confused with soft-leaved sedge (p. 313). See notes on soft-leaved sedge for ways to distinguish these 2 species.

1 cm

General: Perennial sedge; flowering stems erect, 10–40 cm tall, exceeding leaves; densely tufted.

Leaves: Erect to drooping, narrow, 1–2 mm wide, with rough margins on upper half.

Inflorescence: Slender, 1–5 cm long cluster of 5–8 egg-shaped spikes, with lower spikes separated and upper spikes close together; spikes with 3–9 female flowers above and male flowers at base, 3–7 mm long; bract at base of inflorescence short and bristle-like; seed sacs (perygynia) egg-shaped to elliptic, 2–2.5 mm long, 1–1.5 mm wide, green turning brown, loosely spreading, with rough-margined beak; June–August.

Habitat: Wet organic conifer hardwood swamps, and moist mineral sites.

Notes: A similar species, grey sedge (*Carex canescens*), found in similar (though often more open) habitats, has a grey-green cast and has 10–30 female flowers per spike.

Brenda Chambers

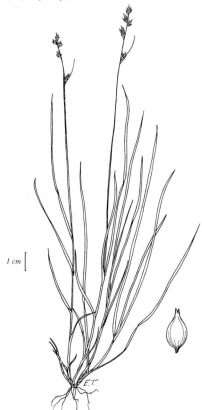

1 cm

General: Perennial sedge; flowering stems 30–80 cm tall, taller than leaves, weak and spreading; in dense clumps.

Leaves: Flat, mostly basal, 2–5 mm wide, light green to yellowish-green.

Inflorescence: Slender, 2–6 cm long cluster of 3–4 spikes with 3–12 female flowers above and few, inconspicuous male flowers at base, 5–12 mm long, pale green or silvery; lowest spike remote, 1–3 cm below upper spikes, surpassed by a bristle-like bract; upper spikes closer together; seed sacs (perigynia) egg- to lance-shaped, 1.3–1.6 mm wide, 4–5.5 mm long, faintly-nerved or nerveless, tapered at base, narrowed above to rough-margined beak about half as long as body; scales egg-shaped, sharp-pointed, silvery, tinged with brown with age, surpassing body of seed sac; May–July.

Brenda Chambers

Habitat: Moist to dry tolerant hardwood stands.

Notes: The characteristic silvery cast of this sedge is due to the pale, translucent, pistillate scales.

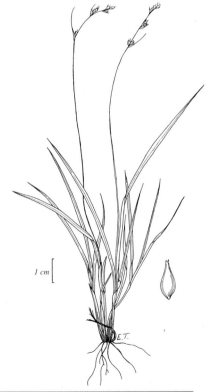

1 cm

General: Perennial sedge; flowering stems slender, erect, sharply rough-angled, 20–40 cm tall; in small leafy clumps with reddish bases, from stout elongate rhizomes.

Leaves: Flat, relatively soft, sometimes slightly rough, ascending, 1–3 mm wide.

Inflorescence: Cluster of 2–3, slightly separated, female spikes with 1 male spike at tip; female spikes rounded to egg-shaped, 3–12 mm long; bract subtending middle female spikelet tapered, scale-like; male spike terminal, club-shaped, stalkless or short-stalked, 0.8–2 cm long, reddish to whitish brown; seed sacs (perigynia) hairy, obtusely 3-angled, 3–4 mm long, 1.3–1.8 mm wide, with slender beak half or more as long as body; May–July.

Bill Crins

Habitat: Thickets, woods and sandy, open acidic sites.

Notes: This species is often dominant, forming large clones in open, sandy acidic habitats in the southern part of the region. It serves an important ecological role in stabilizing sandy substrates.

1 cm

E.T.

Frank Boas

General: Perennial sedge, light green; flowering stems usually 10–50 cm tall, thread-like, hairless, erect or spreading; from thread-like, creeping underground stems; densely tufted.

Leaves: Very narrow, about 1 mm wide, slightly shorter than flowering stem, upright, thin, wiry; margins smooth except near tip.

Inflorescence: Solitary terminal spike, narrowly oblong, 4–16 mm long, with short male portion at tip and overlapping female flowers below; seed sacs (perigynia) green to yellowish-green, 2.4–6.2 mm long, oblong or elliptic, rounded (beakless) at tip, tapered at base, finely nerved; scales membranous, much shorter than seed sacs; lower scales tipped with sharp, firm point; bract at base of inflorescence about length of spike, rough-margined, abruptly pointed, sometimes with a long bristle-tip (awn); June–August.

Habitat: Wet organic conifer swamps.

Notes: The species name *leptalea*, from the Greek *lepto*, 'thin' or 'slender,' refers to the thread-like stems and very narrow leaves of this sedge.

1 cm

General: Perennial sedge, pale green; flowering stems slender, smooth or strongly roughened, hairless, sharply angled, erect, 10–80 cm tall; usually in small clumps, from short or long underground stems (rhizomes) with yellowish-woolly roots.

Leaves: Flat, 2–4 mm wide, usually shorter than stem; sheaths at base of plant light brown to pinkish-tinged; margins slightly rolled under.

Inflorescence: Slender 5–12 cm long cluster of 1–4 long-stalked, nodding female spikes with 1 male spike at tip; male spike 4–12 mm long, long-stalked; female spikes short cylindric, 8–20 mm long, nodding or spreading on thread-like stalks 1–4 cm long; seed sacs (perigynia) green with a bluish-white tinge, brown at maturity, elliptic, 3–4 mm long, somewhat

Bill Crins

flattened, short-stalked, nearly beakless; scales lance-shaped with long, narrow point, brown, often with green midrib, longer and narrower than seed sac; lowermost bract usually overtops inflorescence; July–August.

Habitat: Wet organic conifer swamps.

Notes: Few-flowered sedge is also known as *Carex paupercula*.

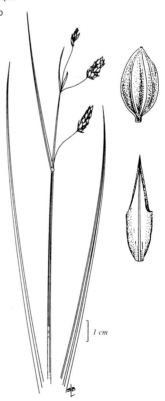

SEDGE FAMILY (CYPERACEAE)

Bill Crins

General: Perennial sedge; flowering stems sharp-angled, 20–50 cm tall, exceeding leaves; in dense clumps, without elongated underground stems (rhizomes).

Leaves: Flat, shorter than flowering stems, 3–5 mm wide, purplish at base.

Inflorescence: Small cluster of 2–3 female spikes with 1 male spike at tip; female spikes short-stalked to stalkless, linear to cylindric, 0.5–1 cm long; lowest 2 spikes separate, with bristle- or leaf-like bract at base; seed sacs (perigynia) hairy, 2.5–4 mm long, rounded-elliptic, tapered to stalk-like base and to beak at tip; scales egg- to lance-shaped, about as long as seed sacs; May–July.

Habitat: Dry to moist tolerant hardwood and intolerant hardwood mixedwood stands; sandy roadsides and clearings.

Notes: The species name *communis* means 'growing in colonies.'

1 cm

SEDGE FAMILY (CYPERACEAE)

General: Perennial sedge; flowering stems curved or spreading, slender, purple-tinged at base, 10–20 cm tall; in loose clumps, from slender rhizomes.

Leaves: Soft, thin, loosely spreading or ascending, usually exceeding flowering stems, 1–3 mm wide.

Inflorescence: Tight cluster of 2–3 female spikes, with 1 male spike at tip; female spikes 2–8-flowered, green or reddish-brown; lowest spike subtended by a green bract 5–20 mm long; male spike 2–5 mm long, may be hidden by female spikes; seed sacs (perigynia) egg-shaped-elliptic, finely hairy, green becoming brownish, 2.5–3 mm long, short-stalked, with 0.4–0.7 mm long beak; scales egg- to lance-shaped, reddish-brown, shorter than seed sacs; May–August.

Habitat: Moist to fresh edges of conifer or mixedwood stands; rocky open ground.

Notes: The species name *deflexa* means 'bent down.'

Brenda Chambers

1 cm

E.T.

LONG-STALKED SEDGE • *Carex pedunculata*
CAREX PÉDONCULÉ

SEDGE FAMILY (CYPERACEAE)

Bill Crins

General: Perennial sedge; flowering stems widely or loosely spreading, reddish at base, 10–30 cm tall; in loose mats.

Leaves: Dark green, flat, stiff, 10–30 cm long, 2–5 mm wide.

Inflorescence: Cluster of 4–5 female spikes with 1 male spike at tip (usually with a few female flowers at its base); female spikes 6–15 mm long, 3- to 8-flowered, short-stalked near top of cluster to long- and slender-stalked near base; seed sacs (perigynia) hairless to slightly hairy, 3-angled, 3.5–5 mm long, with small beak; scales green to purple, ending in a sharp bristle tip (awn); April–May.

Habitat: Fresh tolerant hardwood stands; wet organic hardwood and conifer swamps.

Notes: This species flowers early and has distinctive, reddish-based flowering stems. • The species name *pedunculata* means 'peduncled or stalked,' referring to the long-stalked lower female spike(s).

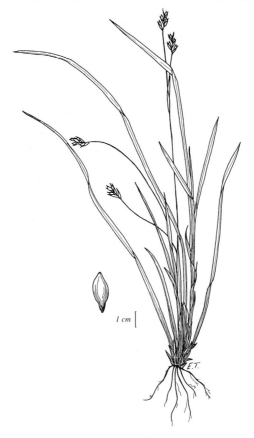

1 cm

E.T.

SEDGE FAMILY (CYPERACEAE)

General: Perennial sedge; flowering stems smooth, arching or ascending with purple bases, 0.2–1 m tall; in dense clumps.

Leaves: Basal leaves usually abundant, dark green, firm, 3–8 mm wide; stem leaves 3–5 mm wide, with 2 prominent lateral veins on the upper side of the leaf.

Inflorescence: Cluster of 3–5 very slender, long-stalked, drooping or spreading female spikes, with 1 male spike at tip; female spikes 2–6 cm long, 3–5 mm wide, linear, loosely flowered; male spike 1–3 cm long; seed sacs (perigynia) egg-shaped, ascending, shiny, short-stalked, with 2 ribs along edges and several nerves, 3–5 mm long, narrowed to a short beak; June–August.

Habitat: Moist to dry, clayey to sandy tolerant hardwood stands.

Brenda Chambers, both

Notes: Drooping wood sedge is a common sedge in tolerant hardwood stands. • Filiform sedge *(Carex gracillima)*, is similar in size and also has purple-based flowering stems, but the seed sacs (perigynia) are beakless and the terminal male spike usually has female flowers near its tip.

1 cm

E.T.

FRINGED SEDGE • *Carex crinita*
CAREX CRÉPU

SEDGE FAMILY (CYPERACEAE)

Bill Crins

General: Perennial sedge; flowering stems rough, sharp-angled, erect, 0.3–1.3 m tall, exceeding main leaves; in large or small clumps.

Leaves: Flat, more or less rough on nerves and margins, 4–12 mm wide; sheaths hairless.

Inflorescence: Loose, erect or arching cluster of 2–5 female spikes with 1–3 stalked male spikes at tip; female spikes narrow, cylindric, 3–10 cm long, on slender and curved stalks; male spikes stalked, slender, up to 5 cm long; seed sacs (perigynia) often inflated, egg- to lance-shaped, 2–4 mm long, 2-ribbed; scales coppery brown with rough flat bristle-tip (awn) up to 10 mm long; May–August.

Habitat: Wet organic hardwood swamps; in moist pockets and at moist edges of tolerant hardwood stands; wet ditches.

Notes: A similar species, nodding sedge (*Carex gynandra*), has rough hairy leaf sheaths. • The species name *crinita* means 'long-haired,' referring to the long bristle (awn) at the tip of each scale.

1 cm

E.T.

General: Smooth, hairless perennial sedge, dark green, sometimes red at shoot bases; flowering stems slender, erect, up to 80 cm tall; clumped, with or without short underground stems (rhizomes).

Leaves: Soft, drooping, flat or folded, up to 80 cm long, 4–8 mm wide, equal to or shorter than flowering stem; margins rough to base.

Inflorescence: 3–5 cm long cluster of 1–3 crowded, stalkless female spikes with a terminal male spike; male spike narrow 15–25 mm long, long-stalked; female spikes essentially stalkless, rounded, 10–20 mm across, 5–10-flowered; seed sacs (perigynia) crowded, spreading, bladder-like, shiny, many-nerved, rounded at base, tapered to rough beak at tip; scales narrowly lance-shaped, half as long as seed sacs; leaf-like bracts at spike bases overtop inflorescence; May–October.

Habitat: Wet organic hardwood and conifer swamps; moist sandy to fine loamy upland tolerant hardwood stands.

Notes: Bladder sedge can be recognized by its large, rounded female spikes with their inflated, bladder-like seed sacs (perigynia).

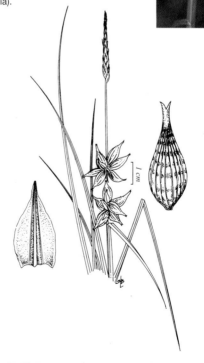

SEDGE FAMILY (CYPERACEAE)

General: Perennial sedge; flowering stems stiff, 3-angled, rough at tip, 15–70 cm tall; densely tufted.

Leaves: Basal leaves stiff, very slender, 1 mm wide; basal sheaths brown, long, persistent, with fine fibres; stem leaves bladeless, conspicuously inflated sheaths, with network of veins and dark, membranous tip.

Inflorescence: Single, erect cottony head (spike), usually silky-white, inversely egg-shaped to rounded, 0.8–1.5 cm long; scales lead-coloured to blackish with whitish margins; fruit tiny, black, 3-sided, dry, hard, single-seeded (achene), tipped with and hidden by long, white bristles; April to mid July.

Habitat: Open sphagnum bogs and boggy thickets.

Notes: The bright white bristles of these seeds look like cotton, hence the name 'cottongrass.' The genus name *Eriophorum* means 'wool-bearing.'

Bill Crins

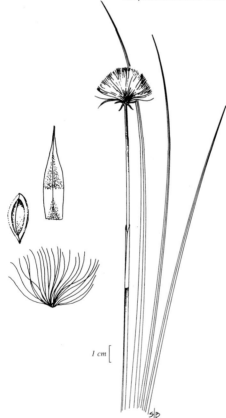

1 cm

SEDGE FAMILY (CYPERACEAE)

General: Perennial sedge; flowering stems thick or slender, smooth, obscurely triangular to nearly rounded in cross-section, leafy, 1–1.5 m tall; in dense tufts with many curving basal leaves, from fibrous roots.

Leaves: Long, flat, rigid, 3–10 mm wide; margins rough.

Inflorescence: Dense to loose, reddish-brown cluster 3–10 cm long; branches ascending to nodding, tipped with loose clusters of slender-stalked spikelets; spikelets numerous, egg-shaped, 3–6 mm long, soft white-hairy with long, protruding bristles at maturity; bracts at base of cluster leaf-like, spreading, usually nodding at tips, often longer than inflorescence; August–October.

Habitat: Shorelines, marshes, swamps, wet ditches and clearings.

Notes: Wool-grass is a native perennial.

Karen Legasy

] *1 cm*

TOAD RUSH • *Juncus bufonius*
JONC DES CRAPAUDS

General: Annual rush; flowering stems low, slender, erect to spreading, unbranched or branching at base, 3–20 cm tall; tufted from fibrous roots.

Leaves: Very slender, rolled inward, 1 mm wide, few, usually shorter than flowering stem; sheaths gradually taper to blade.

Inflorescence: Open, branching, wide, appears laterally flattened, with remote flowers scattered along branches, about 1/2 as tall as plant; flowers whitish, greenish or pale brown, 3–7 mm long, with 6 lance-shaped, scale-like petals and sepals; June–November.

Linda Kershaw

Habitat: Moist to wet open sandy areas.

Notes: The scientific species name *bufonius* means 'pertaining to the toad.' • Toad rush grows in almost every country of the world.

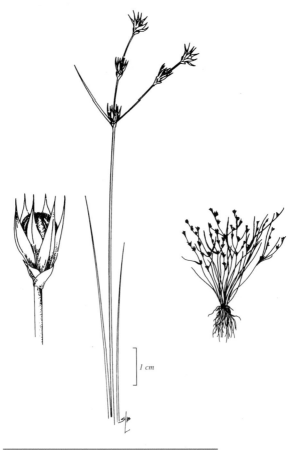

1 cm

Les fougères et alliées

FERNS AND ALLIES

A simple pictorial key is provided to illustrate differences among ferns and rattlesnake ferns (pp. 341–362) and their allies—the horsetails (pp. 331–335) and club-mosses (pp. 336–340). Illustrations of common plants in these groups are included to highlight features that will aid in identification.

PARTS OF A FERN
(pp. 341–362)

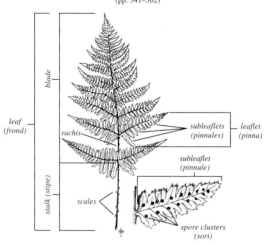

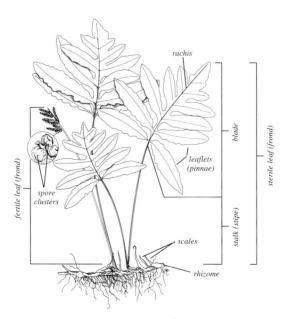

PARTS OF A HORSETAIL
(pp. 331–335)

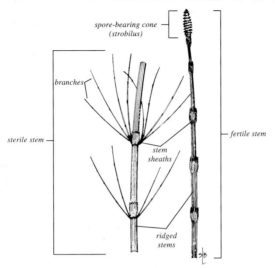

spore-bearing cone
(strobilus)

branches

sterile stem

fertile stem

stem
sheaths

ridged
stems

PARTS OF A CLUB-MOSS
(pp. 336–340)

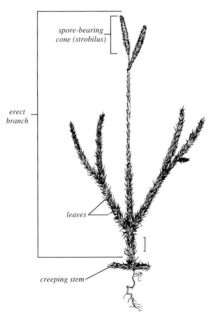

spore-bearing
cone (strobilus)

erect
branch

leaves

creeping stem

HORSETAIL FAMILY (EQUISETACEAE)

General: Delicate, lacy perennial horsetail, erect, up to 50 cm tall; from deep, creeping underground stems (rhizomes).

Stems: Mainly solitary, hollow with central cavity almost 1/2 diameter of stem; fertile stems brownish, with relatively long sheaths, unbranched at first, eventually become green and branched like sterile stems, appear in early spring before sterile stems; sterile stems green, with whorls of feathery branches; branches branched, thin, delicate, solid, ascending to spreading or drooping; sheaths green at base, chestnut brown at top, 4–5-toothed.

Fruiting Structures: Terminal spore-bearing cones, 1.5–3 cm long, on 2–6.5 cm long stalks, soon wither away.

Habitat: Moist clayey to sandy upland cedar mixedwoods; wet organic hardwood and conifer swamps.

Notes: Woodland horsetail differs from field horsetail (p. 332) and meadow horsetail (p. 333) in the way its branches are further branched; those of field and meadow horsetail are not.

OMNR

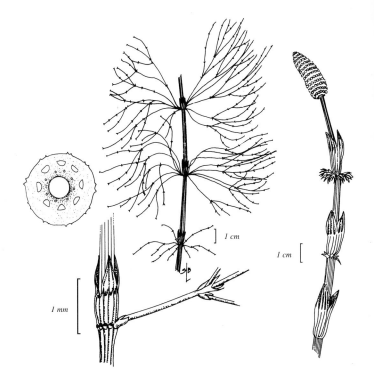

1 cm

1 mm

1 cm

FIELD HORSETAIL • *Equisetum arvense*
PRÊLE DES CHAMPS

HORSETAIL FAMILY (EQUISETACEAE)

Karen Legasy

General: Perennial horsetail, usually erect, commonly up to 25 cm tall, occasionally up to 40 cm; from extensively creeping underground stems (rhizomes).

Stems: Single or clustered, typically 10–12 ridged; fertile stems light brown, unbranched, their whitish sheaths 14–20 mm long and tipped with 5–9 mm long, dark, lance-shaped teeth, appear in early spring before sterile stems and soon wither away; sterile stems slender, green with whorls of bushy, ascending to spreading, unbranched branches and green sheaths with 4 brown to blackish teeth; first branch segment equal to or longer than nearest stem sheath.

Fruiting Structures: Terminal spore-bearing cones, 1.7–4 cm long, on 2.2–5.5 cm long stalks.

Habitat: Wet organic to moist conifer and hardwood swamps; open sandy sites, wet ditches.

Notes: Aboriginal peoples used field horsetail in a remedy for bladder and kidney ailments. See caution in Introduction. • Horsetail species have apparently been used as indicators of gold deposits. Aboriginal peoples used field horsetail as an indicator of where to find water. They also used the plants for smoothing and polishing surfaces. • See notes on meadow horsetail, p. 333.

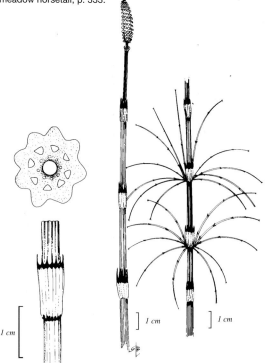

1 cm

1 cm

1 cm

HORSETAIL FAMILY (EQUISETACEAE)

General: Slender, erect, feathery perennial horsetail, 20–40 cm tall; from slender, solid, black, creeping underground stems (rhizomes).

Stems: Usually single; fertile stems unbranched at first, tipped with spore-bearing cone, appear in spring before sterile stems, eventually develop whorls of branches and resemble sterile stems; sterile stems slender, with whorls of short, straight, spreading branches; sheaths have short, dark, sharp-pointed, white-margined, brown teeth; branches unbranched, very slender, 3-angled, 4–15 cm long; first branch segment shorter than nearest stem sheath.

Fruiting Structures: Terminal spore-bearing cones, 2–2.5 cm long, on 2–4.8 cm long stalks.

Habitat: Wet organic to moist hardwood and conifer swamps.

OMNR

Notes: You can distinguish meadow horsetail from the similar-looking field horsetail (p. 332) by its branches and sheaths. The first segment of each branch of meadow horsetail is shorter than the stem sheath near it, whereas the first sheath on each branch of field horsetail is equal to or longer than the stem sheath near it. Meadow horsetail is usually more delicate-looking than field horsetail and has conspicuous, white-margined teeth in its stem sheaths.

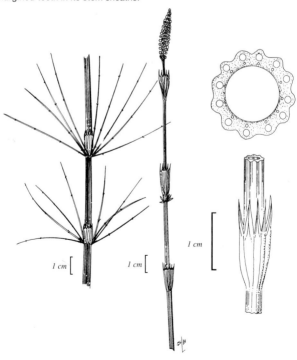

1 cm

1 cm

1 cm

SWAMP HORSETAIL • *Equisetum fluviatile*
PRÊLE FLUVIATILE

Jim Pojar

General: Highly variable perennial horsetail, up to 90 cm tall; from hollow, shiny, creeping underground stems (rhizomes).

Stems: Usually solitary, erect, 3–8 mm thick, 5–10-grooved; fertile and sterile stems similar; stem sheaths have about 18 dark-brown to blackish, short, pointed, rigid teeth with narrow, white margins; central stem cavity large (about 4/5 of diameter); branches none or in irregular to regular whorls, ascending to spreading, 2–15 cm long, nearly smooth, 4–6-ridged, greenish, slender, hollow, unbranched.

Fruiting Structures: Terminal, spore-bearing cones, 9–35 mm long, on 5–32 mm long stalks.

Habitat: In shallow water along edges of streams; in swamps, ponds and ditches.

Notes: One of swamp horsetail's distinguishing features is the large central cavity of its hollow stem. • The species name *fluviatile* means 'relating to or occurring in a river.'

Food Use by Wildlife: plant. • **Mammals:** moose, black bear.

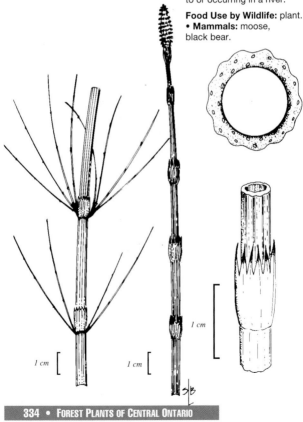

1 cm

1 cm

1 cm

S/B

HORSETAIL FAMILY (EQUISETACEAE)

General: Our smallest horsetail, perennial, up to 20 cm tall, curling and matted; often in dense tufts or mats from shallow, slender, creeping, branching underground stems (rhizomes).

Bill Crins

Stems: Clustered, prostrate, ascending and arched to down-curved or zigzag, solid, dark green, 3-ridged, deeply grooved, slender, 0.5–1 mm thick, 3–20 cm long, unbranched (sometimes with small, irregular branches); sheaths have 3 (sometimes 4) sharp-pointed, triangular teeth with dark centres and light edges; sterile and fertile stems similar, but fertile stems usually more erect.

Fruiting Structures: Short, black, terminal, spore-bearing cones, 2–3 mm long, with a short, pointed tip, stalkless.

Habitat: Wet organic to moist upland sites; wet banks and shorelines.

Notes: You can recognize dwarf scouring rush in the field by its slender, zigzag stems that often form dense tufts.

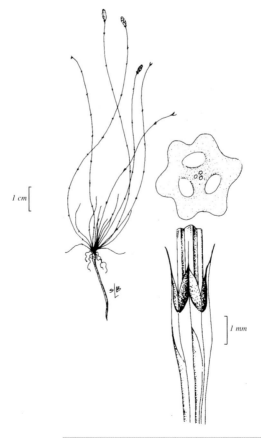

1 cm

1 mm

INTERRUPTED CLUB-MOSS • *Lycopodium annotinum*
LYCOPODE INNOVANT

CLUB-MOSS FAMILY (LYCOPODIACEAE)

General: Low, trailing, prostrate, perennial club-moss, evergreen; stems creeping, up to 2 m long, forked, often slightly covered by thin layer of humus; branches erect, stiff, unbranched or forked 1–2 times, 6–30 cm tall, bristly with stiff, spreading leaves.

Leaves: Needle-like, stiff, narrowly lance-shaped, widest near middle or above, sharp-pointed at tip, narrowed at base, dark green, shiny, 2.5–11 mm long, spreading to some-times bent downward; margins slightly to coarsely toothed toward tip.

Fruiting Structures: Oblong to cylindrical, spore-bearing cones, 0.6–4.5 cm long, solitary, terminal, stalkless.

Habitat: Fresh to moist, sandy to loamy upland tolerant hardwood and intolerant hardwood mixedwood stands; wet organic hardwood and conifer swamps.

OMNR

Notes: The annual growth of interrupted club-moss is indicated by interruptions or indentations in the leaf pattern of its branches. • Interrupted club-moss has been used to make Christmas decorations. • Aboriginal peoples used the spore powder as a drying agent for wounds, diaper rash and nosebleeds. See caution in Introduction.

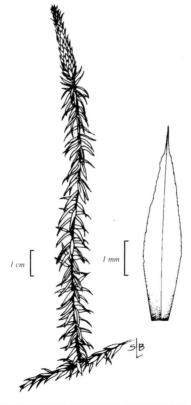

1 cm

1 mm

S|B

CLUB-MOSS FAMILY (LYCOPODIACEAE)

General: Low, trailing, prostrate perennial club-moss, evergreen; stems extensively creeping, branching, rooting at intervals; branches erect, widely forked or branched 1–4 times, up to 25 cm tall, densely covered with stiff, spreading leaves.

Leaves: Needle-like, lance-shaped or linear, 3.5–7 mm long, 0.5–0.8 mm wide, bright green, crowded, ascending to spreading, hair-like at tip; margins untoothed to minutely toothed.

Karen Legasy

Fruiting Structures: Terminal, linear to cylindrical, spore-bearing cones, single or in groups of 2–3, long-stalked.

Habitat: Dry to moist, sandy to silty upland intolerant hardwood mixedwood and tolerant hardwood stands; occasional in wet organic conifer swamps.

Notes: You can recognize wolf's claw club-moss in the field by its widely forking, erect stems with their long-stalked cones. • The spore powder is flammable and has historically been used in fireworks and for flash explosions in photography. The powder was also used to treat skin disorders. See caution in Introduction. • The name *Lycopodium* was taken from the Greek *lycos*, 'wolf' and *podos*, 'foot,' because of the fancied resemblance of the leafy stem tip to the paw of a wolf.

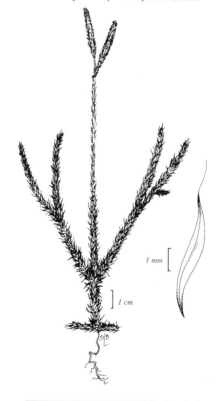

1 mm

1 cm

GROUND PINE • *Lycopodium dendroideum*
LYCOPODE FONCÉ

CLUB-MOSS FAMILY (LYCOPODIACEAE)

Karen Legasy

General: Erect, bushy, 'tree-like,' perennial club-moss, usually less than 25 cm tall; branchless near base, heavily branched above; branches upright, irregular, densely covered with leaves; from deep, creeping underground stems (rhizomes).

Leaves: Needle-like, lance-shaped, tapered to sharp point at tip, narrowed at base, about 5 mm long, dark green, shiny; margins smooth.

Fruiting Structures: Yellow, cylindrical, spore-bearing cones, up to 3.5 cm long, stalkless, erect, at branch tips.

Habitat: Dry to moist, sandy to clayey upland tolerant hardwood, intolerant hardwood mixedwood, cedar mixedwood and pine stands.

Notes: Distinguishing features of ground pine include its distinctive, tree-like shape and its needle-like leaves. • Aboriginal peoples used ground pine in a medicine for stiff joints. See caution in Introduction.

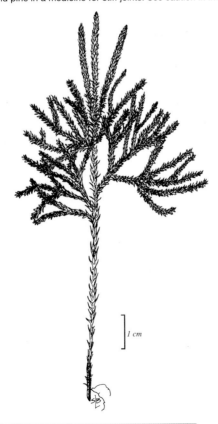

1 cm

CLUB-MOSS FAMILY (LYCOPODIACEAE)

General: Erect, bushy, 'tree-like' perennial club-moss; main stems creeping on ground or slightly below surface and erect at tips, 3–40 cm tall, irregularly forked 3–4 times, flattened.

Leaves: Scale-like (resembling cedar leaves), tiny, less than 1 mm long, in 4 rows (4-ranked), overlapping, with bases extending downward on and fused to stem (except for lower leaves); upper leaves narrow, curved inward; side leaves wider, spreading at tips; lower leaves minute, triangular.

Fruiting Structures: Cylindrical, spore-bearing cones, 1–3 cm long, solitary or in groups of 2–5; stalks simple or forked, 3–6 cm long, at branch tips.

Karen Legasy

Habitat: Dry to moist, sandy to silty upland pine, intolerant hardwood mixedwood and tolerant hardwood stands.

Notes: Also known as *Lycopodium digitatum*. • You can recognize ground cedar by its flattened branches, which resemble the branches of eastern white cedar (p. 22).

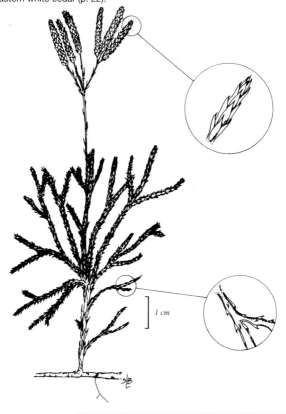

1 cm

CLUB-MOSS FAMILY (LYCOPODIACEAE)

General: Perennial club-moss; stems horizontal, creeping, 10–40 cm long, often covered with humus and rooting, bearing many withered brown leaves, ascending to erect at tips; erect stems 10–25 cm tall, densely covered in leaves, forked or branched 1–3 times.

Leaves: Needle-like, lance-shaped, usually widest towards tip, dark green, shiny, 0.6–1.5 cm long, 0.8–2.6 mm wide, widely spreading to irregularly bent downward; margins sparsely toothed at tip; dead leaves remain on plant.

Brenda Chambers

Fruiting Structures: Kidney-shaped, orangish-yellow spore cases (sporangia) in upper leaf axils.

Habitat: Dry to moist, sandy to clayey upland tolerant hardwood and intolerant hardwood mixedwood stands; wet organic conifer swamps.

Notes: Also known as *Lycopodium lucidulum*. • Shining club-moss may be confused with interrupted club-moss (p. 336) in the field, but shining club-moss does not have cones at its branch tips as does interrupted club-moss. Also, the erect portion of shining club-moss's creeping stem is at the stem tip, whereas the erect stems of interrupted club-moss are branches, produced at intervals along the main creeping stem.

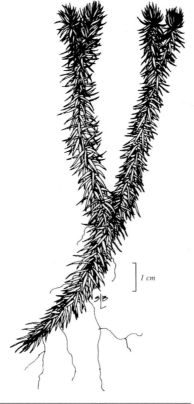

1 cm

UNIQUE

leaves dark green, thick, shiny; leaflets eared at base

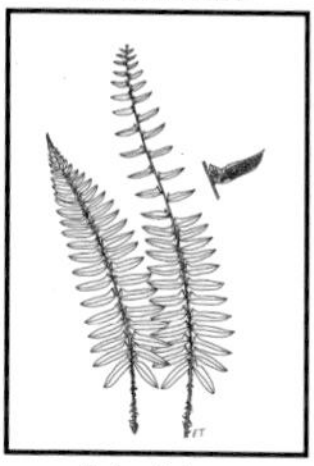

Polystichum acrostichoides (p. 344)

horseshoe-shaped leaves

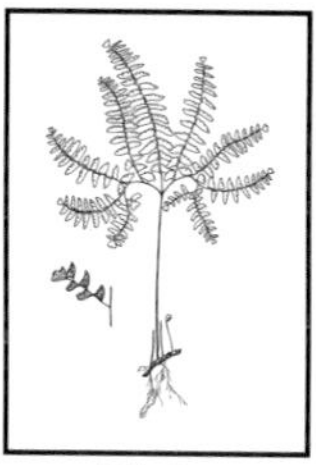

Adiantum pedatum (p. 350)

stems succulent; leaves triangular, with a separate fertile blade

Botrychium virginianum (p. 362)

ONCE DIVIDED

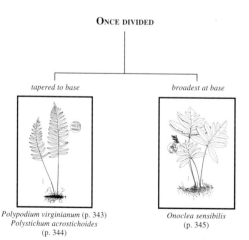

tapered to base

Polypodium virginianum (p. 343)
Polystichum acrostichoides (p. 344)

broadest at base

Onoclea sensibilis (p. 345)

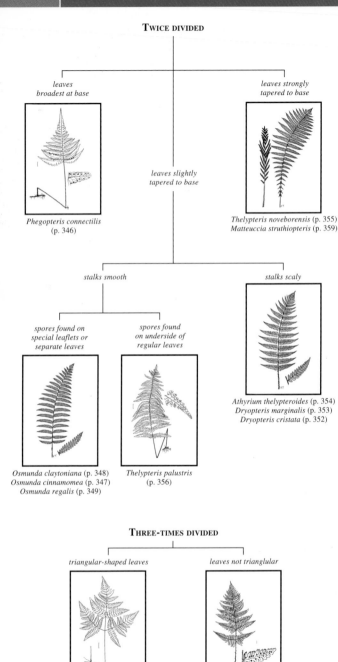

TWICE DIVIDED

*leaves
broadest at base*

*Phegopteris connectilis
(p. 346)*

*leaves strongly
tapered to base*

*leaves slightly
tapered to base*

Thelypteris noveborensis (p. 355)
Matteuccia struthiopteris (p. 359)

stalks smooth

stalks scaly

*spores found on
special leaflets or
separate leaves*

*spores found
on underside of
regular leaves*

Osmunda claytoniana (p. 348)
Osmunda cinnamomea (p. 347)
Osmunda regalis (p. 349)

Thelypteris palustris
(p. 356)

Athyrium thelypteroides (p. 354)
Dryopteris marginalis (p. 353)
Dryopteris cristata (p. 352)

THREE-TIMES DIVIDED

triangular-shaped leaves

leaves not trianglular

Pteridium aquilinium (p. 361)
Gymnocarpium dryopteris (p. 360)
Botrychium virginianum (p. 362)

Dryopteris carthusiana (p. 357)
Athyrium filix-femina (p. 358)
Cystopteris bulbifera (p 351)

POLYPODY FAMILY (POLYPODIACEAE)

General: Evergreen perennial fern, 8–30 cm tall; in mats from spongy, rope-like, scaly, shallow, 2–7 mm thick underground stems (rhizomes).

Leaves: Loosely clustered, erect to spreading, compound, with 10–20 alternate to nearly opposite pairs of leaflets, leathery, deep green, often golden above, lance-shaped to almost oblong, pointed at tip, squared (truncate) at base, 5–25 cm long, 3–6 cm wide; leaflets (pinnae) 3–5 mm wide, 3–5 times longer than wide, rounded to pointed at tip, smaller toward leaf tip; stalk smooth and slender.

Brenda Chambers

Fruiting Structures: Large, dot-like, reddish-brown spore clusters (sori), in 2 rows on underside of leaflets.

Habitat: Dry rocky outcrops in tolerant hardwood and pine stands.

Notes: Common polypody can be cultivated in moist, subacidic potting soil in a sunny location, but it is difficult to establish in woodland gardens.

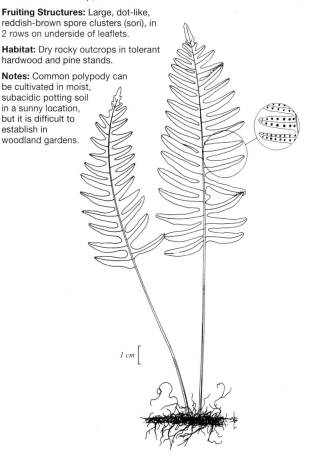

1 cm

Brenda Chambers, both

General: Hardy perennial fern, 10–60 cm tall; in clusters from short, scaly underground stems (rhizomes) covered with old stalk bases and wilted leaves.

Leaves: Clustered, arching, evergreen, compound, divided once (pinnate) into 20–40 pairs of leaflets, dark green, thick, shiny above, 10–60 cm long, 5–18 cm wide, lance-shaped, slightly narrowed at base; leaflets (pinnae) lance-shaped to oblong, short-stalked, with minute bristle-tipped teeth, prominently eared at base; upper, fertile leaflets much smaller, with lowest fertile leaflets noticeably shorter than sterile leaflets immediately below; stalk grooved, very scaly, green (brown at base).

Fruiting Structures: Spore clusters (sori) dot-like, in two or more rows near midrib on undersides of leaflets, red-brown.

Habitat: Fresh to moist, sandy to clayey tolerant hardwood stands; in the southern part of the region, uncommon.

Notes: Braun's holly fern (*Polystichum braunii*) (inset photo), a close relative, is rare in our region. It is spiny-toothed like Christmas fern, but is scalier and more finely cut. It can be found in isolated, cool, moist habitats adjacent to the east coast of Lake Superior.

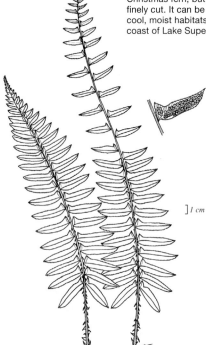

] *1 cm*

E.T.

WOOD FERN FAMILY (DRYOPTERIDACEAE)

General: Somewhat coarse perennial fern, 50–70 cm tall; in loose patches from stout, brown, extensively creeping underground stems (rhizomes).

Leaves: Loosely clustered, compound, once-divided (pinnate); sterile leaves widely triangular with 5–11 pairs leaflets; leaflets lance-shaped to oblong, 3–18 cm long, 1–5 cm wide; margins wavy to deeply cut; wing along central axis progressively wider toward leaf tip; stalks stiff, brittle, naked or with few scattered scales near base.

Karen Legasy

Fruiting Structures: Dark brown, spore-bearing (fertile) leaves, less than 40 cm long, often not developed, lance-shaped to oblong, compound, twice-divided (bipinnate); leaflets (pinnae) erect; sub-leaflets (pinnules) rolled into tight, bead-like balls around spore clusters.

Habitat: Wet organic and moist sandy to clayey hardwood and conifer swamps; wet roadsides.

Notes: The sterile leaves are very sensitive to cold and usually turn black after the first frost. The fertile leaves often persist over winter. Sensitive fern can be cultivated in moist garden soil in partial sun, but its creeping underground stems tend to make it a bit weedy for a garden.

]*1 cm*

MARSH FERN FAMILY (THELYPTERIDACEAE)

Karen Legasy

General: Somewhat coarse perennial fern, up to 40 cm tall; from long, slender, creeping underground stems (rhizomes).

Leaves: Single, erect, narrowly triangular, 6–25 cm long, 4–15 cm wide, compound, once- to nearly twice-divided (pinnate to nearly bipinnate) into 10–25 pairs of opposite leaflets; leaflets long, narrow, pointed, spreading at right angles to axis (except for lowest pair, which bend sharply downward); margins deeply lobed so there appear to be rounded subleaflets; needle-like hairs 0.5–1 mm long on veins and margins of central axis; stalk brown and slightly scaly toward base, straw-coloured above, hairy, up to twice as long as blade.

Fruiting Structures: Small, circular spore clusters (sori) near margins on underside of lower leaflets.

Habitat: Wet organic conifer and hardwood swamps; fresh to moist, sandy to clayey tolerant hardwood (with yellow birch and eastern hemlock) and cedar mixedwood stands.

Notes: Also known as *Thelypteris phegopteris*. • The 2 bottom leaflets, which are distinctly separate from the others and bend downward and outward, distinguish Northern beech-fern.

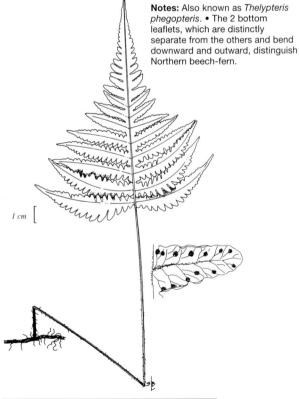

1 cm

ROYAL FERN FAMILY (OSMUNDACEAE)

General: Coarse perennial fern, up to 1.2 m tall; in clusters from stout, stubbly underground stems (rhizomes), with remains of old, withered broken stalks.

Leaves: In clusters, erect to arching, compound, divided 2 times (bipinnate); sterile leaves oblong to lance-shaped, up to 1.6 m long; leaflets (pinnae) in 20 or more pairs, narrow lance-shaped, 15 cm long, 3 cm wide, with tuft of brownish hairs at base, deeply cut into oblong, pointed lobes; stalks leaves green, slightly grooved, coated with cinnamon-brown wool when young, some of which persists.

Brenda Chambers

Fruiting Structures: Shorter, narrower, cinnamon-coloured, erect, club-like, fertile leaves with leaflets (pinnae) tightly hugging woolly stems; spore cases (sporangia) 0.5 mm wide, short-stalked, in clusters on undersides of leaflets, green turning cinnamon; appear before sterile leaves, and wither and die early.

Habitat: Wet organic conifer and hardwood swamps; wet ditches.

Notes: The species name *cinnamomea* means 'cinnamon-coloured,' referring to the distinctive colour of the fertile leaves.

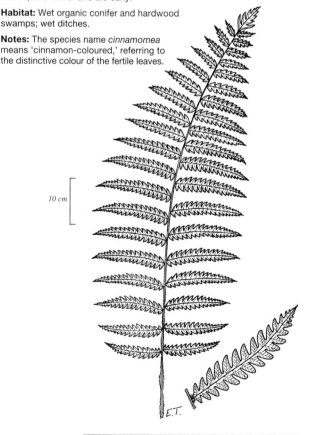

10 cm

E.T.

ROYAL FERN FAMILY (OSMUNDACEAE)

General: Coarse, arching, often over 1 m tall; from stout, creeping underground stems (rhizomes) covered with bases of old leafstalks.

Leaves: Clustered, erect, greenish, spreading from a central point, oblong, widest near middle, tapering to tip and base, compound, almost twice-divided (bipinnate) with leaflets deeply cut into blunt lobes, woolly when young, becoming hairless; sterile leaves usually outer, spreading, up to 1 m tall and 30 cm wide, distinctly arching outward.

Fruiting Structures: Fertile leaves usually inner, erect, taller than sterile leaves, with 2–4 pairs of fertile leaflets near middle; fertile leaflets relatively small, greenish at first, later blackish with dense, dark clusters of spore cases (sporangia).

Habitat: Moist sandy to clayey upland tolerant hardwood stands; wet organic conifer and hardwood swamps; wet forest edges and ditches.

Notes: The common name refers to the fertile leaves, which are 'interrupted' in the middle by the small fertile leaflets.

Brenda Chambers, both

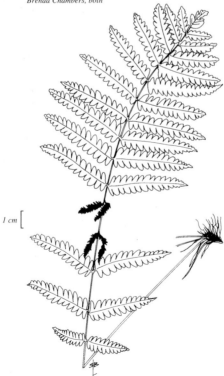

1 cm

General: Delicate perennial fern, 0.6–1.8 m tall; in large crowns of leaves from thick, long-lived underground stems (rhizomes).

Leaves: Clustered from a central point, erect, dull green, egg-shaped to widely so, 30–130 cm long, 10–55 cm wide, compound, twice-divided (bipinnate) with 5–7 pairs of sub-opposite leaflets (pinnae); leaflets oblong-oval to lance-oblong, short-stalked, up to 30 cm long and 14 cm wide; sub-leaflets (pinnules) alternate, 7–10 per side, narrow, oblong, tapered slightly at tip, rounded and oblique at base, hairless, very short-stalked; stalks pinkish, hairless, 20–50 cm long.

Fruiting Structures: Brown, erect, spore-bearing (fertile) leaflets with dense, dark clusters of spore cases (sporangia), in branched, terminal clusters on sterile blades.

Habitat: Wet shorelines of lakes and streams; wet ditches.

Notes: The roots were historically used in a remedy for jaundice and in an ointment for wounds, bruises and dislocations. See caution in Introduction.

Brenda Chambers

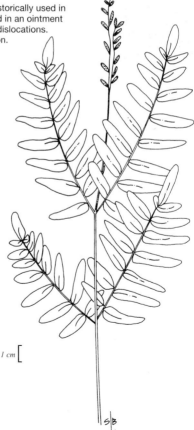

1 cm

S|B

MAIDENHAIR FERN • *Adiantum pedatum*
ADIANTE PÉDALÉ

MAIDENHAIR FERN FAMILY (PTERIDACEAE)

Brenda Chambers

General: Nearly erect, circular or horseshoe-shaped perennial fern, up to 60 cm tall; from extensively creeping, greyish-brown underground stems (rhizomes).

Leaves: Loosely clustered, compound, divided 2 times (bipinnate), bluish-green, thin, very graceful, up to 50 cm wide; leaflets (pinnae) 5–6 on each branch of stalk, each up to 35 cm long and 5 cm wide; sub-leaflets (pinnules) fan-shaped to oblong, up to 25 pairs, deeply cut on upper edge; stalks slender, forked at summit, usually into 2 widely divergent, arched branches; deep red-brown to blackish, shiny, with light brown scales at base,

Fruiting Structures: 1–5 spore clusters (sori), linear to oblong, along upper margin of sub-leaflets.

Habitat: Fresh to moist, calcareous loamy to clayey upland tolerant hardwood stands, with sugar maple and other hardwoods, including basswood.

Notes: Fiddleheads of this species, picked early in the spring before they unfold, are excellent cooked as greens. Historical medicinal uses include the treatment of coughs, asthma, jaundice and kidney ailments. See caution in Introduction. • The species name *pedatum* means 'palmately forking.'

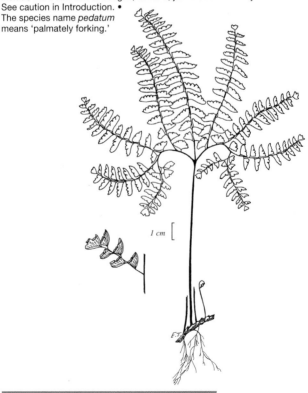

1 cm

WOOD FERN FAMILY (DRYOPTERIDACEAE)

General: Delicate, narrow-leaved perennial fern, 30–90 cm tall; usually in masses over rocks, from creeping scaly underground stems (rhizomes).

Emma Thurley

Leaves: In loose clusters, spreading over ground (not upright), soft, pale green, long triangular, prolonged into very long taper to tips, up to 90 cm long and 20 cm wide at base, compound, divided 3 times (tripinnate), with 20–40 pairs of leaflets (pinnae); leaflets long triangular to linear, drooping at ends; sub-leaflets (pinnules) cut to middle, linear, lowest pair noticeably parallel to stalk; rounded bulblets produced on lower surface drop off and produce new plants; stalks yellow, grooved, swollen and blackish at base.

Fruiting Structures: Few, dot-like spore clusters (sori), scattered away from margins on undersides of sub-leaflets.

Habitat: Damp rocky calcareous drainage areas in tolerant hardwood stands; calcareous rocky slopes, shaded ravines and steep banks.

Notes: The species name *bulbifera* means 'bearing bulbs.'

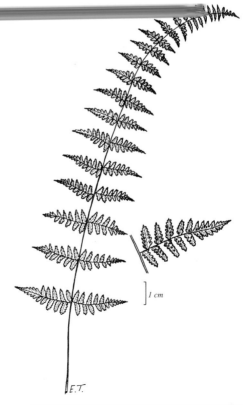
1 cm

E.T.

WOOD FERN FAMILY (DRYOPTERIDACEAE)

Emma Thurley

General: Erect perennial fern, up to 80 cm tall; in clusters from stout underground stems (rhizomes) closely covered with old stalk bases.

Leaves: Clustered, erect, compound, divided 2 times (bipinnate), of 2 types; sterile leaves evergreen, yellow-green, 10–80 cm long, 6–30 cm wide, lance-shaped, widest above middle, tapering to base and tip, with 14–34 pairs of leaflets (pinnae); leaflets triangular to oblong above, widely spaced and tilting horizontally below; lowest pair triangular, distinctly tilted; sub-leaflets (pinnules) oblong, blunt, with slightly toothed margins; stalks green above, brown below, densely covered with brown scales; fertile leaves usually taller, narrower and not evergreen.

Fruiting Structures: Spore clusters (sori) dot-like, on underside of upper leaflets of fertile leaves, midway between margin and midvein of sub-leaflets.

Habitat: Wet organic hardwood and conifer swamps.

Notes: A distinctive fern of wet, forested habitats, this species is quickly recognized by its narrow leaves with widely spaced, ladder-like leaflets.

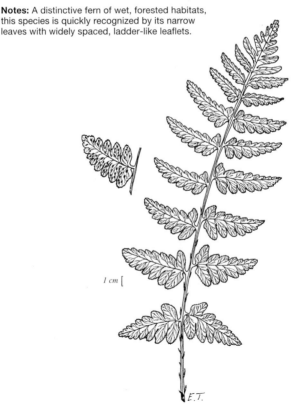

1 cm

E.T.

WOOD FERN FAMILY (DRYOPTERIDACEAE)

General: Evergreen perennial fern, up to 75 cm tall; in clusters from stout ascending underground stems (rhizomes) covered with old stalk bases and scales.

Leaves: Clustered, erect to spreading, dark green to bluish-green above, light green to grey-green below, leathery, 25–75 cm long, 6–30 cm wide, oblong to egg-shaped, evergreen, compound, divided 2 times (bipinnate), with 12–30 pairs of leaflets (pinnae); leaflets lance-shaped, rapidly tapered to point and arching upwards at tip, widely spaced near stalk; sub-leaflets (pinnules) blunt-tipped, toothed or lobed; stalk stout, grooved in front, brownish-green, with golden-brown scales, swollen and dark red-brown at base.

Fruiting Structures: Spore clusters (sori) dot-like, whitish to grey-brown, marginal or nearly so on underside of sub-leaflets.

Habitat: Dry to moist, sandy to loamy upland tolerant hardwood stands; typically in dry open rocky microsites.

Brenda Chambers

Notes: The species name *marginalis* means 'marginal,' and refers to the position of the spore clusters on the underside of the sub-leaflets.

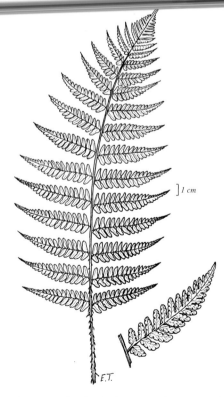

SILVERY SPLEENWORT • *Athyrium thelypterioides*
ATHYRIUM FAUSSE-THÉLYPTÉRIDE

Emma Thurley

General: Soft, pale perennial fern, 30–90 cm tall; in clusters from thick creeping black underground stems (rhizomes).

Leaves: Clustered, erect to spreading, yellowish-green, 30–90 cm long, 10–30 cm wide (at middle), lance-shaped, narrowed at base and tip, compound, divided 2 times (bipinnate), with up to 18 pairs of leaflets (pinnae); leaflets linear to lance-shaped, with yellow-green hairs beneath, lowest pair usually pointing downwards; sub-leaflets with wavy margins; stalks green, very hairy, dark and slightly scaly at base.

Fruiting Structures: Spore clusters (sori), linear or slightly curved, whitish or silvery (turning light brown), 2–8 along lateral veins on underside of sub-leaflets.

Habitat: Moist to fresh upland tolerant hardwood stands.

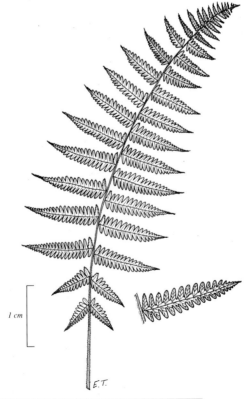

1 cm

E.T.

MARSH FERN FAMILY (THELYPTERIDACEAE)

General: Delicate, pale, shiny perennial fern, up to 60 cm tall; in small clusters from slender, creeping and forking, dark brown underground stems (rhizomes); in spreading colonies.

Leaves: Clustered, erect to spreading, yellow-green, thin, 20–60 cm long, 5–12 cm wide, elliptic to lance-shaped, tapered from middle to pointed tip, gradually narrowed to base, usually minutely hairy (especially beneath), compound, divided 2 times (bipinnate), with 23–46 pairs of leaflets (pinnae); leaflets long, pointed, narrow, lance-shaped, cut to middle into narrow lobes or sub-leaflets (pinnules); lowest leaflets 0.2–1.3 cm long (many times shorter than middle ones); stalks green, scaly, brown at base.

Fruiting Structures: Spore clusters (sori) few, small, near margins on undersides of leaflets.

Habitat: Moist to fresh upland tolerant hardwood stands.

Notes: This fern often develops large colonies at the forest edge or in forest openings. Its yellow-green colour is distinctive.

Brenda Chambers, both

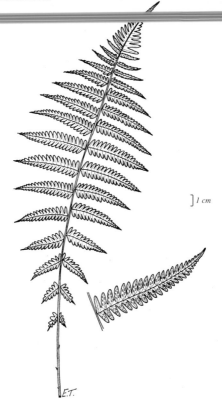

] *1 cm*

E.T.

MARSH FERN • *Thelypteris palustris*
DRYOPTÉRIDE THÉLYPTÉRIDE

MARSH FERN FAMILY (THELYPTERIDACEAE)

Bill Crins

General: Delicate, thin, perennial fern, 10–70 cm tall; from slender, black, long, creeping underground stems (rhizomes).

Leaves: Single, oblong to lance-shaped, pointed, widest near base, green or yellowish-green, up to 45 cm long and 15 cm wide, compound, once to nearly twice-divided (pinnate to nearly bipinnate), with 17–40 pairs of leaflets; leaflets spreading at right angles to axis, lance-shaped, pointed, with deep, oblong, blunt-tipped lobes almost forming sub-leaflets; young leaves minutely hairy on both surfaces; stalks pale green above, black at base, smooth or with a few scattered scales, 9–35 cm long on sterile leaves, up to 70 cm long and longer than blade on fertile leaves.

Fruiting Structures: Dot-like spore clusters (sori) on underside of upper leaflets of fertile leaves.

Habitat: Wet edges of creeks and streams; in ditches.

Notes: You can cultivate marsh fern in potting or garden soil for a woodland garden, but you may find this plant difficult to control because of its spreading and weedy growth habit.

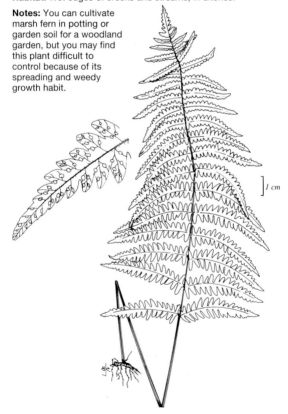

]*1 cm*

General: Delicate perennial fern, 30–70 cm tall, but sometimes up to 1 m; from scaly underground stems (rhizomes).

Leaves: In circular clusters, erect to spreading, dark green, 10–50 cm long, 5–30 cm wide, lance- or egg-shaped to triangular, compound, divided 2 to almost 3 times (bi- to tripinnate), with 17–33 pairs of leaflets (pinnae), progressively smaller upwards, lance- or egg-shaped to triangular; sub-leaflets (pinnules) oblong, blunt, with sharp, pointed teeth on margins; stalk densely scaly below leaflets, and with scattered scales above.

Brenda Chambers

Fruiting Structures: Rounded to kidney-shaped, dot-like spore clusters (sori) near mid-vein on underside of sub-leaflets.

Habitat: In most moisture regimes, soil textures and stand types; generally absent in dry pine and pine-oak stands.

Notes: Historically, the fresh pulp from the underground stems (rhizomes) was applied to cuts and the leaves were soaked for several days to make a liquid used for washing hair. See caution in Introduction. • See notes under lady fern (p. 358).

Food Use by Wildlife: plant. • **Mammals:** snowshoe hare, white-tailed deer. • **Birds:** ring-necked pheasant, spruce grouse, ruffed grouse, woodcock, wild turkey, American crow.

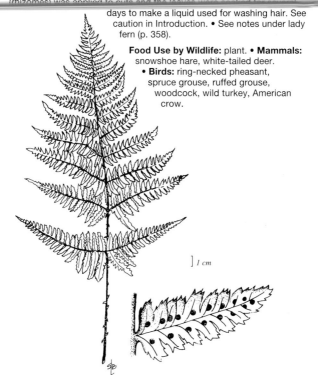

] *1 cm*

LADY FERN • *Athyrium filix-femina ssp. angustum*
ATHYRIUM FOUGÈRE-FEMELLE

Brenda Chambers

General: Perennial fern, up to 1 m tall; from creeping to erect, scaly underground stems (rhizomes).

Leaves: In circular clusters, erect, narrowly to widely lance-shaped, tapered at both ends, 40–90 cm long, 10–35 cm wide, pale green, compound, divided 2 to almost 3 times (bipinnate to sub-tripinnate); leaflets (pinnae) lance-shaped, pointed, short-stalked to stalkless; sub-leaflets (pinnules) oblong to lance-shaped, toothed, merging near leaflet tips; stalks straw-coloured to brownish or reddish, fragile, grooved and scaly near base, shorter than blades.

Fruiting Structures: Dot-like, slightly arching to kidney- or almost horseshoe-shaped spore clusters (sori) on underside of leaflets.

Habitat: Wet organic and moist sandy to clayey hardwood and conifer swamps; fresh to dry upland tolerant hardwood stands; wet ditches.

Notes: Lady fern may be confused with spinulose wood fern (p. 357) in the field. Lady fern can be distinguished by the way its leaves taper, with smaller leaflets near the base. On spinulose wood fern, the leaflets near the base are longer than the leaflets near the middle. The spore clusters are also different. On lady fern, the clusters are almost horseshoe-shaped, whereas on spinulose wood fern they are rounded. • Aboriginal peoples used the leaves to cover food. The fresh fiddleheads were eaten in early spring. See caution in Introduction.

1 cm [

WOOD FERN FAMILY (DRYOPTERIDACEAE)

General: Coarse, erect to arching perennial fern, up to 1.5 m tall; in clusters from stout, scaly underground stems (rhizomes).

Leaves: Clustered, arching; sterile leaves ostrich-plume-shaped, oblong to lance-shaped, widest near top, very gradually narrowed to base, abruptly pointed at tip, up to 1.5 m long, 12–40 cm wide, rich green, compound, once divided (pinnate), with up to 40 (or more) pairs of leaflets (pinnae); leaflets long, narrow-pointed, ascending, deeply cut into up to 30 (or more) pairs of oblong lobes, lowest pair winged or clasping stalk; stalks stout, green, flattened above with deep groove, rounded on back.

Fruiting Structures: Fertile leaves divided once (pinnate), erect, rigid, 20–60 cm tall, green turning brown, with spore clusters on undersides of sub-leaflets; margins fold over spore clusters, forming pod-like structures.

Habitat: Moist sandy to silty pockets in upland tolerant hardwood and intolerant hardwood mixedwood stands; wet organic hardwood swamps; along streams and riverbanks, wet roadsides.

Notes: The dark brown fertile leaves persist over winter.

Brenda Chambers, top;
Emma Thurley, bottom

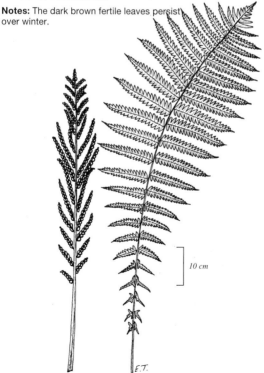

10 cm

E.T.

OAK FERN • *Gymnocarpium dryopteris*
DRYOPTÉRIDE DISJOINTE

WOOD FERN FAMILY (DRYOPTERIDACEAE)

John Seyler

General: Delicate fern, up to 35 cm tall; solitary or in extensive patches from blackish underground stems (rhizomes) that bear scattered leafstalks.

Leaves: Single, widely triangular, thin, lacy, yellowish-green, up to 18 cm long and 25 cm wide, compound, divided 2–3 times (bi- to tripinnate); leaflets 3, triangular, stalked, further divided into round-toothed sub-leaflets; uppermost leaflet longest; stalks scaly at base, shiny and straw-coloured.

Fruiting Structures: Small, circular spore clusters (sori) near margins on underside of leaflets.

Habitat: Wet organic to moist clayey to sandy conifer and hardwood swamps; in fresh to dry upland cedar mixedwoods, and tolerant hardwood stands with a yellow birch and eastern hemlock component.

Notes: Oak fern can be cultivated in potting or rich garden soil, and it makes a nice addition to woodland gardens.

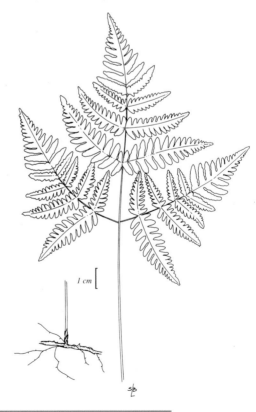

1 cm

BRACKEN FERN FAMILY (DENNSTAEDTIACEAE)

General: Coarse perennial fern, up to 1 m tall; often in large patches from deep, spreading, branching underground stems (rhizomes).

Leaves: Single, widely triangular, up to 90 cm long and wide, held horizontally at the top of erect stalks, compound, divided 2–3 times (bi- to tripinnate); leaflets (pinnae) opposite, oblong, narrowed to blunt tips; lowest pair of leaflets much larger than others and twice-divided

Karen Legasy

(bipinnate); upper leaflets once- to twice-divided (pinnate to bipinnate); sub-leaflets (pinnules) oblong to lance-shaped; underside hairless to hairy; margins rolled under; stalks hairless, straw-coloured to brownish, rigid, swollen at base.

Fruiting Structures: Strips of spore clusters (sori) along edges on underside of leaves, often hidden by curled margins.

Habitat: Dry to moist, rocky to clayey, upland pine, pine-oak, black spruce-pine and intolerant hardwood mixedwood stands; wet organic conifer swamps; cutovers, roadsides.

Notes: Aboriginal peoples used the underground stems and fiddleheads for food. Recent research has reported a **carcinogenic compound in bracken fern that may be harmful to humans and livestock**. If cattle eat mature bracken fern leaves, the hazardous compounds can be passed on to humans through milk. See caution in Introduction.

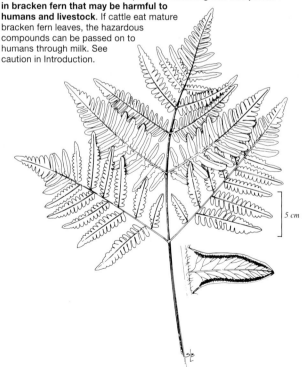

5 cm

RATTLESNAKE FERN • *Botrychium virginianum*
BOTRYCHE DE VIRGINIE

OMNR

General: Erect perennial grape-fern; fruiting stem 5–45 cm tall, smooth, fleshy and pink at base, with single leaf.

Leaves: Single, terminal or near middle of fruiting stem, thin, lacy, triangular, 5–40 cm wide and nearly as long, bright green, compound, 2–3 times divided (bi- to tripinnate); leaflets lance-shaped to oblong, short-stalked, 1–2 times divided (pinnate–bipinnate); sub-leaflets small, oblong; margins sharp-toothed.

Fruiting Structures: Fertile leaves solitary, on 3–30 cm-long stalk from centre of sterile leaf, 2–3 times divided (bi- to tripinnate), 2–20 cm long; leaflets with double rows of stalkless, rounded spore cases (sporangia) 0.5–1 mm in diameter.

Habitat: Moist hardwood swamps; fresh sandy to silty upland tolerant hardwood stands.

Notes: Rattlesnake fern was probably named for the way the fertile spike resembles a rattlesnake rattle.

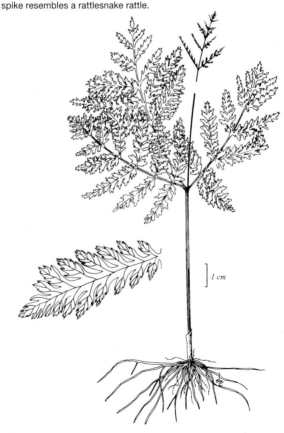

] 1 cm

Les mousses et les hépatiques

MOSSES AND LIVERWORTS

The mosses and liverworts have been divided into 5 groups. These are thallose liverworts (p. 365), leafy liverworts (pp. 366–369), peat mosses (pp. 370–379), upright (acrocarpous) mosses (pp. 380–397), and creeping (pleurocarpous) mosses (pp. 398–414). To aid in identification, diagrams are provided which highlight identifying features of these groups. These are followed by a pictorial key.

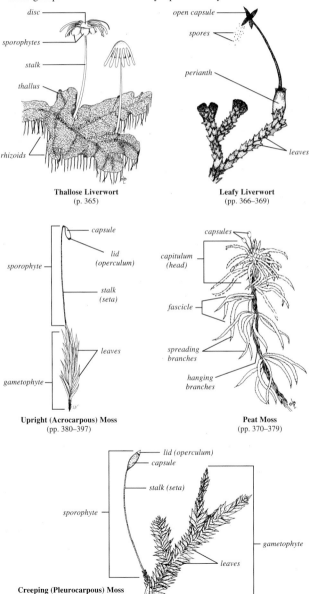

Thallose Liverwort
(p. 365)

disc
sporophytes
stalk
thallus
rhizoids

Leafy Liverwort
(pp. 366–369)

open capsule
spores
perianth
leaves

Upright (Acrocarpous) Moss
(pp. 380–397)

capsule
lid (operculum)
stalk (seta)
sporophyte
leaves
gametophyte

Peat Moss
(pp. 370–379)

capsules
capitulum (head)
fascicle
spreading branches
hanging branches

Creeping (Pleurocarpous) Moss
(pp. 398–414)

lid (operculum)
capsule
stalk (seta)
sporophyte
gametophyte
leaves

KEY TO MOSSES AND LIVERWORTS

plants always leafy;
leaves not in 2 rows,
not lobed; capsules opening
by a lid (operculum)

plants ribbon-like or leafy;
leaves usually in 2 rows,
often lobed; capsules in a
perianth at first, later splitting in 4

Mosses

Liverworts

plants with a flattened,
ribbon-like body (thallus)

plants with
stems and leaves

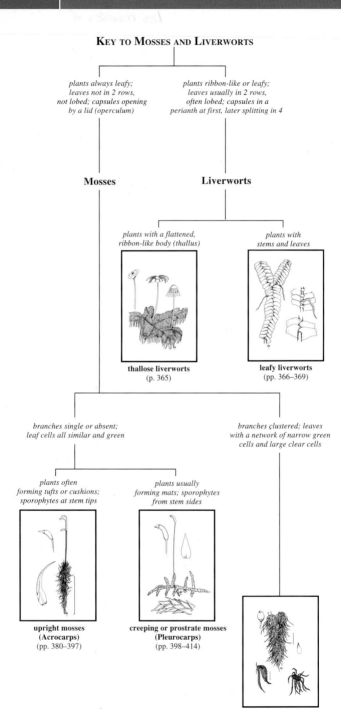

thallose liverworts
(p. 365)

leafy liverworts
(pp. 366–369)

branches single or absent;
leaf cells all similar and green

branches clustered; leaves
with a network of narrow green
cells and large clear cells

plants often
forming tufts or cushions;
sporophytes at stem tips

plants usually
forming mats; sporophytes
from stem sides

upright mosses
(Acrocarps)
(pp. 380–397)

creeping or prostrate mosses
(Pleurocarps)
(pp. 398–414)

peat mosses
(pp. 370–379)

MARCHANTIA FAMILY (MARCHANTIACEAE)

General: Ribbon-like (thallose) liverwort, pale- to dark-green, about 2–10 cm long and 1–2 cm wide, with wavy lobes along margins; upper surface flattened, with barrel-shaped pores, usually with a dark line down the centre; underside purplish, with numerous, yellowish, thin filaments (rhizoids) that anchor plant

Dale Vitt

Fruiting Bodies: Umbrella-like, male or female, stalked, often present; tiny, rounded, bowl-shaped bodies of vegetative reproductive cells (gemmae cups) often present.

Habitat: Wet organic to moist hardwood swamps; on soil, humus, and rotting logs.

Notes: Distinguishing features of green-tongue liverwort are its umbrella-like stalked fruiting bodies, bowl-like gemmae cups, black midrib and smooth upper surface. • Green-tongue liverwort is often confused with snake liverwort (*Conocephalum conicum*) which is larger, has a very coarsely sectored upper surface with white pores, has less noticeable midribs, and emits a distinctive spicy-sweet fragrance when fresh plants are crushed. Green-tongue liverwort has no fragrance, even when crushed.

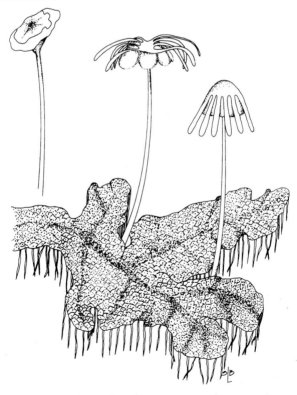

JAMESON'S LIVERWORT • *Jamesoniella autumnalis*

Linda Kershaw

General: Leafy liverwort, green, may be tinged with red; stems usually 1.5–3 cm long; often in flat dark green mats.

Leaves: Oval, entire, arranged in two rows, overlapping from the stem tip downward (like the shingles on a roof); underleaves very small (on young plants) or lacking (on mature plants).

Fruiting Bodies: Large, pouch-like perianths, fringed with thread-like hairs at tip; surrounded by leaf-like bracts fringed with thread-like hairs; capsules circular and black when immature, splitting open into four reddish-brown wings at maturity; stalks transparent, delicate.

Habitat: Fresh to moist, sandy to clayey tolerant hardwood and intolerant hardwood stands; mature to old growth coniferous forest; wet organic conifer swamps; on rotten logs and stumps.

Notes: Jameson's liverwort can be distinguished by the dark green patches it forms. These patches are made up of small plants with large, inflated, perianths fringed with hair-like segments and untoothed leaves that overlap like the shingles of a roof. The perianths are often abundant in the fall (hence the name *autumnalis*), but capsules are not often seen because they mature and disintegrate in the length of a day.

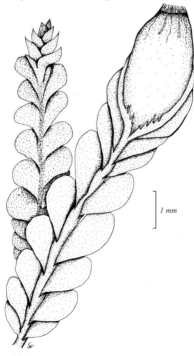

1 mm

LEPIDOZIA FAMILY (LEPIDOZIACEAE)

General: Robust, leafy liverwort, dark green, usually 3–6 mm wide; often forms tufts 6–20 cm thick.

Leaves: 3-toothed at tips, asymmetric, rectangular; in 2 rows, overlapping upward so that leaf at base of branch overlaps next leaf above and so on to branch tip; underleaves much smaller, 4–5-toothed, in 1 row.

Fruiting Bodies: Capsules, rarely produced, emerge from spindle-shaped pouches with 3 corners at the upper end.

Habitat: All moisture regimes and soil textures; on moist microsites in upland eastern white cedar mixedwood and black spruce-pine stands; in wet organic conifer swamps.

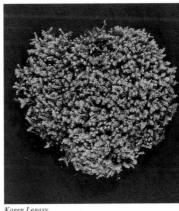

Karen Legasy

Notes: In moist, shaded forest, three-lobed liverwort grows in deep tufts. Otherwise, it forms mats on logs and stumps. The reversed shingle-like appearance of the leaves (think of an arrangement of shingles on a roof, where the base of the stem is the peak of the roof) is a distinguishing feature of three-lobed liverwort. In most other liverworts, the leaves overlap from the tip downward. The 3 small lobes or teeth at the leaf tips are another identifying feature. This is the largest leafy liverwort in Ontario, and is easily noticed.

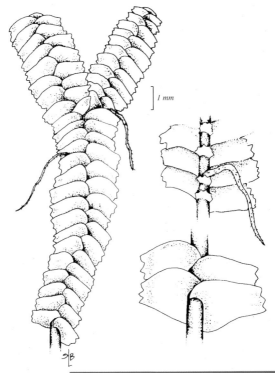

1 mm

Robin Bovey

General: Leafy liverwort, usually 1.5–3 cm long; in deep, loose rusty-brown or red tufts that are easily removed from the substrate.

Leaves: Hand-shaped, divided to 1/2 their length into 3–4 unequal lobes fringed with short, thread-like hairs, wrapped around stem; underleaves small, 2-lobed, with many hair-like segments.

Fruiting Bodies: Large inflated perianths with mouths fringed with thread-like hairs; capsules black and circular when immature, split open into four brown wings at maturity; stalks transparent, delicate.

Habitat: Wet organic conifer swamps; upland fresh to moist, sandy to coarse loamy black spruce-jack pine stands; on soil or rock.

Notes: This species is distinguished by its lobed leaves, fringed with hair-like segments, and its loose, rusty-brown, tufted growth. Naugehyde liverwort (*P. pulcherrimum*, p. 369) is smaller, grows in mats that are tightly appressed to logs, and its leaves are lobed for 3/4 of their length.

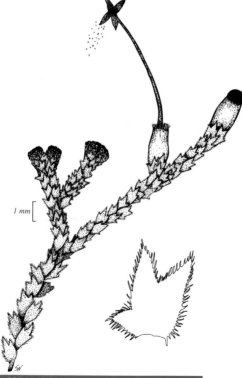

1 mm

PTILIDIUM FAMILY (PTILIDIACEAE)

General: Leafy liverwort, deep green or reddish-brown; in low, flat, dense, fuzzy, often circular tufts, firmly attached to substrate.

Leaves: Hand-shaped, divided to 3/4 their length into 3–4 narrow lobes fringed with thread-like hairs; arranged in 2 rows along upper side of stem, wrapped around stem.

Fruiting Bodies: Large inflated perianths with mouths fringed with thread-like hairs; capsules black and circular when immature, split open into four brown wings at maturity; stalks transparent, delicate.

Robin Bovey

Habitat: All moisture regimes, soil types and stand types; on rotting wood and at tree bases.

Notes: Northern naugehyde liverwort (p. 368) is a similar species that is twice as large and is much more upright. Unlike northern naugehyde liverwort, this species does not grow on rock and it is more firmly attached to its substrate.

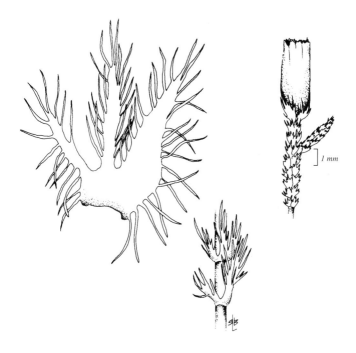

] 1 mm

KEY TO THE PEAT MOSSES (*SPHAGNUM* SPP.)

A. No large terminal buds

stems dark brown to black   *stems green*

stems black and brittle   *stem brown, not brittle*

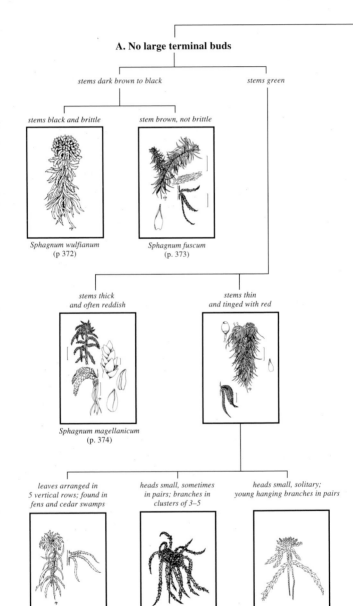

Sphagnum wulfianum
(p 372)

Sphagnum fuscum
(p. 373)

*stems thick
and often reddish*   *stems thin
and tinged with red*

Sphagnum magellanicum
(p. 374)

*leaves arranged in
5 vertical rows; found in
fens and cedar swamps*   *heads small, sometimes
in pairs; branches in
clusters of 3–5*   *heads small, solitary;
young hanging branches in pairs*

Sphagnum warnstorfii
(p. 375)

Sphagnum capillifolium
(p. 376)

Sphagnum angustifolium
(p. 377)

Les mousses

B. Large terminal buds

branches thick; leaves
spreading at right angles

Sphagnum squarrosum
(p. 378)

branches thin; heads
star- shaped, with 5 rays

Sphagnum girgensohnii
(p. 379)

General: Dark- to brownish-green peat moss, occasionally tinged pink to light-red near tip; rounded heads appear woolly or shaggy with densely crowded branches, resemble a clover head; stems erect, wiry, thick, stiff, brittle (snap when broken), dark brown to blackish, about 9 cm tall; branches in clusters of 6–12; forms loose carpets.

Leaves: Stem leaves triangular to tongue-shaped, pointed to slightly rounded and may be fringed at tip, less than 0.8 mm long; branch leaves narrowly egg- to lance-shaped, tapering to long, pointed, toothed at tip only, spreading when dry, 1.0–1.2 mm long; midrib absent.

Fruiting Bodies: Uncommon, rounded, dark brown to black capsules, 1–2 mm in diameter; stalks, short, erect, from tip of plant.

Brenda Chambers, top,
Karen Legasy, bottom

Habitat: Wet organic conifer swamps; moist upland black spruce stands.

Notes: You can recognize this peat moss in the field by its large, distinctive clover-like heads and the 6 or more branches in every branch cluster along its dark, wiry main stem.

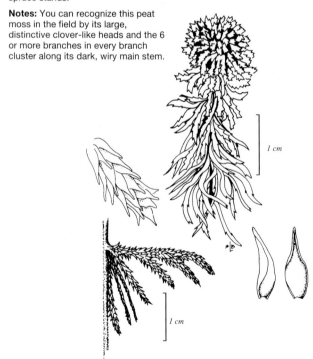

1 cm

1 cm

PEAT MOSS FAMILY (SPHAGNACEAE)

General: Small, slender, delicate, wiry peat moss, normally dark-to rust-brown, occasionally greenish; heads small, compact; stems small, upright, dark reddish-brown, about 10 cm tall; branches slender, in clusters of 3–5; forms dense, brown to brownish-green, rounded, cushion-like patches (hummocks).

Leaves: Stem leaves tongue-shaped, widely rounded at tip, 0.8–1.3 mm long; branch leaves egg- to lance-shaped, 0.9–1.3 mm long, inrolled, tapering to point at finely toothed tip (use hand lens); midrib absent.

Fruiting Bodies: Uncommon, rounded, dark brown to black capsules, 1–1.5 mm in diameter; stalks erect, short, from the head.

Habitat: Wet organic conifer swamps; often on top of large hummocks.

Notes: Common brown peat moss usually grows at the top of old, dry hummocks. Its brownish colour, compact growth, thread-like branches and brown (not brittle) stems help to distinguish this peat moss in the field.

Karen Legasy, both

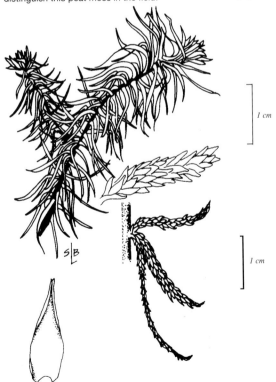

1 cm

1 cm

S|B

PEAT MOSS FAMILY (SPHAGNACEAE)

Brenda Chambers

General: Robust peat moss, green to pinkish, red, or purplish-red; stems erect, red, 8–20 cm tall; branches in groups of 4–6, short, plump, appear thick and swollen; forms hummocks.

Leaves: Stem leaves tongue- or spoon-shaped, widely rounded or minutely fringed at tip, up to 2 mm long; branch leaves widely egg-shaped, curved inward (concave or hood-shaped), minutely toothed at tip, 1.5–2.0 mm long; midrib absent.

Fruiting Bodies: Uncommon, rounded, dark brown to black capsules, 1–2 mm in diameter; stalks short, erect, at tip of plant.

Habitat: Wet organic conifer swamps; moist upland conifer stands.

Notes: Midway peat moss can be distinguished by its robust size, its thick or swollen branches and its stems, which are red when scraped. It is usually green when growing in shaded areas and red when in the open.

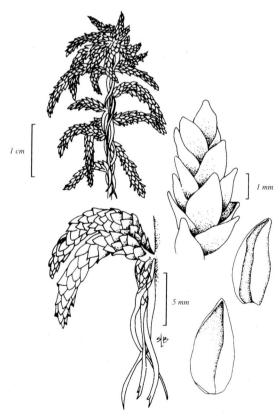

1 cm

1 mm

5 mm

PEAT MOSS FAMILY (SPHAGNACEAE)

General: Delicate, slender peat moss, green in shade or red in sun; stems soft, weak, slender, usually reddish but sometimes brown; branches in clusters of 3–5 with 2 branches spreading away from rest, slender; forms loose carpets and low hummocks.

Robin Bovey

Leaves: Stem leaves tongue-shaped, with shallow notch at wide, rounded tip, toothless, with strong border strongly widened at base, 0.8–1.3 mm long; branch leaves egg- to lance-shaped, curved under near tip when dry, straight when moist, 0.6–1.4 mm long, toothless, with a border; usually spirally arranged in 5 distinct rows; midrib absent.

Fruiting Bodies: Uncommon, rounded, dark-brown or black capsules; stalks short, erect, from the head.

Habitat: Wet calcareous conifer swamps and fens.

Notes: Warnstorf's peat moss is usually red, but even in green forms, the red branch stem is visible among the leaves. This peat moss usually indicates calcium and, therefore, an enhanced wetland nutrient status.

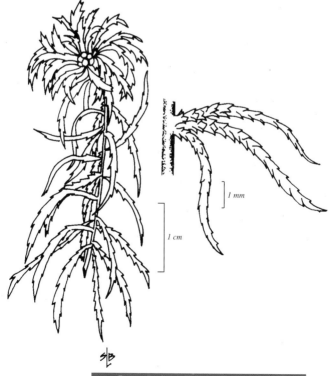

1 mm

1 cm

S|B

Rob Arnup

General: Slender peat moss, usually short, pale or brownish-green to pinkish-red; heads small, rounded; stems erect, stiff, occasionally forked with a double head, 5–8 cm tall; branches in groups of 3–5, long, sweeping, outward-curving; forms tight, carpet-like mounds.

Leaves: Stem leaves tongue-shaped to oblong, slightly notched or fringed at pointed tip, curved inward (use hand lens), 1.0–1.8 mm long; branch leaves egg- to lance-shaped, pointed at tip, curved slightly inward at tip (use hand lens), about 1.2 mm long, closely overlapping; midrib absent.

Fruiting Bodies: Uncommon, rounded, dark brown to black capsules, 1–2 mm in diameter; stalks short, erect, at tip of plant.

Habitat: Wet organic conifer swamps; moist to dry, sandy to fine loamy upland black spruce-pine stands.

Notes: Also known as *Sphagnum nemoreum*. • Small red peat moss can be recognized by its sweeping, outward-curving branches that resemble long hair or tresses. • This species could be confused with wide-tongued peat moss (*Sphagnum russowii*) in the field. However, that species has pores in its outer stem cells, which are not found in small red peat moss. • When dry, peat mosses (*Sphagnum* spp.) are capable of absorbing a large amount of liquid. They have historically been used for personal hygiene, in diapers and to dress wounds (some peat mosses have antibiotic properties). See caution in Introduction.

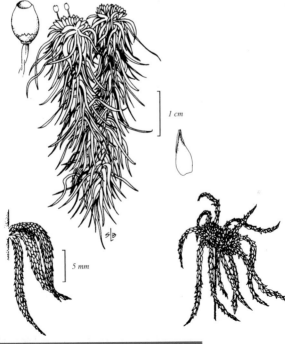

1 cm

5 mm

PEAT MOSS FAMILY (SPHAGNACEAE)

General: Slender peat moss; heads small, flat or rounded, slightly star-like in appearance, greenish to yellowish or brownish; stems upright, pale green to yellowish, occasionally pinkish near branch tips, about 9 cm tall; branches in clusters (fascicles) of 5; hanging branches normally thread-like, almost white, up to 10 mm long, loosely spaced, in distinctive pairs (note hanging branches just below heads); forms dense mats and small hummocks.

Leaves: Stem leaves triangular, blunt-tipped, less than 0.8 mm long, occasionally have minute teeth on margins; branch leaves egg- to lance-shaped, up to 1 mm long, occasionally wavy and curved under on dry branches; midrib absent.

Karen Legasy

Fruiting Bodies: Uncommon, rounded, dark brown or black capsules, 1–2 mm in diameter; stalks short, erect, from the head.

Habitat: Wet organic conifer swamps; on sides of hummocks formed by other peat mosses such as common brown peat moss (p. 373) and midway peat moss (p. 374).

Notes: You can recognize poor-fen peat moss by its slender shape, its often pinkish stem and its distinctive pairs of hanging branches.

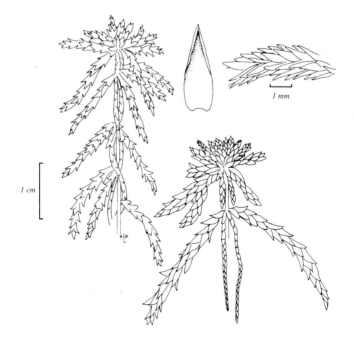

1 mm

1 cm

Frank Boas

General: Robust, tall, stiff peat moss, bright-to pale-green or yellowish; heads with prominent terminal buds surrounded by short, loosely clustered branches; stems thick, green to reddish-brown; branches in groups of 5, loosely clustered, with hanging branches closely covering stem; forms loose mats.

Leaves: Stem leaves oblong to tongue-shaped, slightly jagged at wide, rounded tip, indistinctly bordered, 1.5–2.5 mm long, curved slightly inward; branch leaves egg- to arrowhead-shaped, abruptly narrowing and distinctly rolled inward at pointed, toothed tip, 2–2.8 mm long, wide-spreading or spreading at right angles from base, toothless (except at tips); midrib absent.

Fruiting Bodies: Rounded, dark brown to black capsules on short stalks, from top of head.

Habitat: Wet organic to moist conifer and hardwood swamps and sedge-dominated fens.

Notes: After you identify shaggy peat moss once, you'll find it easy to recognize in the field by its branch leaves (which spread at right angles), its bright green colour, its large size and its large terminal bud.

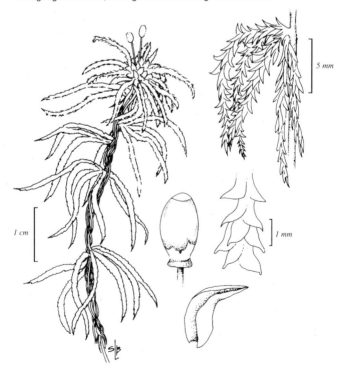

5 mm

1 cm

1 mm

PEAT MOSS FAMILY (SPHAGNACEAE)

General: Slender, wiry peat moss, green to slightly yellow (never red); heads flat, star-shaped with 5 radiating points and a distinct, shiny terminal bud at the centre; stems erect, stiff, woody, snap crisply when broken, green to pale green or yellow, about 8 cm tall; branches in groups of 3–5, spreading; forms large, loose carpets or mounds.

Leaves: Stem leaves widely tongue-shaped, only slightly longer than wide, blunt or flat at tips with a ragged or toothed margin, 1.0–1.3 mm long, tightly pressed against stem; branch leaves egg- to lance-shaped, tapering and strongly rolled inward to pointed, toothed tip, toothless (except at tip), 1.0–1.4 mm long, closely overlapping; midrib absent.

Brenda Chambers

Fruiting Bodies: Uncommon, rounded capsules, 1–2 mm in diameter, dark brown to black; stalks erect, short, from the head.

Habitat: Wet organic to moist, sandy to clayey conifer and hardwood swamps.

Notes: Common green peat moss is one of the most common peat mosses. It can be recognized by its flat, star-shaped tips, spreading branches, and widely tongue-shaped stem leaves with toothed, flat tips.

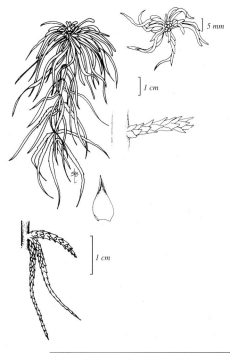

5 mm

1 cm

1 cm

KEY TO THE UPRIGHT (ACROCARPOUS) MOSSES

A. Leaves long and narrow; stems simple or rarely forked

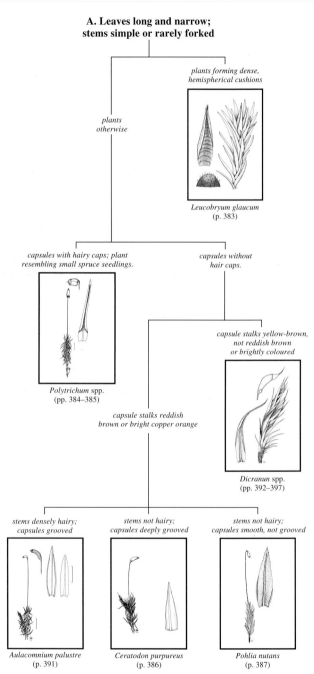

plants forming dense, hemispherical cushions

plants otherwise

Leucobryum glaucum
(p. 383)

capsules with hairy caps; plant resembling small spruce seedlings.

capsules without hair caps.

capsule stalks yellow-brown, not reddish brown or brightly coloured

Polytrichum spp.
(pp. 384–385)

capsule stalks reddish brown or bright copper orange

Dicranum spp.
(pp. 392–397)

stems densely hairy; capsules grooved

stems not hairy; capsules deeply grooved

stems not hairy; capsules smooth, not grooved

Aulacomnium palustre
(p. 391)

Ceratodon purpureus
(p. 386)

Pohlia nutans
(p. 387)

B. Leaves wide, round to egg-shaped or stems distinctly branched, sometimes tree-like

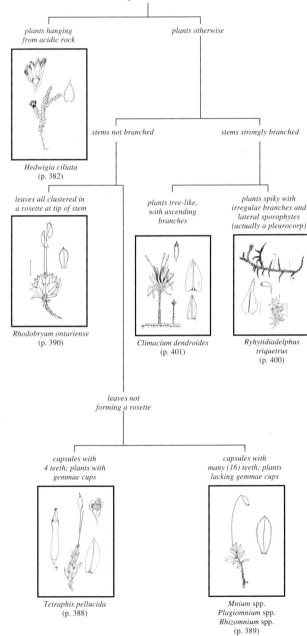

plants hanging from acidic rock

Hedwigia ciliata (p. 382)

plants otherwise

stems not branched

stems strongly branched

leaves all clustered in a rosette at tip of stem

Rhodobryum ontariense (p. 390)

plants tree-like, with ascending branches

Climacium dendroides (p. 401)

plants spiky with irregular branches and lateral sporophytes (actually a pleurocorp)

Ryhytidiadelphus triquetrus (p. 400)

leaves not forming a rosette

capsules with 4 teeth; plants with gemmae cups

Tetraphis pellucida (p. 388)

capsules with many (16) teeth; plants lacking gemmae cups

Mnium spp.
Plagiomnium spp.
Rhizomnium spp.
(p. 389)

GRIMMIA FAMILY (GRIMMIACEAE)

Karen Legasy

General: Coarse, usually robust moss, olive-green, yellow or brown, usually greyish and very brittle when dry, softer and greener when moist; immature stems spreading or curved and ascending, irregularly branched; forms dull, loose to dense mats.

Leaves: Widely oblong to egg-shaped, 1.3–2.3 mm long, erect or pressed to stem with tips spreading when moist and tightly pressed around stem when dry; tips short to long, usually whitish, pointed and irregularly, minutely toothed; margins rolled under on lower 2/3 or sometimes nearly to tip; midrib absent.

Fruiting Bodies: Rounded capsules, pale brown near base, reddish-brown at rim, smooth, shiny when dry, erect, at stem tips, deeply hidden among leaves with finely hairy margins; stalks very short.

Habitat: Wet organic hardwood and conifer swamps; upland tolerant hardwood and intolerant hardwood stands; on acidic rock.

Notes: This moss is also known as witch's hair. • You can recognize ciliate hedwigia moss in the field by the following characteristics: transparent, whitish leaf tips; rigid and greyish when dry but softer and olive-green with spreading leaves when moist; capsules hidden among the leaves. This moss grows like coarse, green hair, hanging on rock.

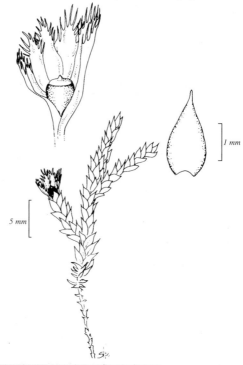

1 mm

5 mm

BRYUM FAMILY (BRYACEAE)

General: Robust, hummock-forming moss, greyish to bluish-green; stems 3–6 cm tall; in dense rounded cushions, usually 2–9 cm tall and 30–50 cm in diameter, but can be much larger.

Leaves: Tubular, lanceolate, often blunt-tipped, thick and fleshy, 3–8 mm long, erect to spreading; midrib occupies most of leaf, giving the distinctive blue-green colour.

Fruiting Bodies: Rare, curved capsules, ribbed, 1.5–2 mm long; stalks red-brown, 9–17 mm long.

Brenda Chambers

Habitat: Dry to moist upland tolerant hardwood, pine and pine-oak stands; in wet organic conifer swamps; on humus, rock or rotten wood.

Notes: This moss is easily recognized by its distinctive, half-circular, greyish to bluish-green cushions. Young colonies of several loose plants can be found growing on rotten logs and humus; identification at this stage usually requires the use of a microscope. There are other mosses that might be confused with young colonies of pin cushion moss, but none that will be confused with the large mature cushions. Vegetative reproduction can occur when brood-leaves, produced in leaf axils, break off and grow into new plants.

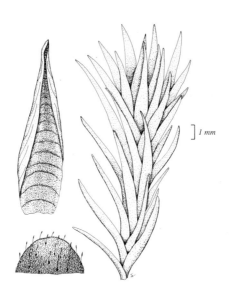

1 mm

Brenda Chambers

General: Robust moss, dark green to brownish; stems stiff, coarse, 4–45 cm tall, usually erect from a prostrate and slightly twisted base, unbranched; in loose to somewhat dense mats or tufts.

Leaves: Lance-shaped, pointed at tip, sheathing, yellowish-brown, often shiny at 2–3 mm long base, about 6–10 mm long, erect to spreading (usually near tips), rolled and pressed to stem when dry, spreading and curved downward when moist; margins sharply toothed nearly to base; midrib extends beyond tip, sometimes toothed at back.

Fruiting Bodies: 4-angled capsules, 3–5 mm long, covered by a hairy cap when young, reddish-brown; stalks 5–9 cm long; at stem tips.

Habitat: All moisture regimes, soil types and stand types; on soil, decaying logs, rocks covered with moss, and peat moss hummocks; disturbed ground.

Notes: Common hair cap moss was used for bedding, to stuff pillows and to make small brooms for dusting. • See notes on juniper moss (p. 385) for distinguishing features.

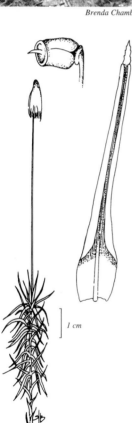

1 cm

General: Robust moss, green or bluish-green to reddish-brown; stems stiff, coarse, 1–13 cm tall, unbranched; often forms extensive, loose mats or tufts.

Leaves: Lance-shaped, with a brown, short to long, toothed, bristle-like point at tip and a 1–2.5 mm long, sheathing base, straight or curved, 4–8 mm long, erect to moderately spreading when moist, pressed to stem when dry; margins toothless, folded inward above base, giving leaves a needle-like appearance; midrib single, extends to tip.

Fruiting Bodies: 4-angled (squarish) capsules, 2.5–5 mm long, longer than wide, dark brown, covered with a pale, hairy cap when young; slightly erect to horizontal on stalks 2–6 cm long; at stem tips.

Brenda Chambers

Habitat: All moisture regimes, soil textures and stand types; on soil, humus or decaying stumps; disturbed ground.

Notes: Both juniper moss and common hair cap moss can be recognized by the 'hairy cap' on their young capsules. Juniper moss can be distinguished from common hair cap moss (p. 384) in that it is usually smaller (1–13 cm tall vs. 4–45 cm tall), its leaves have a brown, bristle-like point at their tips, its upper leaves are inrolled and needle-like (rather than flat) and its margins are toothless (rather than distinctly toothed) (use hand lens). • Awned hair cap moss (*Polytrichum piliferum*) could be confused with juniper moss, but it is a smaller moss, and it has a long, white hair-point at the tip of each leaf.

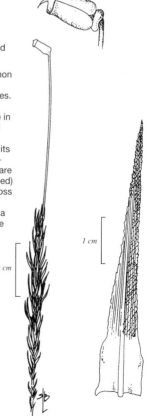

1 cm

1 cm

Dale Vitt

General: Dirty green to yellowish-brown or reddish moss; stems unbranched or forked, 0.5–2.5 cm tall; often in dense, dull tufts or mats.

Leaves: Lance-shaped, 1.8–2 mm long, sharply pointed at tip; margins strongly bent backward from base and nearly to tip where they are irregularly notched or minutely toothed; midrib extends to tip or beyond.

Fruiting Bodies: Cylindrical capsules, deeply furrowed when dry, 2–4 mm long, leaning to horizontal; stalks dark purplish-red, 8–30 mm long; at stem tips.

Habitat: Dry to moist open, disturbed areas; on exposed mineral soil; often appears after fire; occasional on rock or rotting wood.

Notes: Fire moss is quite variable in growth form, and therefore it can be difficult to identify in the field. One distinguishing feature is its deeply furrowed capsules with dark purplish-red stalks. • This moss is considered a weed and is often an early invader of disturbed sites such as burned areas, but it is eventually replaced by other species.

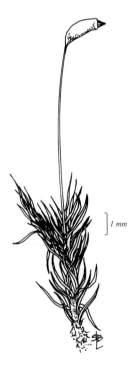

1 mm

General: Moss; stems erect, red, 1–4 cm tall; in loose, green or yellowish, sometimes shiny, upright tufts.

Leaves: Lance-shaped with margins rolled under, weakly toothed at the tip, 2–4 mm long, erect, or twisted around stem when dry; midrib distinct, ending at leaf tip.

Fruiting Bodies: Smooth, orange-brown capsules, symmetrical, tapered at mouth, nodding, 2.5–4 mm long; stalks shiny, orange or red, persistent, 15–30 mm long; at stem tips.

Robin Bovey

Habitat: On rotten logs and stumps, soil, humus, and bark of trees; in rock crevices of cliffs.

Notes: A key identifying feature of this moss is its shiny, orange-red stalks with their nodding, smooth capsules. This species is often confused with fire moss (p. 386), which also has red sporophyte stalks and small, upright tufted plants, but fire moss has upright capsules and grows along open roadsides and in disturbed areas.

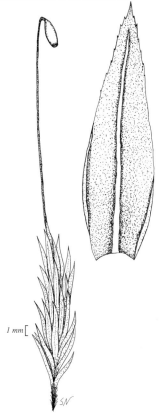

1 mm

SN

FOUR-TOOTH MOSS FAMILY (TETRAPHIDACEAE)

Robin Bovey

General: Green to reddish-brown moss, dull, mnium-like; stems erect, leafy, often end in a bowl-like terminal cluster, 8–15 mm tall, unbranched; in tufts or growing close together.

Leaves: Lower leaves up to 3 mm long egg-shaped, pointed at tip; upper leaves about 1–2 mm long, gradually longer and narrower; margins toothless, flat or inrolled; midrib narrow, single, strong, ending below tip in lower leaves, almost reaching tip in upper leaves.

Fruiting Bodies: Narrowly cylindric capsules, upright, smooth, with 4 teeth at red or brown mouth, 2–3 mm long; stalks erect, 6–14 mm long; at stem tips.

Habitat: All moisture regimes, soil types and stand types; rotten stumps or logs in advanced decay; rarely on soil or rock.

Notes: You can recognize common four-tooth moss by the way its stem often ends in a bowl-like cluster of leaves and by its erect capsule with 4 teeth at the mouth. The sterile stems of this moss frequently end in a cluster of gemmae cups. These are small bodies of a few cells each which serve to vegetatively produce a new plant. If you kick a rotten, moss-covered stump and it collapses, chances are you will find this moss there.

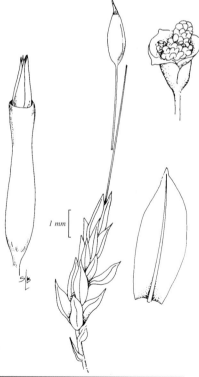

1 mm

MNIUMS • *Mnium* spp.
LEAFY MOSSES • *Plagiomnium* spp.
ROUND MOSSES • *Rhizomnium* spp.

MOSSES

MNIUM FAMILY (MNIACEAE)

General: Loosely tufted, upright-growing mosses, with sterile trailing runners, light to dark green; stems often reddish, sometimes hairy; dry plants tan to brown, with distinctive leaf margins and midribs.

Leaves: Ovate, broadly rounded or pointed at tip; margins thick, singly or doubly toothed, or toothless; midrib distinct, extends to leaf tip and sometimes beyond.

Brenda Chambers

Fruiting Bodies: Cylindric capsules, usually hanging; stalks single or clustered, at stem tips.

Habitat: Dry to moist, rocky to clayey upland eastern white cedar mixedwoods and tolerant hardwood stands with yellow birch and hemlock; wet organic hardwood swamps; on rotting wood or humus.

Notes: *Mnium* species and related genera are distinguished by their leaves, which have a strong border and midrib, and by their cylindrical, hanging capsules. Several common species can be separated using their leaf serrations and number of sporophytes:

Edged lantern mnium (*Mnium marginatum*): Leaf margins with double teeth; sporophytes single.

Red-mouthed mnium (*Mnium spinulosum*): Leaf margins with double teeth; sporophytes in 2s or 3s.

Woodsy mnium (*Plagiomnium cuspidatum*): Leaf margins with single teeth halfway to base.

Common leafy moss (*Plagiomnium medium*): Leaf margins with single teeth from tip to base.

Pointed round moss (*Rhizomnium punctatum*): Leaf margins without teeth.

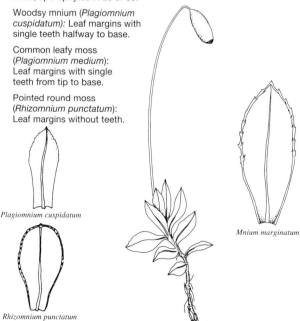

Plagiomnium cuspidatum

Rhizomnium punctatum

Mnium marginatum

General: Robust moss, dark green, commonly reddish-tinged; main stems underground; secondary stems erect, 1–5 cm tall, unbranched; in loose tufts or mats.

Leaves: Oblong to egg-shaped, tipped with an abrupt, rigid, stout point, 5–10 mm long, wide-spreading when moist, erect and irregularly twisted when dry; margins rolled under on lower 2/3 of leaf, toothed on upper 1/3 to 1/2; upper leaves crowded near tip of stem in rosette up to 1.5 cm wide; lower leaves smaller, scale-like, remote; midrib ends at tip or extends shortly beyond.

Fruiting Bodies: Rare, oblong or cylindrical capsules, short-necked, curved, 4–7 mm long, horizontal to hanging, smooth; stalks 2.5–4 cm long, curved or hooked at tip, 1–8 from each rosette; at stem tips.

Habitat: Fresh to moist upland eastern white cedar mixedwoods and tolerant hardwood stands, with hemlock and yellow birch; wet organic hardwood swamps; on soil, humus, decaying logs and rocks.

Notes: Also known as *Rhodobryum roseum*. • In the field, this species and Mnium mosses (p. 389) appear very similar. However, rose moss can be distinguished by its clustered, rose-like formation of upper leaves, its underground primary stem, its relatively long leaf cells, and its indistinct leaf border.

Karen Legasy, top;
Brenda Chambers,
bottom

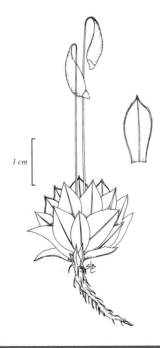

1 cm

BOG MOSS FAMILY (AULACOMNIACEAE)

General: Robust moss, yellowish-green to yellowish-brown, shiny; stems erect, green with a dense woolly or fuzzy reddish-brown to creamy covering, 3–9 cm tall, unbranched or forked; in loose to dense tufts or clusters.

Leaves: Egg- to lance-shaped, blunt to pointed at tip, 2–4 mm long, folded and usually twisted and crisp when dry, erect and spreading when moist; usually crowded and forming a point at tip of stem; margins toothless or minutely toothed (use hand lens); midrib prominent, extends almost to tip.

Linda Kershaw

Fruiting Bodies: Cylindrical capsules, curved, leaning or horizontal to almost erect, reddish-brown, 2.5–4 mm long, solitary, distinctly ribbed when dry; stalks erect, 2.5–4.5 cm long, twisted when dry; at stem tips.

Habitat: Wet organic conifer swamps, fens; on moist rock surfaces.

Notes: Ribbed bog moss may appear greenish or brownish, but it always has a yellowish tinge.

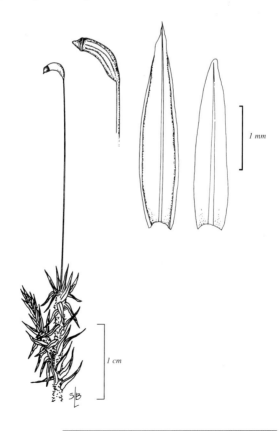

1 mm

1 cm

DICRANUM FAMILY (DICRANACEAE)

Linda Kershaw

General: Dull green to brownish moss; stems 1–3 cm tall, woolly below, usually with clusters of stout, stiff, rounded branchlets with minute leaves pressed together in axils of upper leaves; in small compact tufts.

Leaves: Curved, turned to 1 side or spreading, lance-shaped with slenderly tapered tip or lance-egg-shaped with blunt to pointed tip, 3–5 mm long, nearly tubular; margins strongly curved upward; margins minutely toothed; midrib extends to tip or beyond.

Fruiting Bodies: Urn-shaped capsules, irregularly furrowed when dry, 2–3 mm long, erect or almost so; stalks 10–20 mm long, yellow-brown; at stem tips.

Habitat: All moisture regimes, soil textures and stand types; on rotten logs and stumps.

Notes: Spiky dicranum has almost tubular leaves and miniature branchlets in the axils of its upper leaves. This moss reproduces vegetatively using its brittle, whip-like branchlets, which can easily be broken off by rubbing a tuft of the moss in the palm of your hand. Fragments of broken branchlets are then visible.

1 mm

DICRANUM FAMILY (DICRANACEAE)

General: Dull dark green moss, 2–4 cm tall; stems reddish brown; often forming woolly tufts or mats.

Leaves: Narrowly lance-shaped, 4–7 mm long, slightly keeled in upper part, not wavy when moist, curved to one side, contorted when dry; margins with minute teeth half way to base; midrib extends beyond long-pointed tip.

Fruiting Bodies: Curved capsules, inclined to horizontal, somewhat furrowed, 1.8–3 mm long; stalks yellow, 8–20 mm long; at stem tips.

Frank Boas

Habitat: All moisture regimes, soil types and stand types; on rotten logs or stumps.

Notes: Curved, contorted dry leaves that are all swept to one side on plants that are growing on rotten logs separate this species from other *Dicranum* species. Broom moss (pg. 397) is much more strongly swept to one side, and its leaves are not contorted when dry.

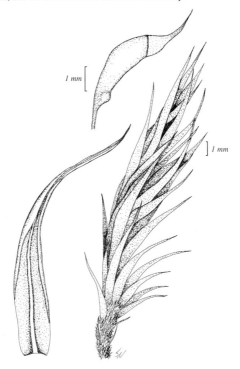

1 mm

1 mm

DICRANUM FAMILY (DICRANACEAE)

Karen Legasy

General: Dark green to brownish or yellow-brown moss; stems erect, 0.5–3 cm tall, unbranched; forms dense, dull, woolly mats or tufts.

Leaves: Narrowly lance-shaped with long, pointed tip, 2–3.5 mm long, very wavy when dry, curved inward but usually not tubular above; margins minutely toothed about 1/2 way to base; midrib extends to tip or slightly beyond.

Fruiting Bodies: Urn-shaped capsules, 1.5–1.75 mm long, erect, usually slightly curved, slightly grooved when dry; stalks smooth, erect, yellow to reddish-brown, 6–14 mm long, solitary or in clusters at stem tips.

Habitat: All moisture regimes, soil textures and stand types; on decaying logs, stumps and tree bases or exposed roots.

Notes: Lawn moss is smaller than most *Dicranum* species. It forms dull, dense, woolly mats that look like small green lawns on the bark at the base of trees or on rotting wood.

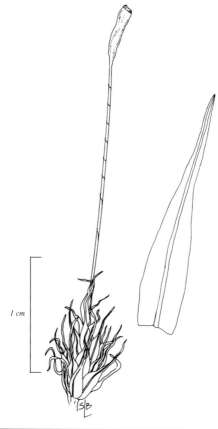

1 cm

DICRANUM FAMILY (DICRANACEAE)

General: Dull green or yellowish-brown moss; stems erect, unbranched, covered with thick, orange or yellowish 'wool,' 4–8 cm tall; in compact tufts.

Leaves: Lance-shaped, 5–9 mm long, folded above, spreading, tapered to long, pointed tip, crisped and curled when dry, dull; margins toothed in upper 1/2 or 1/3; midrib narrow, well-developed.

Fruiting Bodies: Urn-shaped capsules, 2–2.5 mm long, horizontal, strongly curved, grooved; stalks 20–27 mm long, usually clustered in 3s at stem tips.

Karen Legasy

Habitat: All moisture regimes and soil textures; in upland pine and black spruce-pine stands, and wet organic conifer swamps; on thin, sandy soil over rock.

Notes: You can recognize Ontario dicranum by its woolly stems, clustered capsules, and very wide-spreading leaves that are twisted and contorted when dry.

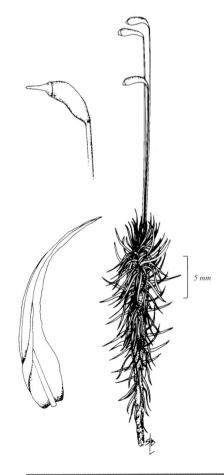

5 mm

General: Showy moss, light- to yellow-green or golden, usually shiny; stems erect, covered with thick, whitish fuzz (rhizoids), 2–15 cm tall, unbranched; forms dense, woolly mats or tufts.

Leaves: Lance-shaped with long, slender, pointed tip, 8–10 mm long, spreading at right angles to stem, obviously wavy (undulate); margins sharply toothed about 1/2 way to base, widely curved downward below middle; midrib very narrow, ends below tip, has 2 toothed ridges along lower side.

Fruiting Bodies: Urn-shaped capsules, strongly curved, 2.5–4 mm long, grooved when empty and dry, horizontal to slightly erect; stalks 2–3.7 cm long, in clusters of 1–5 at stem tips.

Brenda Chambers

Habitat: All moisture regimes and soil textures; in upland pine, black-spruce pine stands and wet organic conifer swamps; on soil, decayed wood, humus and rock.

Notes: Wavy-leaved moss can be distinguished by its wavy leaves, the whitish wool or fuzz on its stem and its clusters of 1–5 capsules (most *Dicranum* species have only 1 capsule at the tip). This is the largest and showiest of our *Dicranum* spp. See notes under broom moss, p. 397.

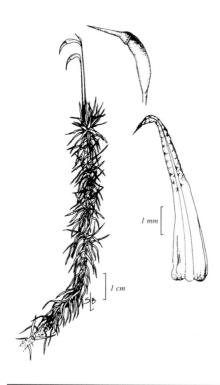

1 mm

1 cm

DICRANUM FAMILY (DICRANACEAE)

General: Fairly robust, usually coarse moss, green to dirty green or yellowish to brownish; stems densely woolly, 2–8 cm tall; forms shiny (sometimes dull) tufts or mats.

Leaves: Highly variable, narrowly to widely lance-shaped with long, slender to broad point, mainly curved, swept to 1 side (as if blown by a strong wind) wet or dry, or occasionally erect to spreading, sometimes wavy (undulate) but not contorted when dry, 3.5–8 mm long; margins usually strongly toothed in upper 1/3; midrib extends to tip or slightly beyond, usually has 4 ridges along lower side.

Karen Legasy

Fruiting Bodies: Urn-shaped capsules, curved, 2.3–5 mm long, smooth to grooved when old, horizontal to almost erect; lid (operculum) often longer than capsule; stalks 18–35 mm long; at stem tips.

Habitat: All moisture regimes, soil textures and stand types; tree bases, humus, rotting logs and rock.

Notes: Distinguishing features of broom moss are its leaves, which are strongly curved toward 1 side and its 1 (rarely 2) capsules at the tip of each stem.

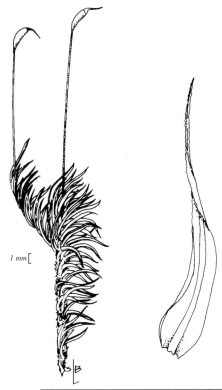

1 mm

KEY TO THE PROSTRATE OR CREEPING (PLEUROCARPOUS) MOSSES

A. Leaves in many rows; branches rounded

branches feather-like

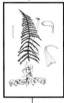

branches irregular, not feather-like

tree-like

Climacium dendroides (p. 401)

creeping

Bracythecium spp. (p. 407)
Callicladium haldanianum (p. 408)

leaves not curved

leaves curved

Ptilium crista-castrensis (p. 403)
Hypnum pallescens (p. 409)
Sanionia uncinata (p. 410)
Callicladium haldanianum (p. 408)

stems 2-3 times divided

Thuidium delicatulum (p. 402)
Hylocomium splendens (p. 404)

stems once divided

Rhytidiadelphus triquetrus (p. 400)
Pleurozium schreberi (p. 405)
Tomenthypnum nitens (p. 406)
Bracythecium spp. (p. 407)
Callicladium haldanianum (p. 408)

**B. Leaves in two rows;
branches appearing flattened**

leaves not shiny

Neckera pennata (p. 413)
Fissidens spp. (p. 414)
Callicladium haldanianum (p. 408)

*leaves very
shiny, green*

leaves strongly curved

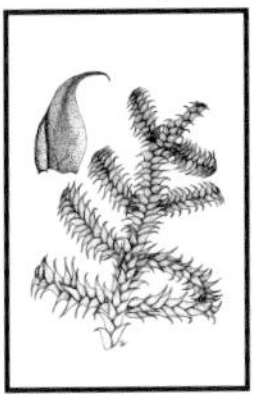

Brotherella recurvans
(p. 411)

leaves not curved

Plagiothecium laetum
(p. 412)

HYLOCOMIUM FAMILY (HYLOCOMIACEAE)

Brenda Chambers

General: Coarse, robust moss; dark- to bright-green or yellowish-green; stems reclining to erect, orange-red, up to 20 cm long; branches irregular, unequal, wide-spreading, orange-red, on all sides of stems, horizontal, usually tapered and curved downward at tips; forms loose mats or tufts.

Leaves: Stem leaves egg- to heart-shaped, with a long, tapered and pointed tip, clasping at base, 3.5–5 mm long, erect and spreading, irregularly wrinkled or pleated, densely toothed in upper 1/2, remotely and minutely toothed toward base; branch leaves oblong to egg-shaped, long-pointed at tip, clasping at base, 1.5–3 mm long; erect or erect and spreading, usually distantly and minutely toothed, but sometimes densely toothed near tip; midribs 2, prominent, extend almost 2/3 of distance to tip.

Fruiting Bodies: Cylindrical, curved capsules, horizontal to hanging, reddish-brown, 1.5–3 mm long; stalks glossy, reddish-brown, 1.5–4.5 cm long, somewhat twisted when dry.

Habitat: Fresh to moist, sandy to fine loamy upland eastern white cedar mixedwoods; on soil, humus and decayed logs.

Notes: Shaggy moss resembles the tail of a cat that has been electrocuted, and it is also commonly known as 'electrified cat's tail moss.'

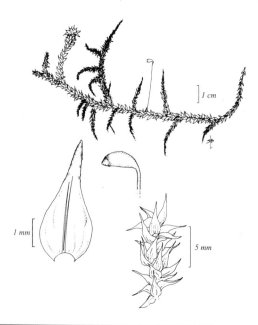

] *1 cm*

1 mm [

5 mm

TREE MOSS FAMILY (CLIMACIACEAE)

General: Tree-like moss, yellowish-green or dark green, dull, but shiny when dry, soft; stems erect, branched, usually 3–9 cm (sometimes up to 13 cm) tall, from horizontal underground stems.

Leaves: Egg- to slightly lance-shaped with blunt to short, pointed tip, 2–3 mm long, pleated; margins coarsely toothed near tip; midrib extends almost to tip; leaves of horizontal stems erect, pressed to stem, not pleated.

Brenda Chambers

Fruiting Bodies: Rare, oblong to cylindrical capsules, 1.5–3 mm long, erect; lid (operculum) beaked and split; stalks 1.8–4.5 cm long.

Habitat: Wet organic, moist sandy to fine loamy hardwood and eastern white cedar swamps.

Notes: You can recognize tree moss by its tree-like shape. Horizontal underground stems connect these small 'trees.'

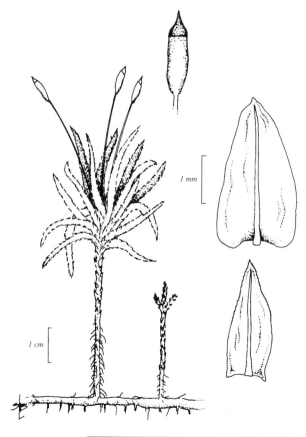

FERN MOSS FAMILY (THUIDIACEAE)

Karen Legasy

General: Robust, fern-like, lacy feathermoss, green to yellowish; stems spreading or ascending and arched, 3–8 cm long, branched 2–3 times; branches on both sides of stem, divided 2–3 times (bi- to tripinnate), resembling a fern leaf; forms mats.

Leaves: Stem leaves triangular to egg-shaped, pointed at tips, minutely toothed and rolled under on margins, 0.6–1.4 mm long, folded or pleated; branch leaves egg-shaped, pointed at tips, erect and spreading, tiny, less than 0.5 mm long, smaller on secondary branches; midrib 1/2 to 2/3 leaf length.

Fruiting Bodies: Cylindrical, curved capsules, 1.8–4 mm long, slightly erect to horizontal; lid (operculum) with a long (1–2 mm) beak; stalks smooth, reddish, 1.5–4.5 cm long.

Habitat: Wet organic hardwood swamps; moist, sandy to fine loamy upland tolerant hardwood stands, with yellow birch and eastern white cedar; on soil, humus, decayed wood or rocks.

Notes: You may encounter a similar species, fern moss (*Thuidium recognitum*), in our region. It is difficult to distinguish the 2 in the field, but you can distinguish common fern moss by its lacy appearance. Common fern moss may be confused with stair-step moss (p. 404), but the vertical stems of common fern moss do not give the image of steps.

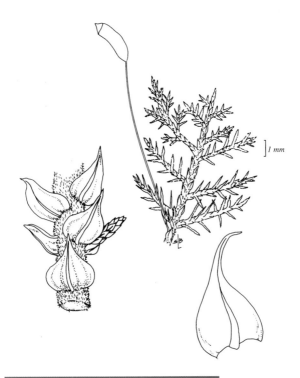

] *1 mm*

General: Robust, striking feathermoss, bright green to goldish, flat, feather-like; stems semi-erect, greenish, 3.5–11 cm tall, once pinnately branched; branches taper to top of stem, curved at tips; densely overlapping in patches to form shiny mats.

Brenda Chambers

Leaves: Stem leaves egg-shaped to triangular with long, pointed tips prominently curved and turned to 1 side, concave, with folds or pleats, 2–3 mm long, flat and minutely toothed toward tip along margins; branch leaves similar but 1.2–2 mm long; midrib double, short or absent.

Fruiting Bodies: Oblong to cylindrical capsules, curved and tapered at base, reddish-brown, dull, 2–3 mm long, horizontal; stalks glossy, reddish-brown, 25–45 mm long.

Habitat: Wet organic conifer swamps; fresh to moist upland black spruce-jack pine and intolerant hardwood mixedwood stands; on logs, stumps, rocks, and mineral soil, in association with Schreber's moss.

Notes: Its distinctive feather- or plume-like appearance gives this moss its common name.

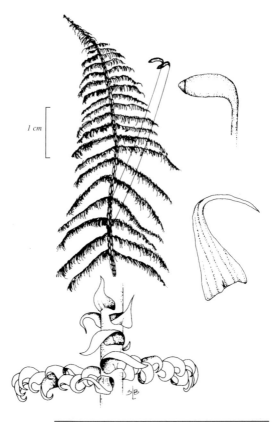

1 cm

HYLOCOMIUM FAMILY (HYLOCOMIACEAE)

General: Delicate feathermoss, yellowish to brownish-green; stems arched, wiry, reddish to brownish, covered with green scales, up to 15 cm long; branches further branched 2–3 times (bi- to tripinnate) to give a lacy appearance; new growth starts near centre of previous year's shoot, giving plant a step-like appearance, with annual, somewhat triangular, step-like layers; forms loose, silk-like mats.

Leaves: Stem leaves egg-shaped, sharply pointed and curled at tips, curled inward in a spoon-like fashion, with tiny teeth along margins (use hand lens), about 3 mm long, loosely overlapping; branch leaves similar but tightly overlapping, tips not curled, and about 1.2 mm long; midrib double, from less than 1/3 to nearly 1/2 of leaf length.

Fruiting Bodies: Egg-shaped capsules, brown to reddish-brown, 1.5–2.7 mm long, horizontal; lid (operculum) long-beaked; stalks smooth, reddish, 12–30 mm long, scattered along stem.

Habitat: Wet organic conifer and hardwood swamps; fresh to moist upland black spruce-pine and eastern white cedar mixedwoods; on soil, humus or decayed wood.

*Linda Kershaw, top;
Brenda Chambers, bottom*

Notes: You can recognize stair-step moss by its distinctive, step-like appearance.

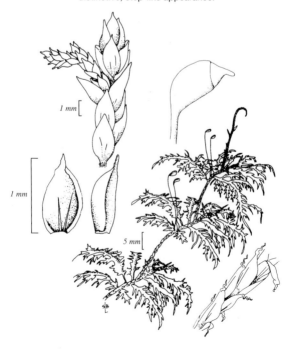

1 mm

1 mm

5 mm

HYLOCOMIUM FAMILY (HYLOCOMIACEAE)

General: Bright, glossy, robust feathermoss, light green to yellowish; stems ascending to erect from prostrate base, red, 7–16 cm tall, once pinnately branched, with a feather-like appearance; forms large, shiny mats.

Leaves: Stem leaves egg-shaped to widely oval, pointed to blunt or rounded at tips, curved inward to give a spoon-like appearance, with tiny teeth on margins at tip (use hand lens), about 1.5 mm long; branch leaves similar, about 1.3 mm long; midrib double, very short, not always apparent.

Karen Legasy

Fruiting Bodies: Cylindrical, curved capsules, brown to reddish-brown, 2–2.5 mm long; stalks glossy, reddish-brown, 20–43 mm long, scattered along stem.

Habitat: All moisture regimes, soil textures and stand types, although more typical of pine and black spruce-pine stands.

Notes: Schreber's moss has commonly been called 'big red stem moss' because of its red stems.

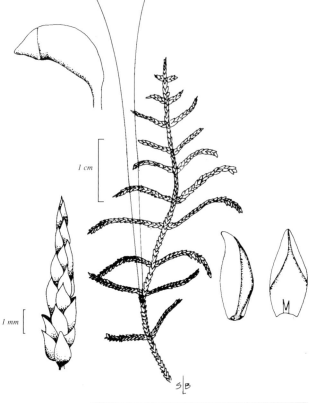

1 cm

1 mm

S|B

General: Robust, rigid moss, yellowish-green to golden-brown, usually glossy or shiny when dry; stems ascending, brown, woolly or felt-like, 5–15 cm tall, once pinnately branched; branches normally straight, horizontal; in dense to loose mats or tufts.

Leaves: Stem leaves slenderly triangular, tapering to long, pointed tip, toothless, 3–4 mm long, stiff, erect and spreading, occasionally turned toward 1 side, strongly folded or pleated lengthwise; branch leaves similar to stem leaves; midrib slender, difficult to see, 3/4 or more of leaf length.

Karen Legasy

Fruiting Bodies: Oblong to cylindrical capsules, curved, orangish-brown, 2–3 mm long, horizontal to slightly erect; stalks smooth, reddish, 2–5 cm long.

Habitat: Wet organic conifer swamps, fens.

Notes: This species is also known as fuzzy brown moss. It can be recognized by its yellowish-golden colour, its erect stems with their brown, woolly covering and its erect, pleated leaves, which are usually shiny when dry.

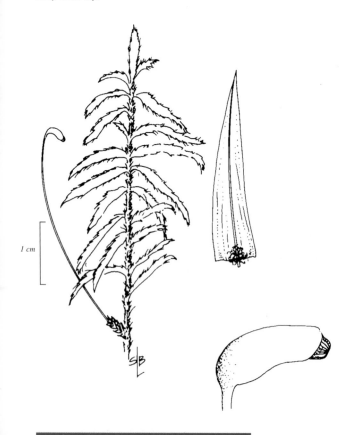

1 cm

RAGGED MOSS FAMILY (BRACYTHECIACEAE)

General: Slender to somewhat robust moss, green to whitish-green or yellowish-golden; stems curved or creeping, with ascending tips, irregularly pinnately branched; spreading, slender, runner-like stems often present; usually in flat, small to large, slightly shiny mats.

Leaves: Stem leaves egg- to lance-shaped, with long, pointed tips, often pleated, and finely toothed, erect or ascending, crowded; branch leaves similar, but usually shorter and narrower; single midrib usually ends below leaf tip.

Karen Legasy

Fruiting Bodies: Oblong to egg-shaped capsules, with wider end toward tip, relatively short and thick, curved, leaning to horizontal; stalks elongated, smooth or rough.

Habitat: Dry to moist, sandy to clayey upland sites; all stand types; on rock, soil, decaying wood and tree bases.

Notes: You can recognize ragged mosses by their greenish to yellowish colour and their strong single midribs, and by the flat, slightly shiny mats they form. The different species of this genus are difficult to distinguish in the field.

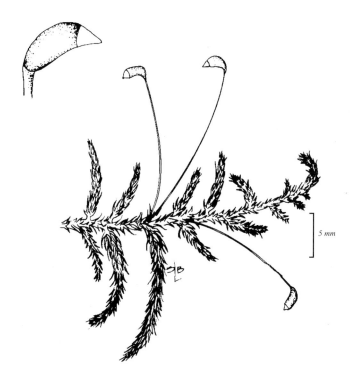

5 mm

HYPNUM FAMILY (HYPNACEAE)

Brenda Chambers

General: Moderately robust, bright green to yellowish or brownish moss; stems creeping, once-branched, with short, unequal, tapered, flat, straight or curved branches; forms flat, shiny, often extensive mats.

Leaves: Stem leaves egg-shaped, short- or long-pointed at tip, toothless, 1–1.7 mm long, often arranged slightly toward 1 side (especially near branch ends, which may appear flat), erect and spreading; branch leaves similar to stem leaves; midrib double, very short or absent.

Fruiting Bodies: Narrowly cylindrical, curved capsules, erect to inclined, 1.7–3 mm long; lid (operculum) long-beaked; stalks slender, erect, reddish to brownish, 16–32 mm long.

Habitat: All moisture regimes, soil textures and stand types; on logs, rocks and at base of trees.

Notes: This moss can be recognized by its short, flattened branches, which often resemble small swords, and by its erect, curved capsules. The appearance of its leaves does not change much, whether they are wet or dry. • This moss is commonly confused with the ragged mosses (p. 407), but the leaves of those species have midribs and are often toothed on their margins.

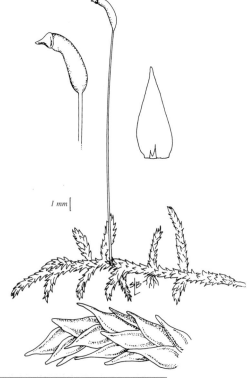

1 mm

HYPNUM FAMILY (HYPNACEAE)

General: Dark- to yellow-green moss; stems creeping, once pinnately branched, resembling a feather; forms dense, slightly shiny mats.

Leaves: Oblong to egg-shaped, pointed at tip, tiny, 0.6–1 mm long, crowded, erect and spreading, curved and turned to 1 side; margins minutely toothed; midrib short, double.

Fruiting Bodies: Cylindrical, curved capsules, 1.5–2 mm long, almost erect; stalks 6–13 mm long.

Habitat: Dry to moist upland tolerant hardwood and intolerant hardwood mixedwood stands; on tree bases and logs.

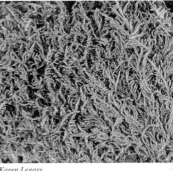

Karen Legasy

Notes: You can recognize stump pigtail moss by its very small size (which makes this a difficult genus to identify in the field) and by the way it usually grows on bark at the base of hardwood trees.

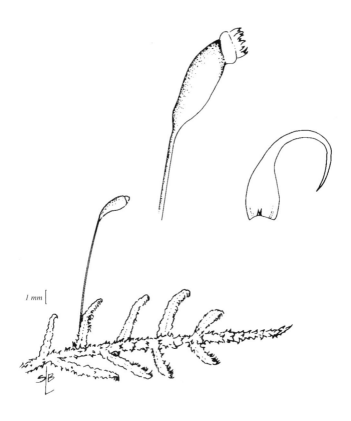

1 mm

AMBLYSTEGIUM FAMILY (AMBLYSTEGIACEAE)

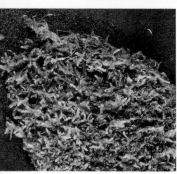

Karen Legasy

General: Slender, green, yellowish or brownish moss; stems creeping to crowded and ascending or erect, irregularly pinnately branched; forms loose, slightly shiny mats or tufts.

Leaves: Lance-shaped, long-pointed at tip, 2.5–4 mm long, strongly curved and turned toward same side (giving stem tip a hooked, sickle-shaped appearance), strongly folded or pleated lengthwise (especially when dry); margins remotely and minutely toothed toward tip; midrib single, extends to tip.

Fruiting Bodies: Oblong to cylindrical capsules, curved, usually horizontal, brown, 1.7–3 mm long; stalks twisted, red, 1.5–3.2 cm long.

Habitat: Wet organic conifer swamps; in many upland stand types; on decaying wood, humus, bark at tree bases and rocks.

Notes: Also known as *Drepanocladus uncinatus*. • You can recognize sickle moss by the way its hooked leaves curve or curl toward 1 side of the stem or branch. • Sickle moss was named for its curled leaves, which are said to resemble a sickle (a tool with a curved blade used for cutting tall grass, etc.).

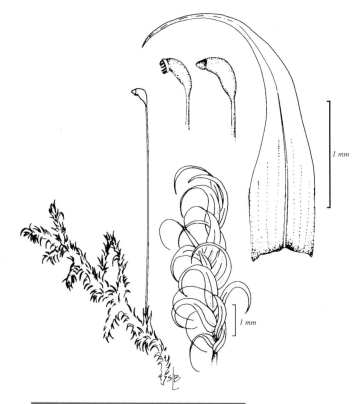

1 mm

1 mm

General: Very shiny, yellowish or golden green moss with its leaf tips strongly curved to one side; stems flat, pinnately branched; grows pressed to the ground in mats.

Leaves: Broad at base, tapered to a curved tip, 1–1.4 mm long, flattened into one plane, crowded, yet loosely overlapping; margins toothed at tip; midrib lacking.

Fruiting Bodies: Curved capsules, smooth, erect or inclined, brown, 1–1.5 mm long; stalk orange-brown, 7–17 mm long.

Habitat: Dry to moist, sandy to loamy upland tolerant hardwood and pine stands; in wet organic conifer and hardwood swamps; on logs, soil, humus, tree bark or exposed roots.

Notes: This shiny, mat-forming moss is quite distinctive and can only be matched in brightness by glossy moss (p. 412), which does not have curved leaf tips. • The name *recurvans* refers to the curved capsules and leaf tips.

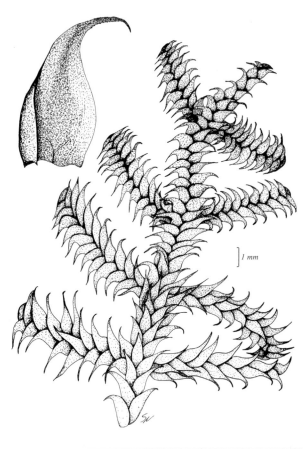

] *1 mm*

COTTON MOSS FAMILY (PLAGIOTHECIACEAE)

General: Moderately robust, glossy, yellow-green or gold; stems creeping, forming flat shiny mats.

Leaves: Oblong, pointed at tip, 1.2–1.5 mm long, separate to slightly overlapping, spreading, flattened together into one plane; margins sometimes rolled under; midrib short, double.

Fruiting Bodies: Symmetrical capsules, almost erect, pale yellow-brown, 1.5–2 mm long; stalk 8–16 mm long.

Habitat: All moisture regimes, soil types and stand types; on humus and on the base of trees.

Notes: Glossy moss resembles brotherella moss (p. 411), but that species has curved leaf tips. • Toothed cotton moss (*Plagiothecium denticulatum*) is very similar in shape, but it is larger and dark green, and is commonly found in swamps and sedge mats.

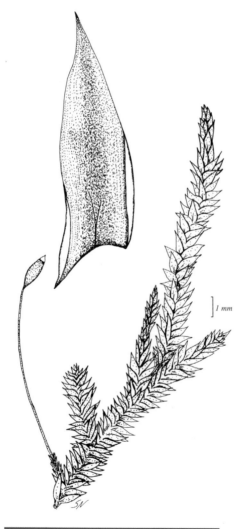

1 mm

NECKERA FAMILY (NECKERACEAE)

General: Light- to yellow-green moss, soft, shiny; stems creeping,
5–10 cm long, once to twice pinnately branched; branches horizontal
or somewhat hanging, feather-like, with numerous flat, wide, spreading
branches; forms mats.

Leaves: Oblong-lance-shaped to oblong-egg-shaped, pointed at
tip, 2–2.5 mm long, soft, strongly undulate, shiny; margins distinctly
wavy and minutely toothed to middle or below; midrib short and double
or lacking.

Fruiting Bodies: Oblong-egg-shaped capsules, brown, about 1.5 mm
long, completely covered by leaves; stalks about 1 mm long (shorter
than capsule).

Habitat: On tree trunks, logs or rocks.

Notes: You can recognize feathery neckera moss by its flat, outward-
spreading branches and its soft, shiny, wavy leaves. It eventually
becomes faded and shaggy. This moss is very common in older, moist
coniferous forest, where it can form an impressive cover over a large
portion of the trunks of live trees.

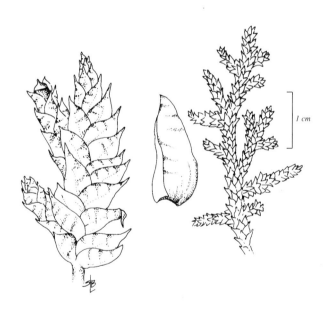

1 cm

Karen Legasy

General: Minute to fairly large moss; stems mainly erect, flattened, unbranched or sparsely branched, growing close together or tufted.

Leaves: Narrow, tongue- to lance-shaped, flattened together in 2 rows, crowded, split at the base, sheathing the stem below and the next leaf above; midrib single, usually well-developed, rarely absent, ends near tip or extends shortly beyond.

Fruiting Bodies: Symmetrical or curved capsules, often abruptly narrowed below mouth, smooth, erect to inclined; stalks terminal or lateral, elongated, straight or wavy, often abruptly bent at base.

Habitat: On soil, rocks, logs and tree bases; in wet areas, often submerged.

Notes: There are approximately 15 species of the genus *Fissidens* in Ontario. This genus is very difficult to identify to the species level in the field. • *Fissidens* means 'split tooth' and refers to the usually forked teeth at or near the capsule mouth.

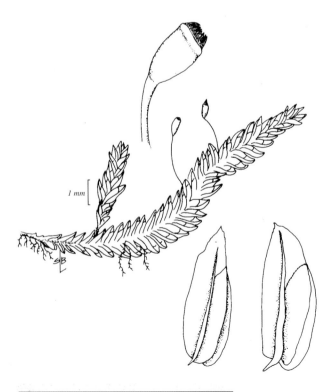

1 mm

Les lichens

LICHENS

This group has been divided into 4 subgroups based on differences in growth form. These are leaf lichens, club/cup lichens, shrub lichens, and hair (pendant) lichens. The following illustrations highlight identifying features of some common lichens, followed by a pictorial key to the lichen subgroups.

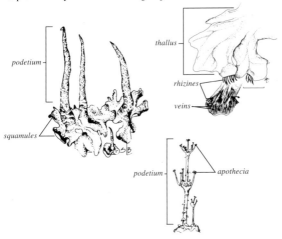

KEY TO LICHEN SUBGROUPS

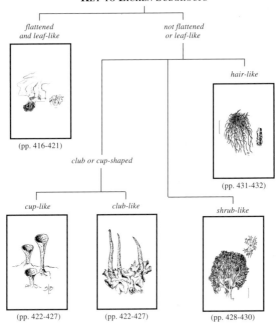

PELT LICHEN FAMILY (PELTIGERACEAE)

General: Prostrate to sub-erect leaf lichen; lobes range from wide, shallow and flat to small and semi-erect; upper surface green to greyish, bright-green when moist, with scattered, dark-green to dark-brown wart-like markings; underside greyish-brown to dirty white near margins, with indistinct veins, densely covered with matted, wool-like or cottony hairs; loosely attached to ground.

Fruiting Bodies: Large reddish or black structures (apothecia) on the upper surface of extended lobes.

Habitat: All moisture regimes, soil textures and stand types; among feathermosses such as Schreber's moss and stair-step moss, or on soil.

Notes: Spotted dog lichen can be recognized by its bright green colour (when wet) and its dark, wart-like markings.

Linda Kershaw

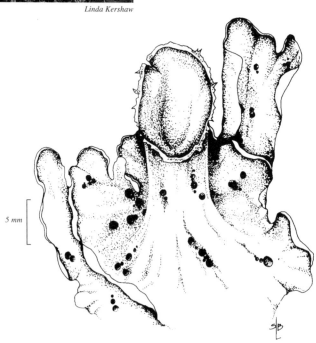

5 mm

PELT LICHEN FAMILY (PELTIGERACEAE)

General: Lobed leaf lichen up to 30 cm or more across; lobes up to 8 cm across; upper surface light brown to greyish brown with whitish tinge when dry, darker brown when wet, has fine hairs; underside tan to whitish, with long, slender, root-like rhizines; margins wavy, usually turned downward; pressed to ground.

Fruiting Bodies: Very common; conspicuous, distinct, brown-black, folded structures (apothecia), 6–10 mm across, on vertical tips of lobes; grey soredia found in circular patches on upper surface.

Habitat: Wet organic hardwood swamps; moist to dry intolerant hardwood stands; on rocks, rotten wood and tree bases.

Linda Kershaw

Notes: Dog's tooth lichen is distinguished by the distinct veins on the underside of its lobes, its distinctive, flaring, root-like rhizines, the grey felt on its upper surface, and its down-curved margins. • Dog's tooth lichen was once believed to be a cure for rabies, which is how it got its common name. See caution in Introduction.

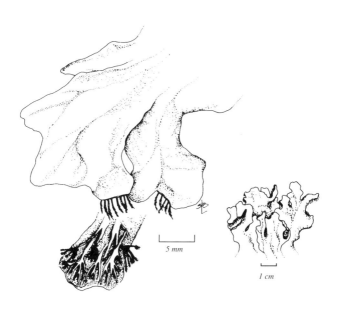

5 mm

1 cm

LUNGWORT FAMILY (LOBARIACEAE)

Brenda Chambers

General: Distinctive leaf lichen with pattern of ridges and hollows on surface, light brownish green, 5–15 cm across; lower surface pale tan, lightly covered with fuzzy hair; loosely attached to its substrate.

Fruiting Bodies: Apothecia rare; isidia and/or powdery soredia commonly found along margins and on ridges.

Habitat: Fresh tolerant hardwood stands; on the trunks of large sugar maple trees; uncommon.

Notes: This species was widely used in the Middle Ages to treat lung diseases because of its strong resemblance to lung tissue. It is extremely sensitive to air pollution, as are many lichens.

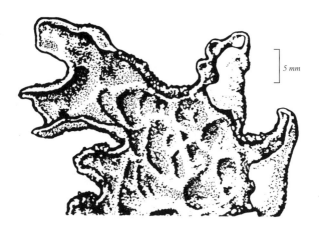

5 mm

ROCK TRIPE FAMILY (UMBILICARIACEAE)

General: Circular leaf lichen, up to 30 cm in diameter, leathery when wet, brittle when dry, flat to undulating, with smooth margins; upper surface brown to green when fresh; lower surface black with numerous hair-like, ball-tipped rhizines; attached to the substrate by a central stalk (stipe).

Fruiting Bodies: Rare, apothecia, convex, 4 mm wide.

Habitat: Shaded rock cliffs, boulders of granite or sandstone; often found along rocky lakeshores.

Brenda Chambers

Notes: Rock tripe was used by the aboriginal people and the *coureurs de bois* in soups and stews. After soaking, it was cooked to a mush and used as a thickener for soups and stews. • It is not recommended that any lichens be eaten due to the **toxic pollutants** accumulated in their thalli. See caution in introduction.

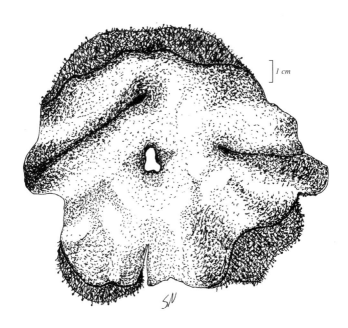

1 cm

Robin Bovey

General: Leaf lichen with a distinctive net-like pattern of white cracks on the pale grey surface, 3–10 cm broad; lobes 1–6 mm wide; lower surface black, with dense, branched, root-like rhizines.

Fruiting Bodies: Apothecia uncommon; powdery soredia very common, along cracks on upper surface.

Habitat: Wet organic hardwood and conifer swamps; dry to moist, sandy to fine loamy upland tolerant hardwood, intolerant hardwood mixedwood and pine stands; on trees, logs and occasionally rock.

Notes: This species is also known as 'cracked lichen.' It often covers large portions of trees. The narrow lobes and soredia-filled cracks are distinguishing features of waxpaper lichen. • Waxpaper lichen can be confused with the less common *Myelochroa aurulenta* (previously called *Parmelina aurulenta*), which has circular spots of powdery soredia instead of soredia-filled cracks. • *Parmelia* spp. are used by ruby-throated hummingbirds and wood pewees in nest construction. • See notes for monk's hood lichen (p. 421).

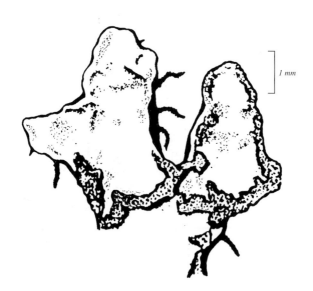

1 mm

PARMELIA FAMILY (PARMELIACEAE)

General: Lobed leaf lichen 6–12 cm across; lobes 1–2 mm wide, hollow, burst at the tips revealing powdery clumps of soredia; upper surface pale grey-blue or greenish, wrinkled; lower surface black, lacks root-like rhizines.

Fruiting Bodies: Apothecia uncommon; powdery soredia very common, whitish grey-green, easily seen inside bursting tube-like lobes.

Habitat: Wet organic conifer swamps; dry to moist, sandy to loamy upland tolerant hardwood, intolerant hardwood, pine and black spruce-pine stands; on twigs or bark of trees; amongst moss and occasionally on rocks or soil.

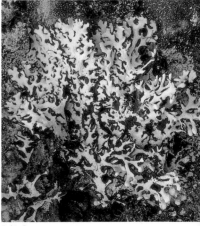
Robin Bovey

Notes: The raised lobe tips, split with copious powdery soredia, are a key identifying feature. • Monk's hood lichen could be confused with the very common waxpaper lichen (p. 420), but the lobes of that lichen are flattened and solid (not rounded and hollow), with white cracks in their grey surface. • This species is one of the few lichens that is pollution-tolerant. It can be found in metropolitan landscapes.

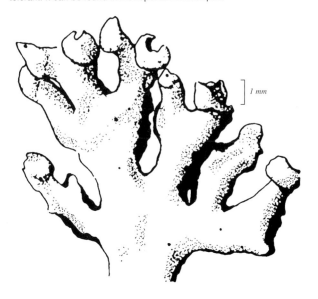
1 mm

Robin Bovey

General: Erect cup lichen with narrow, hollow, cup-like growths (podetia); cups narrow, whitish mineral-grey, 1–4 cm tall, sparingly branched in upper parts, jagged, inrolled and often bearing secondary cups at margins; leaf-like lobes at base (squamules) 1–3 mm long, 1 mm thick, pale- to olive-green or brownish above, white beneath, irregularly lobed, ascending, flat or curved inward, scattered or in tufts, scalloped along margins.

Fruiting Bodies: Apothecia small, brown, around cup margins; soredia fine-powdery, on cups (podetia).

Habitat: Soil or humus, rotten stumps and wood.

Notes: You can recognize powdered funnel cladonia in the field by the inrolled margins of its cups, and by its hollow stalks and powdery (sorediate) mostly unbranched clubs.

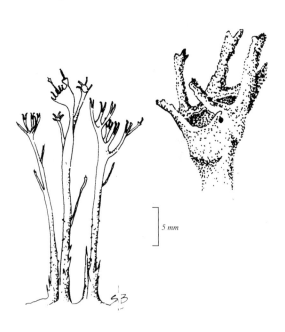

5 mm

CLADONIA FAMILY (CLADONIACEAE)

General: Erect cup lichen with hollow, cup-like growths (podetia); cups goblet-shaped, greenish mineral-grey, 0.5–1.5 cm tall, unbranched, round-toothed and more or less curled inward on margins; leaf-like lobes at base (squamules) persist or disappear, 4–7 mm long and almost as wide, whitish or olive-whitish and dull to rarely shiny above, white and darkened toward base beneath, ascending, scalloped on margins.

Fruiting Bodies: Apothecia brown, along cup margins; pycnidia found *Karen Legasy* either on cup margins or on primary, basal, leafy lobes (squamules); powdery soredia abundant on cups (podetia).

Habitat: All moisture regimes, soil types and stand types; on soil, rocks, tree bases, rotting wood, and amongst mosses.

Notes: False pixie cup can be recognized by its goblet-shaped cups, copious powdery soredia, and brown apothecia. It is our most commonly collected cup lichen. • Fringed pixie cup (*Cladonia fimbriata*) is similar to false pixie cup, but its cups are taller and narrower (trombone-shaped).

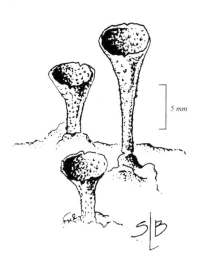

5 mm

General: Club lichen with hollow, awl-shaped clubs (lacking cups), lacking powdery soredia; clubs weakly branched in upper parts, 1–2 cm tall, yellowish or grey; small lobes (squamules) at base of plant.

Fruiting Bodies: Apothecia and pycnidia both very common, red, and at club tips.

Habitat: Dry to fresh, sandy upland black spruce-pine and pine stands; on soil, humus, rotting logs, stumps and rocky open areas.

Notes: British soldiers is easily recognized, and thus is one of the first lichens learned by nature enthusiasts. There are other club-shaped species of *Cladonia* in our region with red fruiting structures, but they have powdery clumps of soredia, or cups are present • Another common lichen, red pixie cup (*Cladonia borealis*) is often mistaken for British soldiers, but that lichen has red apothecia along the rim of its cups, whereas British soldiers has no cups.

Brenda Chambers

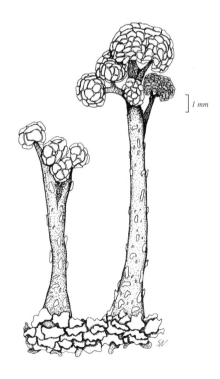

] *1 mm*

CLADONIA FAMILY (CLADONIACEAE)

General: Club lichen with pointed, club-like growths (podetia), or occasionally with small, narrow, irregular cups; clubs whitish-green, 0.5–3 cm tall, unbranched; large leaf-like lobes at base (squamules) 2–5 mm long, thick, dense, rounded or hollowed, olive-green or powdery-white to brownish above, white beneath, toothless or round-toothed on margins.

Fruiting Bodies: Apothecia uncommon, produced at tips of clubs or on cup rims; clubs and cups covered with powdery soredia.

Karen Legasy

Habitat: All moisture regimes, soil types and stand types; on humus, rotting wood, tree bases; over rocks in moist areas.

Notes: This is a very common pointed club lichen throughout this region. However, several other, very similar species are also found in this area. These include yellow toothpick lichen (*Cladonia bacilliformis*), which has yellow clubs (podetia), scarlet toothpick lichen (*Cladonia bacillaris*), which has red apothecia, and horn lichen (*Cladonia cornuta*), which has powdery soredia in round patches on the upper end of its clubs (podetia).

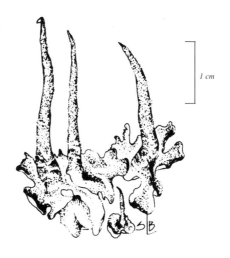

1 cm

Derek Johnson

General: Club/cup lichen with hollow, cup-like growths (podetia); cups variable, with sieve-like openings in interior, greenish to brownish mineral-grey, smooth (lacking soredia), 3–5 cm tall, often with secondary ranks of cups or branched clubs; leaf-like lobes at base (squamules) 1–4 mm long, up to 0.5 mm wide, olive-green or brownish above, white and darkened toward base beneath, ascending, flat to curved inward.

Fruiting Bodies: Apothecia at branch tips, rare, dark brown; pycnidia more common, on margins of cups; powdery soredia absent.

Habitat: On soil, along roadsides, among mosses and on rotting logs in moist areas.

Notes: You can recognize perforated cladonia by the sieve-like openings on the interior surfaces of its cup. However, when these distinctive cups are absent, perforated cladonia is easily confused with fork lichen (*Cladonia furcata*) or slender cup lichen (*Cladonia gracilis*). Without cups, close inspection and chemical tests are needed to separate these species.

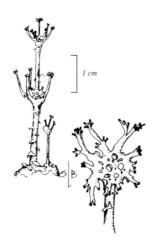

1 cm

CLADONIA FAMILY (CLADONIACEAE)

General: Cup lichen, short, wide, goblet-shaped, greenish, 1–2 cm tall; cup margins smooth to toothed or bearing small clubs (rarely small cups); leaf-like lobes at base (squamules) 1–7 mm long, up to 5 mm wide, persistent or disappearing, irregularly toothed or lobed on margins.

Fruiting Bodies: Apothecia on cup margins or at tips of stout clubs on cup margins, red; pycnidia usually on cup margins; soredia granular, on upper and inside suface of cups.

Karen Legasy

Habitat: Wet organic conifer and hardwood swamps; dry to moist, rocky to loamy upland pine and black spruce-pine stands.

Notes: You can recognize red-fruited goblet lichen by the deep red reproductive structures (apothecia) on its cup margins and by its short, wide cups. • Red-fruited goblet lichen could be confused with the taller, false pixie cup (p. 423), but the apothecia of that species are brown. Red pixie cup (*Cladonia borealis*) is also very similar, but it does not have powdery soredia.

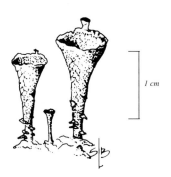

1 cm

Linda Kershaw

General: Upright shrub lichen with distinctive coral- or cauliflower-like, compact heads, pale greenish or yellowish-grey, 6–10 cm tall; branches interwoven, lack main stem; in separate or loosely clumped colonies.

Fruiting Bodies: Apothecia small, at branch tips, brown, rare; pycnidia more common, at branch tips; powdery soredia not usually present.

Habitat: Dry to fresh, rocky to sandy upland pine and black spruce-pine stands.

Notes: Coral lichen can be recognized by its lack of a main stem and its multiple branching, which give it a distinctive cauliflower shape. The branches end in star-shaped whorls of 4–6 tiny branches surrounding a central hole. The coral- or cauliflower-like heads of coral lichen distinguish it from yellow-green lichen (*Cladina mitis*) and reindeer lichen (p. 429). These species both have distinctive main stems, which coral lichen lacks. • Coral lichen is often used to make trees and shrubs in model train-set layouts.

1 cm

CLADONIA FAMILY (CLADONIACEAE)

General: Ascending shrub lichen with a distinct, round, stem with numerous branches and a fibrous, ash-grey surface, 6–10 cm tall, usually branching in 4s; scattered but often forms extensive colonies.

Fruiting Bodies: Apothecia dark brown, at branch tips, rare; pycnidia at branch tips, common.

Habitat: Wet organic conifer swamps; dry to moist, rocky to loamy pine, black spruce-pine and intolerant hardwood mixedwood stands.

Karen Legasy

Notes: Yellow-green lichen (*Cladina mitis*) is very similar in appearance, but it is greenish-yellow, whereas reindeer lichen is ash-grey. Reindeer lichen can also be confused with coral lichen (p. 430), but reindeer lichen has a main stem and does not form distinctive rounded, coral- or cauliflower-like heads like those of coral lichen.

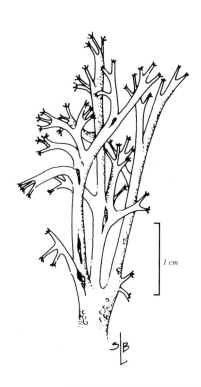

1 cm

CORAL LICHEN FAMILY (STEREOCAULACEAE)

Brenda Chambers

General: Prostrate to sub-erect shrub lichen, whitish mineral-grey, 4–8 cm tall; branches with thick woolly coating, numerous, short, and curved under near top, few at base; underside with thick hairs.

Fruiting Bodies: Apothecia at branch tips, numerous, dark brown; cephalodia small, in black tufts, inconspicuous, hidden in dense hairs on underside of branches.

Habitat: On soil over rocks, on humus and in open places.

Notes: Foam lichen can be recognized by its foam-like appearance and mineral-grey colour. • This species could be confused with rock foam lichen (*Stereocaulon saxitile*), which grows firmly attached to rocks, or with common foam lichen (*Stereocaulon paschale*), which has a sparse woolly coating and distinct, dark brown cephalodia.

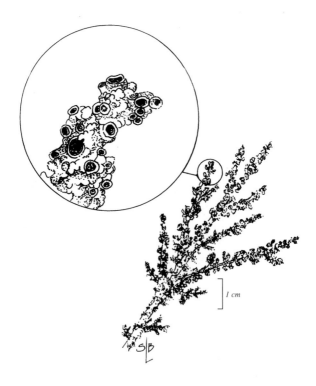

1 cm

OLD MAN'S BEARD FAMILY (USNEACEAE)

General: Erect or hanging hair lichens, usually attached to tree branches by a holdfast, greenish-yellow, 4–40 cm long (depending on species), usually have a central stem with longer or shorter lateral branches projecting; main stem often dark at base, minutely warty, knobby or powdery; branches lack the inner cartilaginous strand found in the main stem; surface of main branches sometimes pitted with grooves or depressions.

Fruiting Bodies: Apothecia lateral or terminal, rare; isidia and powdery soredia abundant on some species.

Habitat: On coniferous or deciduous trees and shrubs (arboreal).

Karen Legasy

Notes: Old man's beard can often be seen hanging from tree branches. These lichens can be recognized by their main central stems, with side branches, and by the strong, cartilaginous strand at the centre of their main stems. You can expose this resilient strand by carefully pulling the stem lengthwise between your finger and thumbnail.

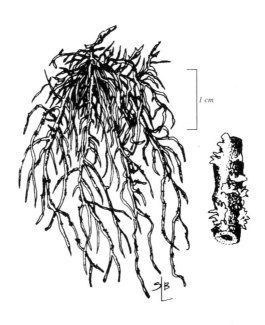

1 cm

OLD MAN'S BEARD FAMILY (USNEACEAE)

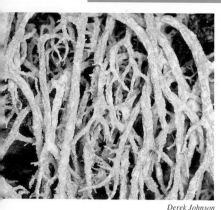

Derek Johnson

General: Hanging shrub lichen, with a distinctive wrinkled appearance, pale yellowish, 7–8 cm long, very soft and pliable to touch; branches somewhat flattened and angular with many sub-branches; powdery clumps of cells (soredia) on surface.

Fruiting Bodies: Apothecia very rare; soredia granular, very abundant.

Habitat: Wet organic conifer swamps; dry to fresh, sandy to coarse loamy upland black spruce-pine stands; on bark and twigs of trees and shrubs.

Notes: Spruce-moss is easy to recognize, and can be found quite easily on spruce, tamarack or birch trees. This lichen differs from other hanging lichen species because it has branches that are dull, angular or flat, pliable and powdery (use hand lens).

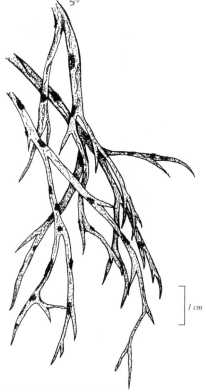

1 cm

Use of Forest Plants
in this Guide as Food by Wildlife

An estimated 300 terrestrial vertebrates spend all or part of the year in central Ontario. Of these, 248 species live in forest or riparian habitats (37 amphibian and reptiles, 54 mammals and 157 birds). Food habit studies are difficult and time consuming to conduct and complete information for all species does not exist. Some uses of plants as food are presented for the following birds (55 species) and mammals (25 species).

Exclusion of a species does not necessarily mean that that species does not eat plants in this region. It may simply mean we do not have sufficient information on the diet of that particular species. Similarly, some of the animals included in the text may eat more species of plants than are indicated. The use of plants as food by wildlife is often presented for a genus [e.g. maples (*Acer* species), blueberries (*Vaccinium* species)].

Food habit studies often report that a particular animal feeds on a type of plant (e.g. a maple or blueberry), but do not indicate which species of the genus is being eaten. Some of the species in a genus may be used and others may not. The species is indicated whenever specific information was available.

Common and Scientific Names
of Mammals Referred to in Text

Virginia opossum	*Didelphis virginiana*
Eastern cottontail	*Sylvilagus floridanus*
Snowshoe hare	*Lepus americanus*
Eastern chipmunk	*Tamias striatus*
Least chipmunk	*Eutamias minimus*
Gray squirrel	*Sciurus carolinensis*
Red squirrel	*Tamiasciurus hudsonicus*
Flying squirrels	*Glaucomys spp.*
Beaver	*Castor canadensis*
Deer mouse	*Peromyscus maniculatus*
White-footed mouse	*Peromyscus leucopus*
Red-backed vole	*Clethrionomys gapperi*
Heather vole	*Phenacomys intermedius*
Rock vole	*Microtus chrotorrhinus*
Meadow jumping mouse	*Zapus hudsonicus*
Woodland jumping mouse	*Napeozapus insignis*
Porcupine	*Erethizon dorsatum*
Red fox	*Vulpes vulpes*
Black bear	*Ursus americanus*
Raccoon	*Procyon lotor*
American marten	*Martes americana*
Fisher	*Martes pennanti*
Striped skunk	*Mephitis mephitis*
White-tailed deer	*Odocoileus virginianus*
Moose	*Alces alces*

Common and Scientific Names of Birds Referred to in Text

Ring-necked pheasant	*Phasianus colchicus*
Spruce grouse	*Dendragapus canadensis*
Ruffed grouse	*Bonasa umbellus*
Wild turkey	*Meleagris gallopavo*
American woodcock	*Scolopax minor*
Red-headed woodpecker	*Melanerpes erythrocephalus*
Red-bellied woodpecker	*Melanerpes carolinus*
Yellow-bellied sapsucker	*Sphyrapicus varius*
Downy woodpecker	*Picoides pubescens*
Hairy woodpecker	*Picoides villosus*
Black-backed woodpecker	*Picoides articus*
Northern flicker	*Colaptes auratus*
Pileated woodpecker	*Dryocopus pileatus*
Eastern phoebe	*Sayornis phoebe*
Great Crested flycatcher	*Myiarchus crinitus*
Eastern kingbird	*Tyrannus tyrannus*
Tree swallow	*Tachycineta bicolor*
Blue jay	*Cyanocitta cristata*
American crow	*Corvus brachyrhynchos*
Black-capped chickadee	*Parus atricapillus*
Red-breasted nuthatch	*Sitta canadensis*
White-breasted nuthatch	*Sitta carolinensis*
Brown creeper	*Certhia americana*
Veery	*Catharus fuscescens*
Swainson's thrush	*Catharus ustulatus*
Hermit thrush	*Catharus guttatus*
Wood thrush	*Hylocichla mustelina*
American robin	*Turdus migratorius*
Gray catbird	*Dumetella carolinensis*
Brown thrasher	*Toxostoma rufum*
Cedar waxwing	*Bombycilla cedrorum*
European starling	*Sturnus vulgaris*
Yellow-rumped warbler	*Dendroica coronata*
Pine warbler	*Dendroica pinus*
Northern cardinal	*Cardinalis cardinalis*
Rose-breasted grosbeak	*Pheuticus ludovicianus*
Indigo bunting	*Passerina cyanea*
Rufous-sided towhee	*Pipilo erythrophthalmus*
Clay-coloured sparrow	*Spizella pallida*
Song sparrow	*Melospiza melodia*
Swamp sparrow	*Melospiza georgiana*
White-throated sparrow	*Zonotrichia albicollis*
Dark-eyed junco	*Junco hyemalis*
Rusty blackbird	*Euphagus carolinus*
Common grackle	*Quiscalus quiscula*
Pine siskin	*Carduelis pinus*
American goldfinch	*Carduelis tristis*
Evening grosbeak	*Coccothraustes vespertinus*
Orchard oriole	*Icterus spurius*
Northern oriole	*Icterus galbula*
Pine grosbeak	*Pinicola enucleator*
Purple finch	*Carpodacus purpureus*
House finch	*Carpodacus mexicanus*
Red crossbill	*Loxia curvirostra*
White-winged crossbill	*Loxia leucoptera*

GLOSSARY

achene: a small, dry, hard, single-seeded fruit that does not open

annual: having a life cycle (from seed to maturity to seed) of one year

apothecium (apothecia): a cup-shaped structure of most lichens in which the disc-shaped or elongated fruiting body is found

ascending: growing on an upward slant

awn: a slender bristle, usually terminal

berry (berries): a fleshy fruit with a pulpy interior, usually containing several seeds

biennial: having a life cycle (from seed to maturity to seed) of 2 years

blade: the flat, expanded part of a leaf

bloom: a whitish, powdery coating that is often waxy

bog: an open or sparsely treed wetland habitat poor in mineral nutrients (supplied primarily by precipitation) and characteristically acidic

bract: a reduced or modified leaf, usually at the base of a flower or inflorescence

branchlet: a small branch, usually growth of the most recent year

callus: a hard swelling out or bulging from the surrounding surface; the firm, thickened base of a lemma

calyx (calyxes): a whorl of sepals, at first enclosing the flower bud

capsule: a dry fruit, usually containing 2 or more seeds and splitting into sections at maturity; the spore-bearing structure of a moss or liverwort

catkin: a dense spike of small male or female flowers without petals

cephalodium (cephalodia): a warty outgrowth with blue-green algal cells on or within a lichen

clasping: partially surrounding another structure, usually referring to leaf bases around stems

conifer mixedwoods: mixedwood stands dominated by conifer species, including black spruce, white spruce, balsam fir, larch, white cedar or jack pine

conifer swamps: wetland forests of black spruce, larch, balsam fir, eastern white cedar, white spruce and/or red spruce, often with ground cover of peat mosses

corolla: a whorl of petals

cuspidate: with a sharp, firm point at the tip

cutover: an area that has been logged

deciduous: shedding leaves at maturity or at the end of the growing season

decurrent: extending down the stem, usually referring to the base of a leaf that extends down the stem giving the stem a winged appearance

disc flower: a floret with a tubular corolla in a composite flower head of the Aster family

diuretic: a substance that promotes the excretion of urine

drupe: a fleshy or pulpy, 1-seeded fruit in which the seed has a hard, stony covering

ephemeral: short-lived

evergreen: with leaves that stay on the plant throughout the year

fen: an open or sparsely treed wetland habitat, richer in minerals than a bog due to groundwater and surface input of nutrients and ranging from acidic to alkaline

fibrous: made up of or looking like fibres

follicle: a dry fruit developed from a single ovary and splitting open along 1 side only

free end: the unattached end

frond: the leaf of a fern; in mosses, a stem which is closely and regularly branched on 1 plane, thus resembling a fern leaf

gemma (gemmae): a small, asexual reproductive structure of a moss or liverwort that becomes detached from its parent and develops into a new individual

gland: a small structure that secretes a substance or substances such as oil or nectar

glume: a bract at the base of a grass spikelet, usually in a pair

growing degree days: cumulative heat units (in Celsius) above daily mean temperature of 5° Celsius, over the entire growing season

habit: the general appearance of a plant

hardwood mixedwood: a mixedwood stand dominated by trembling aspen, white birch, balsam poplar or black ash

hardwood swamp: a wetland forest of black ash, white elm, poplars, red maple, yellow birch and/or eastern white cedar.

herb: a plant without woody above-ground parts, its stems dying back to the ground each year

GLOSSARY

herbaceous: herb-like

hummock: a mound produced by the build-up of peat mosses, sedges or grasses, often creating drier elevated areas in a bog or swamp

humus: dark-brown to black, decomposed plant remains in the organic layer of a soil; the origin or structure of the original plants is, for the most part, impossible to identify

hypanthium (hypanthia): a cup- or saucer-shaped organ around an ovary and bearing the sepals, petals and/or stamens

inflorescence: a cluster of flowers in various arrangements, including spikes and panicles

internode: the section of stem between 2 adjacent nodes

intolerant: inability of a tree or plant to develop and grow in the shade of (and in competition with) other trees and plants

intolerant hardwood mixedwood: a stand that typically contains a mixture of poplars, white birch, white spruce and balsam fir. Pines, eastern white cedar and tolerant hardwoods may also be present.

involucre: a set of bracts beneath an inflorescence

isidium (isidia): a tiny wart-like outgrowth on a lichen, containing algal cells and serving in vegetative reproduction

lateral: on or from the side of an organ

leaflet: one of the divisions of a compound leaf

lemma: the lower of the 2 bracts immediately enclosing a single grass flower

lens-shaped: 2-sided and convex

lenticel: a raised, cork-like marking or spot on young bark

ligule: in grasses, the thin outgrowth from the inner surface of a leaf at the junction of the sheath and blade

lip: a projection of expansion of a structure, commonly referring to the lower petal of a flower such as an orchid or violet

litter: the top layer of organic matter (e.g. leaves and twigs) on the forest floor

lobe: a partial division of an organ such as a leaf

marsh: an open wetland habitat with submerged and floating grass-like vegetation, with up to 2 m of water (fluctuates seasonally), and typically neutral with high oxygen levels

membranous: thin, soft, pliable and somewhat semitransparent

moisture regime: an 11-point mumeric scale of available water in soil, ranging from theta (very dry) to 9 (very wet)

mucro: a small, short, slender, abrupt tip

node: the point on a stem at which a leaf, bud or branch arises; in grasses there is a noticeable swelling or 'joint' at each node

nutlet: a small nut

obscurely toothed: having teeth that are difficult to see without magnification

operculum (opercula): the lid of a moss capsule

organic: the soil type in which an accumulation of decaying vegetation (peat moss, sedges) occurs in a water-saturated environment.

ovary (ovaries): a structure that encloses young, undeveloped seeds

ovoid: shaped like an egg

overtop: to be taller than

panicle: a branched inflorescence

papilla (papillae): a minute, wart-like swelling or bulge

perennial: growing for 3 or more years, usually flowering and producing fruit each year

perianth: in flowers, the sepals and petals of a flower, collectively; in liverworts, a tube of 2–3 fused leaves, surrounding a developing sporophyte

peristome: the fringe of teeth at the opening of a moss capsule

perigynium (perigynia): a membranous sac that encloses the seed of a sedge

persist: to remain attached after normal function has been completed

petal: a member of the inside ring of modified flower leaves (the corolla), usually white or brightly coloured

pinna (pinnae): the primary division of a pinnately compound leaf

pinnate: divided with segments arranged on 2 sides of a central axis, feather-like (see leaf structure in pictorial glossary)

pinnule: the secondary or smallest division of a pinnately compound leaf

pistil: the female organ of a flower, usually consisting of a stigma, style and ovary

pith: the soft, spongy tissue at the centre of some stems and branches

pod: a dry fruit, especially of the pea family

podetium (podetia): a stalk-like, hollow structure on a lichen, that supports apothecia

pome: a fruit with a core, such as an apple

prickle: a small, slender, sharp outgrowth

proliferation: a small outgrowth or offshoot

pycnidium (pycnidia): a small, flask-shaped reproductive structure on a lichen, in which spore-like cells are produced

reticulate: having a network of veins

rhizine: a thread-like fungal hypha that attaches a lichen to its substrate

rhizoid: a thread-like, branched or unbranched growth on a moss or liverwort, that serves as an anchor or is used for absorption

rhizome: a horizontal underground stem, distinguished from a root by the presence of nodes, buds and/or scale-like leaves

rosette: a circular cluster of leaves, usually at the base of a plant

runner: a very slender, wiry stolon

samara: a winged fruit that does not open

saprophyte: a plant, usually without green colour, that derives its food from dead organic matter

scabrous: rough

scale: a small, thin or flat structure; in cones of conifer trees the scales are woody and enclose the seeds

scape: a leafless flowering stem

scarious: having a thin, dry, membranous texture

sepal: a member of the outside ring of modified flower leaves (the calyx), usually green

sheath: a tubular envelope surrounding another organ (e.g. the lower part of some grass and sedge leaves surround the stem)

sinus: the space or cleft between 2 lobes

soil texture: the proportion of sand, silt and clay in a mineral soil. Soil texture classes generally represent a gradient of decreasing sand and increasing clay content (sandy, coarse loamy, silty, fine loamy, clayey).

soredium (soredia): a powdery clump of algal cells and fungal hyphae on a lichen, serves in vegetative reproduction

sorus (sori): a cluster of small spore cases (sporangia) on the underside of a fern leaf, often circular or horseshoe-shaped

spike: a narrow, elongated flower cluster with flowers that are stalkless or nearly so

spikelet: a small or secondary spike; the smallest flower cluster of a grass or sedge inflorescence

spindle-shaped: long and thin

sporangium (sporangia): a sac or case in which spores are produced

sporophyte: the spore-bearing part or phase of a plant

spur: a slender, tubular or sac-like, usually hollow projection, often on a flower

squamule: a small, scale-like lobe of a lichen

stalk: the stem of a leaf, flower or capsule

stigma: the surface in a flower where the pollen lands, usually at the tip of the female organ

stipe: a stalk-like support; the stalk of a fern leaf from the base of the blade to the point of attachment to the rhizome

stipule: a small, leaf-like growth at the base of a leafstalk

stolon: a creeping, horizontal branch or stem that produces new shoots from the base of a plant

style: the stalk of a pistil, connecting the ovary and the stigma

sub-: a prefix meaning 'slightly'

swamp: a thicket or wooded wetland with standing to gently flowing water (seasonally fluctuating) and strong sub-surface flow, rich in mineral nutrients, neutral to moderately acidic, and with little development of peat

taproot: a primary, downward-growing root

tendril: a slender, twining outgrowth from a stem or leaf, used for climbing or support

thallose: having a thallus

thallus (thalli): a plant body that is not differentiated into stems and leaves, common in lichens and also in some liverworts

thicket: a dense patch of shrubs or small trees

tolerant: ability of a tree or plant to develop and grow in the shade of (and in competition with) other trees and plants. Tree species in this group exhibit a range of tolerance

tolerant hardwoods: stands with tree species including sugar maple, yellow birch, hemlock (a conifer included in this group) and other hardwoods (red oak, basswood, black cherry, white ash, etc.)

truncate: having a blunt or abrupt base or tip that is nearly straight across

tuber: a short, thickened underground stem

tubercle: a small swelling or projection

unarmed: without thorns or prickles

undulate: with a wavy appearance

whorl: a group of 3 or more similar organs radiating around a node (see pictorial glossary)

winged: with a flattened extension from the side, referring to leaf stalks and to some types of fruits

PICTORIAL GLOSSARY

Leaf Arrangement

alternate

opposite

whorled

basal

Leaf Margin

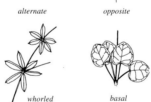

untoothed   *toothed*   *double-toothed*

round toothed

lobed

Leaf Structure

simple *compound (pinnate)* *compound (trifoliate)*

pinnatifid *pinnate-pinnatifid* *bipinnate* *tripinnate*

Leaf Structure

egg-shaped *inversely egg-shaped* *lance-shaped* *inversely lance-shaped* *elliptic* *oval*

linear *oblong* *heart-shaped* *round* *kidney-shaped*

Allen, A.W., P.A. Jordan and J.W. Terrell. 1897. *Habitat Suitability Index Models: Moose, Lake Superior Region*. U. S. Fish & Wildlife Service. Biological Report 82. 47 pp.

Andrus, R.E. 1980. *Sphagnaceae (Peat Moss Family) of New York State*. University of the State of New York, Albany. 89 pp.

Angier, B. 1978. *Field Guide to Medicinal Wild Plants*. Stackpole Books, Harrisburg, Pa. 320 pp.

Angier, B. 1974. *Field Guide to Edible Wild Plants*. Stackpole Books, Harrisburg, Pa. 256 pp.

Anon. 1974. *Seeds of Woody Plants in the United States*. U.S.D.A., Forestry Service, Agric. Handbook 450. 416 pp.

Argus, G.W., and D.J. White. 1977. The rare vascular plants of Ontario. *Syllogeus* no. 14.1-63/1-66.

Bailey, L.H. 1963. *How Plants Get Their Names*. Dover Publications Inc., New York.

Baldwin, D.A., and R.A. Sims. 1989. *Field Guide to the Common Plants in Northwestern Ontario*. Forestry Canada - Ontario Region, Sault Ste. Marie, and Ontario Ministry of Natural Resources, Toronto. 344 pp.

Baldwin, W.K.W. 1958. *Plants of the Clay Belt of Northern Ontario and Quebec*. National Museum of Canada Bulletin No. 156, Ottawa. 304 pp.

Banfield, A.W.F. 1974. *The Mammals of Canada*. National Museum of Natural Sciences, University of Toronto Press, Toronto. 438 pp.

Bell, F.W. 1991. *Critical Silvics of Conifer Crop Species and Selected Competitive Vegetation in Northwestern Ontario*. Forestry Canada, Ontario Ministry of Natural Resources, COFRDA Rep. 3310/NWOFTDU Tech. Rep. 19. 177 pp.

Bentley, C.V., and F. Pinto. 1994. *Autecology of Selected Understory Vegetation in Central Ontario*. Central Region Science & Technology, Technical Report # 31/VMAP Technical Report #93-08. Ontario Ministry of Natural Resources, North Bay, Ont. 169 pp.

Bergen, J.Y. 1908. *Essentials of Botany*. The Athenaeum Press, Boston. 267 pp.

Berglund, B., and C.E. Bolsby. 1974. *The Edible Wild*. Pagurian Press Ltd., Toronto. 1988 pp.

Blouch, R.I. 1984. Northern Great Lakes states and Ontario forests. pp. 391-410. In L.K. Halls, ed., *White-tailed Deer: Ecology and Management*. Wildlife Management Institute, Stackpole Books, Toronto. 870 pp.

Blouin, G. 1984. *Weeds of the Woods: Small Trees and Shrubs of the Eastern Forest*. Goose Lane Editions, Fredericton, N.B. 125 pp.

Bowles, J.M. 1983. *Field Guide to the Common Forest Plants of the Claybelt Region of Northern Ontario*. Ontario Ministry of Natural Resources, Timmins, Ont.

Brown, L. 1979. *Grasses: An Identification Guide*. Houghton Mifflin Co., USA. 240 pp.

Brown. D.T., and G.J. Doucet. 1991. *Temporal changes in winter diet selection by white-tailed deer in a northern deer yard*. Journal of Wildlife Management 55:361-376.

Brunton, D.F., and W.J. Crins. 1992. *Checklist of the Vascular Plants of Algonquin Provincial Park (Revised)*. Algonquin Park Technical Bulletin No. 4. Friends of Algonquin Park and Ontario Ministry of Natural Resources.

Bunney, S. 1992. *The Illustrated Encyclopedia of Herbs: Their Medicinal and Culinary Uses*. Chancellor Press, London, Eng. 320 pp.

Cadman, M.D., P.F.J. Eagles, and F.M. Helleiner. 1987. *Atlas of the Breeding Birds of Ontario*. Federation of Ontario Naturalists, University of Waterloo Press, Waterloo, Ont. 617 pp.

Case, F.W. Jr. 1964. *Orchids of the Western Great Lakes Region*. Cranbrook Institute of Science, Bloomfield Hills, Mich. 147 pp.

Chapman, F.B. 1947. *The Hawthorns for Wildlife Food and Cover*. Ohio Conservation Bulletin 11(8):20-21.

Cobb, B. 1963. *A Field Guide to the Ferns and Their Related Families*. Peterson Field Guide Series. Houghton Mifflin Co., Boston. 281 pp.

Conard, H.S. 1956. *How to Know the Mosses and Liverworts*. Wm. C. Brown Co. Publ., Dubuque, Iowa. 226 pp.

Corns, I.G.W., and R.M. Annas. 1986. *Field Guide to Forest Ecosystems of West-Central Alberta*, Northern Forestry Centre, Canadian Forest Service, Edmonton, Alta. 251 pp.

Crete, M., and J. Bedard. 1975. Daily browse consumption by moose in Gaspé Peninsula, Quebec. Journal of Wildlife Management 39:368-373.

Crum, H.A. 1991. *Liverworts and Hornworts of Southern Michigan*. University of Michigan Herbarium. 233 pp.

Crum, H.A., 1989. *A Focus on Peatlands and Peat Mosses*. University of Michigan Press, Ann Arbor. 306 pp.

Crum, H.A., and L.E. Anderson. 1981. *Mosses of Eastern North America*. Volumes 1 & 2. Columbia University Press, New York. 1327 pp.

Crum, H.E. 1983. *Mosses of the Great Lakes Forest*, 3rd edition. University of Michigan Herbarium, Ann Arbor. 417 pp.

Cumming, H.G. 1987. Sixteen years of moose browse surveys in Ontario. *Alces* 23:125-155.

DeGraaf, R.M., and D.D. Rudis. 1986. *New England Wildlife: Habitat, Natural History, and Distribution*. U.S.D.A. Forest Service, Broomall, Pa. 491 pp.

Densmore, F. 1974. *How Indians Use Wild Plants for Food, Medicine and Crafts*. Dover Publications Inc., New York. 397 pp.

Dore, W.G. and J. McNeill. 1980. *Grasses of Ontario*. Agriculture Canada. Monograph 26. 566 pp.

Erichsen-Brown, C. 1979. *Use of Plants for the Past 500 Years*. Breezy Creeks Press, Aurora, Ont. 510 pp.

Erichsen-Brown, C. 1989. *Medicinal and Other Uses of North American Plants: A Historical Survey with Special Reference to the Eastern Indian Tribes*. Dover Publ. Inc., New York. 512 pp.

Euler, D. 1979. *Vegetation Management for Wildlife in Ontario*. Ontario Ministry of Natural Resources, Toronto. 61 pp.

Fernald, M.L. 1979. *Gray's Manual of Botany*, 8th edition. Van Nostrand Publ. Co., N.Y. 1632 pp.

Fleurbec. 1993. *Fougères, prèles et lycopodes*. Fleurbec, Saint-Henri-de-Lévis, Qué. 511 pp.

———. 1987. *Plantes sauvages des lacs, rivières et tourbières*. Fleurbec, Saint-Augustin, Que. 399 pp.

Foster, S., and J.A. Duke. 1990. *A Field Guide to Medicinal Plants: Eastern and Central North America*. Houghton Mifflin Co., Boston. 366 pp.

Fowells, H.A. (Compiler). 1965. *Silvics of Forest Trees of the United States*. U.S.D.A. Forest Service, Agric. Handbook No. 271. 762 pp.

Frankton, C., and G.A. Mulligan. 1971. *Weeds of Canada*. Canadian Department of Agriculture Publ. 948. 217 pp.

Gibbins, Bryan R. and M. Newton-White. 1978. *Wildflowers of the North*. Highway Book Shop, Cobalt, Ont. 215 pp.

Gleason, H.A. 1974. *The New Britton and Brown Illustrated Flora of the Northeastern United States and Adjacent Canada*. Vols.I, II and III. Hafner Press, New York. 1732 pp.

Glime, J.M. 1993. *The Elfin World of Mosses and Liverworts of Michigan's Upper Peninsula and Isle Royale*. Isle Royale Natural History Association, Houghton, Mich. 148 pp.

Grieve, M., and C.F. Level. 1992 (revised ed.) *A Modern Herbal*. Tiger Books Int., London, Eng. 912 pp.

Grimm, W.C. 1966. *How to Recognize Shrubs*. The Stackpole Company. Harrisburg, Pa. 319 pp.

Hale, M.E. 1979. *How to Know the Lichens*. 2nd edition. William C. Brown Co. Publ., Dubuque, Iowa. 246 pp.

Hills, G.A. 1959. *A Ready Reference to the Description of the Land of Ontario and Its Productivity*. Ontario Department of Lands and Forests, Division of Research, Maple, Ont.

Holling, D., *et al*. 1981. *Medicine for Heroes*. Mississauga South Historical Society. 93 pp.

Hosie, R.C. 1969. *Native Trees of Canada*. Canadian Forest Service, Ottawa.

Hosie, R.C. 1979. *Native Trees of Canada*. 8th ed., Fitzhenry and Whiteside Ltd., Don Mills, Ont. 317 pp.

Hutchins, A.R. 1973. *Indian Herbalogy of North America*. Merco, Windsor 14, Ontario.

Joyal, R. 1976. Winter Foods of Moose in La Verendrye Park, Quebec: An Evaluation of Two Browse Survey Methods. Canadian Journal of Zoology 54:1765-1770.

Klimas, J.E., and J.A. Cunningham. 1974. *Wildflowers of Eastern America*. Random House of Canada Ltd., Toronto. 273 pp.

Knap, A.H. 1975. *Wild Harvest: An Outdoorsman's Guide to Edible Wild Plants in North America*. Pagurian Press Ltd., Toronto. 192 pp.

Kotanen, T.M. 1982. *Vascular Plants of Algonquin Provincial Park*. Compiled by W.J. Crins (1977). 6th Ed. Algonquin Park Museum, Ontario Ministry of Mines and Natural Resources. 39 pp.

Krussmann, G. 1976. *Manual of Cultivated Broad-Leaved Trees & Shrubs*, Vols. I, II & III. Timber Press, Beaverton, Ore. 1403 pp.

Lamoureux, G., et collaborateurs. 1988. *Plantes sauvages printanières*. Fleurbec, Saint-Augustin, Quebec. 7th edition. 247 pp.

Lellinger, D.B. 1985. *A Field Manual of the Ferns and Fern-allies of the United States and Canada*. Smithsonian Institution Press, Washington. 389 pp.

MacKey, B.G., *et al*. 1994. *A New Digital Elevation Model of Ontario*. Department of Natual Resources Canada, Canadian Forest Service, Sault Ste. Marie, Ont. NODA/NFP Technical Report TR-6. 26 pp. plus Appendix I.

REFERENCES

MacKay, B.G., et al. 1996. Site Regions Revisited; A Climatic Analysis of Hill's Site Regions for the Province of Ontario Using a Parametric Method. Canadian Journal of Forest Research Vol. 26 (3): 333-354 (in press).

※ Marie-Victorin, Frère. Flore Laurentienne. Les Presses de l'Université de Montréal, Montréal, Qué. 2nd Edition. 923 pp.

Martin, A.C., H.S. Zim, and A.L. Nelson. 1961. American Wildlife and Plants: A Guide to Wildlife Food Habitats. Dover Publ. Inc., New York. 500 pp.

McCarthy, T., et al. 1994. Field Guide to Forest Ecosystems of Northeastern Ontario. Draft. Field Guide FG-001. Northeast Science & Technology, Ontario Ministry of Natural Resources, Timmins, Ont.

McKay, S., and P. Catling. 1979. Trees, Shrubs and Flowers to Know in Ontario. Alger Press, Oshawa, Ont. 208 pp.

McQueen, C.B. 1990. Field Guide to the Peat Mosses of Boreal North America. University Press of New England, Hanover. 138 pp.

Meyer, J.E., C. Meyer and D.E. Meyer. 1979. The Herbalist. Meyerbooks, Glenwood, Ill.

Morton, J., and J. Venn. 1990. A Checklist of the Flora of Ontario: Vascular Plants. Department of Biology, University of Waterloo, Waterloo, Ont. 218 pp.

Newmaster, S., and A. Lehela. 1995. The Ontario Plant List. Ontario Minstry of Natural Resources, Ontario Forest Research Institute, Sault Ste. Marie. Research Paper 123. 550 pp.

Niering, W., and N. Olmstead. 1979. The Audubon Society Field Guide to North American Wildflowers: Eastern Region. Alfred A. Knopf, Inc., N.Y. 887 pp.

Oldham, M.J. 1994. Natural Heritage Resources of Ontario: Rare Vascular Plants. Natural Heritage Information Centre, Peterborough, Ont. 48 pp.

Peterson, L.A. 1977. A Field Guide to Edible Wild Plants. Houghton Mifflin Co., Boston. 330 pp.

Peterson, R.T., and M. McKenny. 1968. A Field Guide to Wildflowers of Northeastern and North-Central North America. Houghton Mifflin Co., Boston. 420 pp.

Petrides, G.A. 1972. A Field Guide to Trees and Shrubs. The Peterson Field Guide Series. 2nd ed. Houghton Mifflin Co., Boston. 428 pp.

Rollins, J.A. 1974. Viburnums. U.S.D.A. Forest Service. Gen. Tech. Rep. NE-9:140-146.

※ Rouleau, R. 1990. Petite flore forestière du Québec. Les Publications du Québec, Québec. 2nd Edition. 250 pp.

Rowe, J.S. 1972. Forest Regions of Canada. Publication 1300. Canadian Forest Service, Ottawa

Semple, J.C., and S.B. Heard. The Asters of Ontario: Aster L. and Virgulus Raf. (Compositae: Asteraceae). University of Waterloo, Waterloo, Ont. 88 pp.

Semple, J.C., and G.S. Ringius. 1992. Goldenrods of Ontario: Solidago and Euthamia. University of Waterloo, Waterloo, Ont. 82 pp.

Sharp, W.M. 1974. Hawthorns. U.S.D.A. Forest Service Gen. Tech. Rep. NE-9:59-64.

Skeleton, E.G., and E.W. Skeleton. 1977. Haliburton Flora: An Annotated List of the Vascular Plants of the County of Haliburton. Royal Ontario Museum, Life Sciences Miscellaneous Publications, Toronto. 142 pp.

Soper, J.H., and M.L. Heimburger. 1982. Shrubs of Ontario. Royal Ontario Museum, Publications in Life Sciences, Toronto. 495 pp.

Theberge, J.B. Legacy: The Natural History of Ontario. McClelland & Stewart Inc., Toronto. 396 pp.

Van Allen Murphey, E. 1959. Indian Uses of Native Plants. Mendocino County Historical Society, Fort Bragg, Ca. 81 pp.

Vitt, D., J. Marsh, and R. Bovey. 1988. Mosses, Lichens & Ferns of Northwest North America. Lone Pine Publishing, Edmonton, Alta. 296 pp.

Voss, E.G. 1972. Michigan Flora. A Guide to the Identification and Occurrence of the Native and Naturalized Seed-plants of the State. Part I. Gynosperms and Monocots. Cranbrook Institute of Science and University of Michigan Herbarium.

Walker, M. 1984. Harvesting the Northern Wild. The Northern Publishers, Yellowknife, NWT. 224 pp.

Weiner, M.A. 1972. Earth Medicine - Earth Foods. The Macmillan Co. 214 pp.

Wells-Gosling, N. 1985. Flying Squirrels: Gliders in the Dark. Smithsonian Institution Press, Washington. 128pp.

Westbrooks, R.G., and J.W. Preacher. 1986. Poisonous Plants of Eastern North America. University of South Carolina Press, Columbia, S.C. 226 pp.

White, J.H. (revised by R.C. Hosie). 1973. The Forest Trees of Ontario. Ontario Ministry of Natural Resources, Toronto. 119 pp.

Whiting, R.E., and P.M. Catling. 1986. Orchids of Ontario: An Illustrated Guide. Canacoll Foundation, Ottawa. xii and 169 pp.

A

Abies balsamea 13
Acer
  pensylvanicum 107
  rubrum 37
  saccharinum 39
  saccharum 38
  spicatum 106
Achillea millefolium 276
Actaea
  pachypoda 190
  rubra 190
adder's tongue lily 155
Adiantum pedatum 350
Agrimonia
  gryposepala 203
agrimonia, hooked 203
Agropyron repens 297
Agrostis
  gigantea 300
  scabra 300
alder
  green 51
  speckled 52
Allium tricoccum 154
Alnus
  incana 52
  viridis 51
Amelanchier
  arborea 58
  bartramiana 59
  laevis 60
  sanguinea 61
  spicata 62
Anaphalis
  margaritacea 277
Andromeda polifolia 82
Anemone
  canadensis 184
  quinquefolia 185
anemone,
  Canada 184
  wood 185
Antennaria neglecta 278
Apocynum
  androsaemifolium 257
Aquilegia canadensis 183
Aralia
  hispida 234
  nudicaulis 233
  racemosa 235
arbutus, trailing 73
Arctostaphylos
  uva-ursi 72
Arisaema triphyllum 153
Aronia melanocarpa 64
arrow-wood, downy 101
Asarum canadense 179
ash,
  black 40
  green 42
  red 42
  white 41
aspen,
  largetooth 24
  trembling 25
Aster
  ciliolatus 285
  cordifolius 286
  canceolatus 283
  lateriflorus 280
  macrophyllus 287
  nemoralis 281
  puniceus 282
  simplex 283
  umbellatus 284
aster,
  bog 281
  calico 280
  ciliolate 285
  heart-leaved 286
  large-leaved 287
  panicled 283
  purple-stemmed 282
  tall flat-topped 284
Athyrium
  filix-femina 358
  thelypterioides 354
Aulacomnium
  palustre 391
avens,
  large-leaved 202
  water 202
  yellow 202

B

baneberry,
  red 190
  white 190
basil, wild 255
basswood, American 36
Bazzania trilobata 367
bearberry 72
bedstraw,
  fragrant 264
  northern 266
  rough 265
  small 266
beech, American 30
beech-drops 151
beech-fern, northern 346
bellflower,
  large-flowered 160
  marsh 260
bellwort 160
betony, Canada wood 262
Betula
  alleghaniensis 28
  papyrifera 27
  pumila 53
bilberry, oval-leaved 79
bindweed,
  black fringed 174
birch,
  dwarf 53
  white 27
  yellow 28
bishop's cap 194
bitternut 26
blackberry,
  bristly 119
  common 119
  smooth 118
Blephariglottis
  psycodes 144
bloodroot, Canada 222
blue cohosh 192
blue joint, Canada 302
blue-eyed grass,
  common 169
bluebells, northern 258
blueberry,
  low sweet 77
  velvet-leaf 78
bluegrass,
  Canada 305
  Kentucky 305
  swamp 305
bog laurel 85
  bog rosemary 82
Botrychium
  virginianum 362
Brachyelytrum
  erectum 296
Bracythecium spp. 407
British soldiers 424
brome grass,
  fringed 308
Bromus ciliatus 308
Brotherella
  recurvans 411
buckthorn,
  alder-leaved 69
buffalo berry 90
bugleweed, northern 252
bunchberry 241
buttercup,
  kidneyleaf 188
  tall 187

C

Calamagrostis
  canadensis 302
Callicladium
  haldanianum 408
Caltha palustris 189
Campanula
  aparinoides 260
  rotundifolia 259
  uliginosa 260
Capnoides
  sempervirens 224
Cardamine diphylla 225
Carex 312
  arctata 323
  brunnescens 315
  canescens 315
  communis 320
  crinita 324
  deflexa 321
  deweyana 316
  disperma 313
  gracillima 323
  gynandra 324
  intumescens 325
  leptalea 318
  lucorum 317
  magellanica 319
  paupercula 319
  pedunculata 322
  trisperma 314
Carya cordiformis 26
cattail 171
Caulophyllum
  thalictroides 192
cedar,
  eastern white 22
Cerastium
  fontanum 180
  vulgatum 180
Ceratodon purpureus 386

Chamaedaphne
  calyculata 83
cherry,
  black 35
  choke 66
  pin 65
chickweed,
  mouse-ear 180
  northern 181
Chimaphila umbellata 87
choke cherry 66
chokeberry, black 64
Chrysanthemum
  leucanthemum 275
Cinna latifolia 301
cinquefoil,
  marsh 200
  three-toothed 124
Circaea
  alpina 231
  lutetiana 232
  quadrisciculata 232
Cirsium muticum 274
Cladina
  mitis 428
  rangiferina 429
  stellaris 428
Cladonia
  bacillaris 425
  bacilliformis 425
  borealis 424
  cenotea 422
  chlorophaea 423
  coniocraea 425
  cornuta 425
  cristatella 424
  fimbriata 423
  furcata 426
  gracilis 426
  multiformis 426
  pleurota 427
cladonia,
  perforated 426
  powdered funnel 422
Claytonia
  caroliniana 176
  virginica 176
Clematis
  verticillaris 127
  virginiana 127
Climacium dendroides 401
Clinopodium vulgaris 255
Clintonia borealis 156
clintonia, yellow 156
clover,
  alsike 208
  red 209
  white 210
club-moss,
  interrupted 336
  shining 340
  wolf's claw 337
coltsfoot, sweet 279
columbine, wild 183
Comandra livida 178
comandra,
  northern 178
Comptonia peregrina 55
Conocephalum
  conicum 365
Coptis trifolia 182

coral-root,
  early 148
  large 149
  spotted 149
  striped 150
Corallorhiza
  maculata 149
  striata 150
  trifida 148
Cornus
  alternifolia 70
  canadensis 241
  rugosa 92
  stolonifera 91
corpse plant 251
Corydalis
  sempervirens 224
corydalis, pale 224
Corylus cornuta 54
cottongrass, sheathed 326
cow wheat 263
cranberry,
  high-bush 103
  large 81
  small 80
cranesbill, Bicknell's 227
Crataegus 63
  chrysocarpa 63
  douglasii 63
  succulenta 63
cucumber-root,
  Indian 168
currant,
  northern wild black 109
  red 109
  skunk 112
  swamp black 108
  wild black 111
Cypripedium 152
  acaule 138
  calceolus 139
  reginae 138
Cystopteris bulbifera 351

D
daisy, ox-eye 275
Dalibarda repens 201
dandelion 268
Danthonia
  spicata 304
Dentaria diphylla 225
Deschampsia
  flexuosa 309
devil's paintbrush 270
dewberry, northern 116
dewdrop 201
Diaphasiastrum
  digitatum 339
Dicentra
  canadensis 223
  cucullaria 223
Dicranum
  flagellare 392
  fuscescens 393
  montanum 394
  ontariense 395
  polysetum 396
  scoparium 397
dicranum,
  Ontario 395
  spiky 392

Diervilla lonicera 97
Dirca palustris 71
dock 177
dogbane, spreading 257
dogwood,
  alternate-leaved 70
  red osier 91
  round-leaved 92
Drepanocladus
  uncinatus 410
Drosera
  intermedia 173
  linearis 173
  rotundifolia 173
Dryopteris
  carthusiana 357
  cristata 352
  marginalis 353
Dutchman's breeches 223
dwarf-elder 234

E
elder,
  red-berried 129
  wild 234
elderberry, common 128
elm, white 34
Elymus repens 297
enchanter's nightshade,
  smaller 231
  tall 232
Epifagus virginiana 151
Epigaea repens 73
Epilobium
  adenocaulon 230
  angustifolium 229
  ciliatum 230
  leptocarpum 230
Epipactis helleborine 152
Equisetum
  arvense 332
  fluviatile 334
  pratense 333
  scirpoides 335
  sylvaticum 331
Eriophorum
  vaginatum 326
Erythronium
  americanum 155
Eupatorium
  maculatum 273
  purpureum 273
evening primrose 228
everlasting, pearly 277
Evernia
  mesomorpha 432

F
Fagus grandifolia 30
fern,
  bracken 361
  bulblet 351
  Christmas 344
  cinnamon 347
  crested wood 352
  interrupted 348
  lady 358
  maidenhair 350
  marginal wood 353
  marsh 356
  New York 355
  northern beech 346

oak 360
ostrich 359
rattlesnake 362
royal 349
sensitive 345
spinulose wood 357
fir, balsam 13
fireweed 229
Fissidens species 414
flag, multicolored
blue 170
foam lichen, woolly 430
foamflower 195
Fragaria
vesca 198
virginiana 197
Fraxinus
americana 41
nigra 40
pennsylvanica 42

G

Galium
asprellum 265
boreale 266
trifidum 266
triflorum 264
Gaultheria
hispidula 74
procumbens 75
Gaylussacia baccata 76
Geocaulon lividum 178
Geranium
bicknellii 227
robertianum 227
Geum
aleppicum 202
macrophyllum 202
rivale 202
ginger, wild 179
ginseng, dwarf 236
Glyceria striata 306
goldenrod,
bog 290
Canada 288
hairy 291
rough-stemmed 289
goldthread 182
Goodyera
oblongifolia 147
repens 146
tesselata 147
gooseberry,
prickly 113
smooth 110
grape-woodbine 126
grass-of-Parnassus,
marsh 196
ground cedar,
southern 339
ground pine 338
Gymnocarpium
dryopteris 360

H

Habenaria
hyperborea 143
obtusata 141
psycodes 144
hairgrass, common 309

harebell 259
hawkweed,
orange 270
yellow 271
hawthorn 63
black 63
golden-fruited 63
long-spined 63
hazel, beaked 54
heal-all 256
Hedwigia ciliata 382
helleborine 152
hemlock,
eastern 18
ground 44
Hepatica
acutiloba 186
americana 186
triloba 186
heptatica,
blunt-lobed 186
round-lobed 186
Heracleum lanatum 239
herb robert 227
hickory, bitternut 26
Hieracium
aurantiacum 270
caespitosum 271
hobblebush, common 102
holly fern, Braun's 344
holly, winterberry 56
honeysuckle,
bush 97
fly 94
glaucous 96
hairy 93
mountain fly 95
hop hornbeam 29
hop-clover 207
horehound, cut-leaved
water 252
horsetail,
field 332
meadow 333
swamp 334
woodland 331
huckleberry, black 76
Huperzia lucidula 340
Hylocomium splendens 404
Hypericum perforatum 226
Hypnum pallescens 409
Hypogymnia
physodes 421
Hypopitys monotropa 250

I-J-K

ice plant 251
Ilex verticillata 56
Impatiens
biflora 221
capensis 221
Indian pipe 251
Iris versicolor 170
ironwood 29
Jack-in-the-pulpit,
small 153
Jamesoniella
autumnalis 366

jewelweed, spotted 221
Joe-Pye weed,
purple 273
spotted 273
Juglans cinerea 26
Juncus bufonius 328
juneberry,
downy 58
low 62
mountain 59
juniper,
common 45
creeping 45
Juniperus
communis 45
horizontalis 45
Kalmia
angustifolia 86
polifolia 85

L

Labrador tea 84
Lactuca species 272
ladies'-tresses,
hooded 145
nodding 145
northern 145
shiny 145
lady's slipper,
showy 138
pink 138
yellow 139
larch, American 14
Larix laricina 14
Lathyrus ochroleucus 206
laurel,
bog 85
sheep 86
leatherleaf 83
leatherwood 71
Ledum groenlandicum 84
leek, wild 154
lettuce,
white 269
white wild 269
wild 272
Leucanthemum vulgare 275
Leucobryum glaucum 383
lichen,
common foam 430
coral 428
cracked 420
dog's tooth 417
fork 426
fringed pixie cup 423
horn 425
monk's hood 421
powder horn 425
red pixie cup 424
red-fruited goblet 427
reindeer 429
rock foam 430
scarlet toothpick 425
slender cup 426
spotted dog 416
waxpaper 420
yellow toothpick 425
yellow-green 428–9
lily,
blue bead 156
trout 155
lily-of-the-valley, wild 159

Limnorchis hyperborea 143
Linnaea borealis 89
Listera cordata 140
liverwort,
  green-tongue 365
  Jameson's 366
  naugehyde 369
  northern
    naugehyde 368
  snake 365
  three-lobed 367
Lobaria pulmonaria 418
Lonicera
  caerulea 95
  canadensis 94
  dioica 96
  hirsuta 93
  villosa 95
lousewort 262
lungwort 418
  European 258
Lycopodium
  annotinum 336
  clavatum 337
  dendroideum 338
  digitatum 339
  lucidulum 340
Lycopus
  americanus 252
  uniflorus 252
Lysiella obtusata 141

M

Maianthemum
  canadense 159
  racemosum 157
  trifolium 158
mandarin, white 162
manna grass, fowl 306
maple,
  mountain 106
  red 37
  silver 39
  striped 107
  sugar 38
Marchantia
  polymorpha 365
marigold, marsh 189
Matteuccia
  struthiopteris 359
mayflower, Canada 159
meadow-rue, fall 191
meadow-sweet,
  broad-leaved 68
  narrow-leaved 67
Medeola virginiana 168
Melampyrum lineare 263
melic grass, false 303
Mentha arvensis 253
Menyanthes trifoliata 240
Mertensia paniculata 258
Milium effusum 307
millet-grass, wood 307
mint, field 253
Mitchella repens 88
Mitella
  diphylla 194
  nuda 193
mitrewort, naked 193

Mnium 389
  marginatum 389
  spinulosum 389
mnium,
  edged lantern 389
  red-mouthed 389
  woodsy 389
Moneses uniflora 243
Monotropa
  hypopitys 250
  uniflora 251
mooseberry 105
moss,
  awned hair cap 385
  beautiful branch 408
  broom 397
  brotherella 411
  ciliate hedwigia 382
  common fern 402
  common four-
    tooth 388
  common hair cap 384
  common leafy 389
  curly heron's-bill 393
  electrified cat's tail 400
  feathery neckera 413
  fern 402
  fire 386
  fuzzy brown 406
  glossy 412
  golden fuzzy fen 406
  juniper 385
  lawn 394
  leafy 389
  mnium 389
  nodding pohlia 387
  northern tree 401
  peat (see peat moss)
  pin cushion 383
  plume 403
  pointed round 389
  ragged 407
  ribbed bog 391
  rose 390
  round 389
  Schreber's 405
  shaggy 400
  sickle 410
  split-tooth 414
  stair-step 404
  stump pigtail 409
  toothed cotton 412
  wavy-leaved 396
mountain ash,
  American 122
  showy 123
mountain-holly 57
Myelochroa aurulenta 420

N-O

nannyberry 100
Neckera pennata 413
Nemopanthus
  mucronantus 57
nettle,
  slender 177
  stinging 177
oak,
  bur 32
  red 33
  white 31
oat grass, poverty 304
Oenothera biennis 228
old man's beard 431

Onoclea sensibilis 345
orchid,
  blunt-leaf 141
  Hooker's 142
  large round-
    leaved 142
  small purple-
    fringed 144
  tall northern
    green 143
Orthilia secunda 244
Oryzopsis asperifolia 298
Osmorhiza
  claytonii 237
  longistylis 237
Osmunda
  cinnamomea 347
  claytoniana 348
  regalis 349
Ostrya virginiana 29
Oxalis
  acetosella 211
  stricta 211
Oxycoccus microcarpon 80

P

Panax
  quinquefolia 236
  trifolius 236
Parmelia sulcata 420
Parmelina aurulenta 420
Parnassia palustris 196
parsnip,
  cow 239
  water 240
Parthenocissus
  inserta 126
  quinquefolia 126
  vitacea 126
partridgeberry 88
peat moss,
  common brown 373
  common green 379
  midway 374
  poor-fen 377
  shaggy 378
  small red 376
  Warnstorf's 375
  wide-tongued 376
  Wulf's 372
Pedicularis
  canadensis 262
Peltigera
  aphthosa 416
  canina 417
Petasites
  frigidus 279
  palmatus 279
Phegopteris connectilis 346
Phleum pratense 299
Picea
  glauca 15
  mariana 16
  rubens 17
pin cherry 65
pine,
  eastern white 21
  jack 19
  red 20
pinesap 250

Pinus
  banksiana **19**
  resinosa **20**
  strobus **21**
pitcher plant, purple **172**
pixie cup,
  false **423**
  fringed 423
  red 424
Plagiomnium **389**
  cuspidatum 389
  medium 389
  Plagiothecium
  denticulatum **412**
  laetum **412**
Plantago major **267**
plantain, common **267**
Platanthera
  hookeri **142**
  hyperborea **143**
  obtusata **141**
  orbiculata **142**
  psycodes **144**
Pleurozium schreberi **405**
Poa
  compressa 305
  palustris 305
  pratensis 305
  saltuensis **305**
Pohlia nutans **387**
poison ivy **125**
Polygala paucifolia **220**
polygala, fringed **220**
Polygonatum pubescens **163**
Polygonum cilinode **174**
Polypodium
  virginianum **343**
polypody, common **343**
Polystichum
  acrostichoides **344**
  braunii 344
Polytrichum
  commune **384**
  juniperinum **385**
  piliferum 385
poplar, balsam **23**
Populus
  balsamifera **23**
  grandidentata **24**
  tremuloides **25**
Potentilla
  palustris **200**
  tridentata **124**
Prenanthes **269**
  alba 269
  altissima 269
prince's pine **87**
Prunella vulgaris **256**
Prunus
  pensylvanica **65**
  serotina **35**
  virginiana **66**
Pteridium aquilinum **361**
Ptilidium
  ciliare **368**
  pulcherrimum **369**
Ptilium
  crista-castrensis **403**
Pulmonaria officinalis 258
pussytoes, field **278**

Pyrola
  americana **248**
  asarifolia **247**
  chlorantha **249**
  elliptica **245**
  minor **246**
  rotundifolia 248
  secunda 244
  virens 249
pyrola,
  greenish-flowered **249**
  lesser **246**
  pink **247**
  round-leaved **248**

Q-R
quackgrass **297**
Quercus
  alba **31**
  macrocarpa **32**
  rubra **33**
raisin, northern wild **99**
Ranunculus
  abortivus **188**
  acris **187**
raspberry,
  dwarf **115**
  flowering **114**
  wild red **117**
rattlesnake-plantain,
  green-leaved 147
  dwarf **146**
  tesselated **147**
rattlesnake root **269**
redtop **300**
reedgrass,
  broad-leaved **301**
Rhamnus alnifolia **69**
Rhizomnium **389**
  punctatum 389
Rhodobryum
  ontariense **390**
  roseum 390
Rhus radicans 125
Rhytidiadelphus
  triquetrus **400**
Ribes
  americanum **111**
  cynosbati **113**
  glandulosum **112**
  hirtellum **110**
  hudsonianum 109
  lacustre **108**
  nigrum 111
  triste **109**
rice grass, white-grained
  mountain **298**
rock tripe **419**
Rosa
  acicularis **120**
  blanda **121**
rose,
  prickly wild **120**
  smooth wild **121**
Rubus
  allegheniensis **119**
  canadensis **118**
  flagellaris **116**
  idaeus **117**
  odoratus **114**
  parviflorus 114
  pubescens **115**
  setosus 119

Rumex **177**
  acetosella **175**
rush, toad **328**

S
Salix
  bebbiana **46**
  discolor **47**
  humilis **48**
  lucida **49**
  petiolaris **50**
Sambucus
  canadensis **128**
  racemosa **129**
Sanguinaria canadensis **222**
Sanicula marilandica **238**
Sanionia uncinata **410**
Sarracenia purpurea **172**
sarsaparilla,
  bristly **234**
  wild **233**
Satureja vulgaris 255
Schizachne
  purpurascens **303**
Scirpus cyperinus **327**
scouring rush, dwarf **335**
Scutellaria
  galericulata **254**
  lateriflora 254
sedge **312**
  bladder **325**
  bristle-stalked **318**
  brownish **315**
  common **320**
  Dewey's **316**
  distant **317**
  drooping wood **323**
  few-flowered **319**
  filiform 323
  fringed **324**
  grey **315**
  long-stalked **322**
  nodding **324**
  northern **321**
  soft-leaved **313**
  three-fruited **314**
serviceberry,
  red-twigged **61**
  smooth **60**
sheep laurel **86**
Shepherdia canadensis **90**
shinleaf **245**
shorthusk, bearded **296**
silverrod 291
Sisyrinchium
  montanum **169**
skullcap,
  mad-dog 254
  marsh **254**
Smilacina
  racemosa **157**
  trifolia **159**
smilacina,
  three-leaved **158**
snakeroot, black **238**
snowberry **98**
  creeping **74**

Solidago
  bicolor 291
  canadensis 288
  hispida 291
  rugosa 289
  uliginosa 290
Solomon's seal,
  false 157
  hairy 163
Sorbus
  americana 122
  decora 123
sorrel, sheep 175
spear grass,
  bushy pasture 305
speedwell, common 261
Sphagnum
  angusifolium 377
  capillifolium 376
  fuscum 373
  girgensohnii 379
  magellanicum 374
  nemoreum 376
  russowii 376
  squarrosum 378
  warnstorfii 375
  wulfianum 372
spikenard 235
Spiraea
  alba 67
  latifolia 68
Spiranthes
  cernua 145
  gracilis 145
  lucida 145
  romanzoffiana 145
spleenwort, silvery 354
spring beauty,
  Carolina 176
  Virginia 176
spruce,
  black 16
  red 17
  white 15
spruce-moss 432
squawroot 192
squirrel corn 223
St. John's-wort,
  common 226
starflower 242
Stellaria
  borealis 181
  calycantha 181
Stereocaulon
  paschale 430
  saxitile 430
  tomentosum 430
strawberry,
  barren-ground 199
  common 197
  woodland 198
Streptopus
  amplexifolius 162
  roseus 161
sundew,
  round-leaved 173
  slender-leaved 173
  spatulate-leaved 173
sweet cicely 237
  smooth 237

sweet-fern 55
Symphoricarpos albus 98

T-U-V
tamarack 14
Taraxacum officinale 268
Taxus canadensis 44
Tetraphis pellucida 388
Thalictrum
  dasycarpum 191
  pubescens 191
Thelypteris
  noveboracensis 355
  palustris 356
  phegopteris 346
thimbleberry 114
thistle, swamp 274
Thuidium
  delicatulum 402
  recognitum 402
Thuja occidentalis 22
Tiarella cordifolia 195
tickle grass 300
Tilia americana 36
timothy 299
Tomenthypnum
  nitens 406
toothwort,
  broad-leaved 225
touch-me-not, spotted 221
Toxicodendron
  radicans 125
Trientalis borealis 242
Trifolium
  aureum 207
  hybridum 208
  pratense 209
  repens 210
Trillium
  cernuum 167
  erectum 165
  grandiflorum 164
  undulatum 166
trillium,
  nodding 167
  painted 166
  red 165
  white 164
trout lily 156
Tsuga canadensis 18
twayblade,
  heart-leaved 140
twinflower 89
twisted-stalk, rose 161
Typha latifolia 171
Ulmus americana 34
Umbilicaria
  mammulata 419
Urtica dioica
  var. *gracilis* 177
Usnea species 431
Uvularia grandiflora 160
Vaccinium
  angustifolium 77
  macrocarpon 81
  myrtilloides 78
  ovalifolium 79
  oxycoccus 80

Veronica officinalis 261
vetch,
  American 205
  cow 204
vetchling, marsh 206
Viburnum
  acerifolium 104
  alnifolium 102
  cassinoides 99
  edule 105
  lantanoides 102
  lentago 100
  opulus 103
  rafinesquianum 101
  trilobum 103
viburnum,
  maple-leaved 104
Vicia
  americana 205
  cracca 204
Viola
  adunca 212
  blanda 214
  canadensis 218
  incognita 214
  macloskeyi 215
  pensylvanica 219
  pubescens 219
  renifolia 216
  septentrionalis 213
  sororia 217
violet,
  Canada 218
  downy yellow 219
  hooked-spur 212
  kidney-leaved 216
  northern blue 213
  northern white 215
  smooth yellow 219
  sweet white 214
  woolly blue 217
virgin's bower,
  purple 127
  Virginia 127
Virginia-creeper,
  five-leaved 126

W-Y
wake-robin 165
Waldsteinia
  fragarioides 199
willow,
  Bebb's 46
  pussy 47
  shining 49
  slender 50
  upland 48
willow-herb, northern 230
wind flower 185
wintergreen 75
  one-flowered 243
  one-sided 244
wood fern, spinulose 358
wood-sorrel, upright 211
wool-grass 327
yarrow 276